12th International Conference on 3D and Advanced Dosimetry 2022

Journal of Physics: Conference Series Volume 2630

Online
19 - 22 July 2022

ISBN: 978-1-7138-8215-2
ISSN: 1742-6588

Printed from e-media with permission by:

Curran Associates, Inc.
57 Morehouse Lane
Red Hook, NY 12571

Some format issues inherent in the e-media version may also appear in this print version.

Institute of Physics
Dirac House, Temple Back
Bristol BS1 6BE UK

Phone: 44 1 17 929 7481
Fax: 44 1 17 920 0979

techtracking@iop.org

Additional copies of this publication are available from:

Curran Associates, Inc.
57 Morehouse Lane
Red Hook, NY 12571 USA
Phone: 845-758-0400
Fax: 845-758-2633
Email: curran@proceedings.com
Web: www.proceedings.com

TABLE OF CONTENTS

Journal of Physics: Conference Series **2630** (2023) 011001 doi:10.1088/1742-6596/2630/1/011001

IC3DDose 2022 - 12th International Conference on 3D and Advanced Dosimetry

Preface

The 12th IC3DDose conference on advanced radiation dosimetry was held in person in Quebec City, Quebec, Canada from June 20 to 23, 2022. The meeting was scheduled back-to-back with the Canadian Organization of Medical Physicists' (COMP) annual scientific meeting, offering unique opportunities for scientific knowledge sharing and extensive networking. It was a refreshing restart for the IC3DDose community after the cancelled meeting in 2020 and a reduced scope virtual meeting in 2021.

A total of 71 participants from 11 countries met for 3.5 days in a mix of refresher/review sessions and proffered presentations covering state-of-the-art dosimetry encompassing all major dosimetry techniques used for standard external beam and brachytherapy. Presenters also covered advanced applications addressing the challenges of radiation measurement in small field, MR-Linac, microbeam and FLASH dosimetry. A total of 14 invited presentations and 25 contributed talks were given the at meeting including papers in a Rising Star Competition and in a single hybrid session with virtual participants from Australia, China and Japan. The associated papers presented in these proceedings add to an already extensive collection of 118 review and 574 proffered papers in 9 previous volumes of the Journal of Physics Conference Series from the past conferences.

A feature for this year's meeting was that all contributed speakers on a given day were asked to mount their presentation slides on poster boards for a dedicated closing discussion session entitled "Meet the speakers of the day". This allowed attendees to engaged in deeper discussions, beyond the question periods at the end of their talks. The daily sessions reinforced the IC3DDose philosophy to foster strong mingling of all participants, from trainees to more experienced attendees. A keynote address and the final morning of sessions were open to COMP registrants also. This made the topics of the IC3DDose meeting available to a much wider audience.

The meeting was a great success and the attendees developed new expertise and new networks through the many interactions over the meeting's four days. Colleagues from Aarhus Denmark announced at the end of the meeting that they will now undertake to organize a 13th meeting in 2024.

The local organizing committee sincerely thanks the sponsors for our meeting: Elekta and Varian as bronze sponsor and MedScint in the start-up category. We further acknowledge support from COMP, l' Association Québécoise des Physicien(ne)s Médicaux Cliniques, l' Université Laval and CRCHU de Quebec-Université Laval.

Louis Archambault, Luc Beaulieu and L. John Schreiner
(for the organizing and scientific committee)

International Scientific Committee

Louis Archambault, Canada; Sven Bäck, Sweden; Peter Balling, Denmark; Clive Baldock, Australia; Luc Beaulieu, Canada (chair); Sam Beddar, USA; Sofie Ceberg, Sweden; Yves de Deene, Australia; Simon Doran, UK; Geoffrey Ibbott, USA; Andrew Jirasek, Canada; Kevin Jordan, Canada; Boyd McCurdy, Canada; Mark Oldham, USA; L. John Schreiner, Canada; Hiroshi Yasuda, Japan

Journal of Physics: Conference Series **2630** (2023) 011001 doi:10.1088/1742-6596/2630/1/011001

Local Organizing Committee

Louis Archambault, Luc Beaulieu, L. John Schreiner

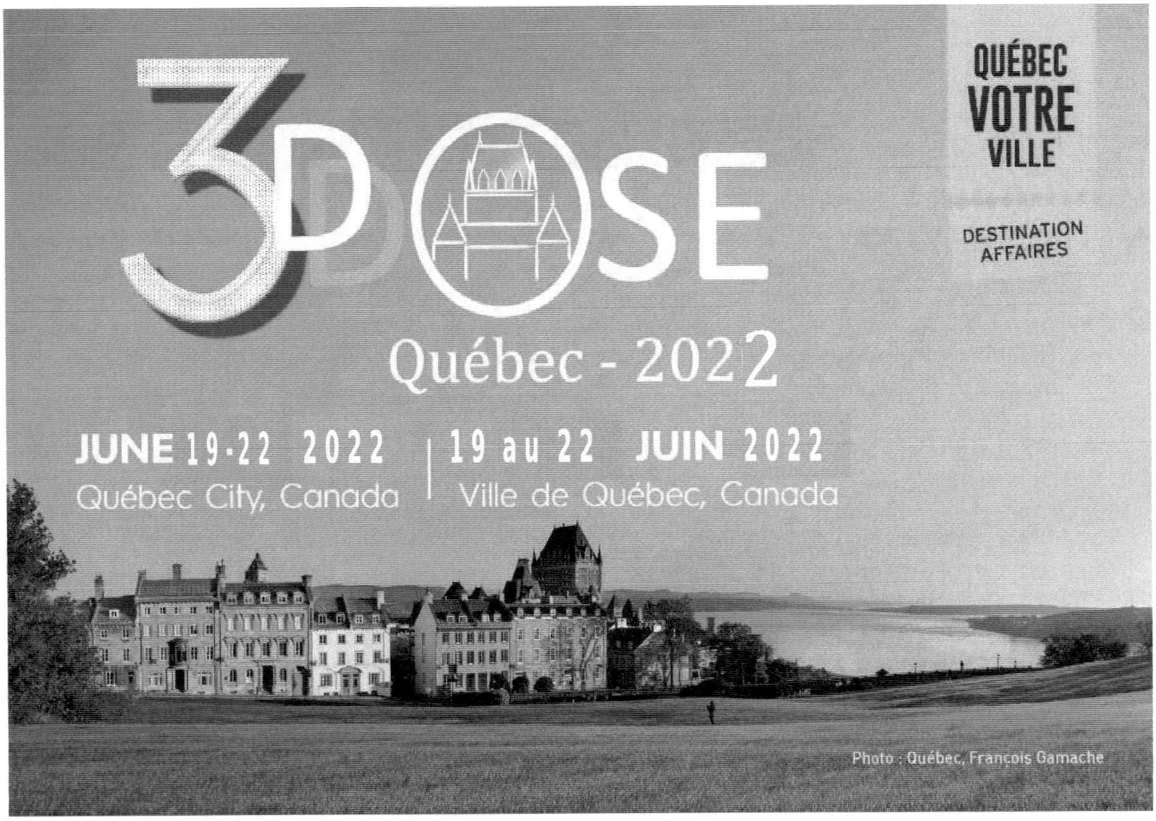

Detailed Program of the

12th INTERNATIONAL CONFERENCE ON 3D and ADVANCED DOSIMETRY (IC3Ddose)

June 2022, Château Frontenac, Quebec City.

Journal of Physics: Conference Series **2630** (2023) 011001 doi:10.1088/1742-6596/2630/1/011001

Contents

Journal of Physics: Conference Series **2630** (2023) 011001 doi:10.1088/1742-6596/2630/1/011001

About

The meeting is the 12th edition of the DosGel/IC3Ddose biennial gathering of scientists on the topic of advanced dose measurements, including 3D and 4D dose imaging. Attendance is usually limited to 120-130 participants in order to create a unique opportunity for all those interested, including graduate students and medical physics residents, to have one-on-one exchanges with the world-leading experts in the field. In addition to regular contributed talks, vendor exhibit, and a poster session, the 2022 edition will feature over a dozen refresher courses, a Rising Star oral talk competition as well as panel discussions and debates.

Topics (but not limited to)

- 3D and 4D dosimetry;
- GEL dosimetry;
- Film dosimetry;
- EPID dosimetry;
- Scintillation and Cherenkov;
- End-to-end QA;
- Dosimetry in extreme conditions:
  - Small field;
  - MR-Linac;
  - FLASH.

You will find last year's presentations here: `https://youtube.com/playlist?list=PLOwwIxJBoN1AiczRciTFBO-wFA28zvlKu`

Organizing committee

Louis Archambault (ULaval, Canada)
Luc Beaulieu (ULaval, Canada)
Kevin Jordan (Western Ontario, Canada)
Andrew Jirasek (UBC-O, Canada)
Boyd McCurdy (UManitoba, Canada)
John Schreiner (Queen's, Canada)

International Committee

Sven Bäck (Skane, Sweden)
Clive Baldock (Western Sydney, Australia)
Peter Balling (Aarhus, Denmark)
Sam Beddar (MD Anderson, USA)
Sofie Ceberg (Lund, Sweden)
Yves De Deene (Macquarie, Australia)
Simon Doran (ICR, UK)
Geoffrey Ibbott (Vermont, USA)
Mark Oldham (Duke, USA)

A WARM THANK YOU TO OUR SPONSORS

Commercial Sponsors

Elekta

varian
A Siemens Healthineers Company

Start-up Sponsor

medscint

Association Support

COMP OCPM
Canadian Organization of Medical Physicists
Organisation canadienne des physiciens médicaux

AQPMC
Association québécoise des physicien(ne)s médicaux cliniques

CERCLE DES AMBASSADEURS DE QUÉBEC

Academic Sponsors

UNIVERSITÉ LAVAL
Département de physique, de génie physique et d'optique

CENTRE DE RECHERCHE SUR LE CANCER

CHU de Québec

AXE ONCOLOGIE

12th International Conference on 3D and Advanced Dosimetry IOP Publishing
Journal of Physics: Conference Series **2630** (2023) 011001 doi:10.1088/1742-6596/2630/1/011001

AT A GLANCE

See detailed program for accurate session times

Time	Sunday June 19th	Monday June 20th	Tuesday June 21st	Wednesday June 22nd	Thursday June 23rd
08:30 - 09:10		On-site registration / Badge pickup 09:00 Opening	On-site registration / Badge pickup	On-site registration / Badge pickup / COMP pre-meeting day	**COMP Panel (<u>in COMP meeting room</u>): FLASH in Radiotherapy - Mechanisms,Tec hnology, Dosimetry and Delivery**
09:10 - 09:30		Invited talks	Invited talks	Invited talks	
09:30 - 10:30					Invited talk
10:30 - 10:50		**Pause santé / Health Break / Exhibit**			
10h50-11h30		Proffered session	Proffered session	Proffered session	Invited talk
11:30 - 12:00					vendor talks, Awards, future meetings
12:00 - 13:00		Lunch and exhibit	Free to roam the old City and enjoy a terrasse!	Lunch and exhibit	Lunch
13:00 - 14:20		**Rising Stars Competition**		**Dosimetry in Extreme Conditions**	**COMP Young Investigators Symposium: IC3Ddose participants invited!**
14:25 - 15:15					
15:15 - 15:35		**Pause santé / Health Break**		**Pause santé / Health Break**	
15:35 - 16:15		Invited talk	Invited talk and Proffered **(start at 15h20)**	**Dosimetry in Extreme Conditions (cont...)**	
16:15-17:00		**Meet the Speakers / Posters of the day**		**Meet the Speakers / Poster Session**	
17:00 - 17:30	Welcome Registration	**Virtual session**	**Meet the Speakers / Posters of the day**		
17:30 - 18:00				**Joint IC3Ddose-COMP Keynote - by Dr John Schreiner**	
18h30-					
Evening	Free to explore the city	Free to explore the city	TBD	Joint Social / Exhibit	

DAY 1 – JUNE 20th

CT: Contributed Talk, IS: Invited Speaker, KL: Keynote Lecture, IT: Invited Talk.

8h15	**On site Registration / Badge pickup**		
9h00	**Opening of the 12th IC3Ddose conference**		
9h10-10h30	**Refresher #1** Chair : Michelle Hilts, UBC-O, Canada		
9h10-9h50	IS	**L. John Schreiner** Queen's University, Canada	End-to-End QA
9h50-10h30	IS	**Simon Doran** Division of Radiotherapy and Imaging, ICR, UK	Gel dosimetry: Optical reading
10h30-10h50	**Health Break**		
10h50-12h00	**Proffered session #1** Chairs : Sven Bäck and Andrew Jirasek		
10h50-11h05	CT	**Nathaniel Woolvett and Kevin Jordan** London, Ontario, Canada	Modified optical computed tomography scanner for low temperature scanningr
11h05-11h20	CT	**Andy Ogilvy, S Collins, M Guenter, D Huang, W Hare, M Hilts, and A Jirasek** UBC-O, Canada	Preliminary commissioning of a prototype solid tank optical CT scanner for 3D radiation dosimetry
11h20-11h35	CT	**Mélissa Cinq-Mars, J-D Jutras and L Beaulieu** Université Laval, Canada	On the comparison of two NIPAM gel dosimeters with high resolution 3D MRI sequences
11h35-11h50	CT	**Lia Barbosa Valdetaro, Ludvig Paul Muren, Peter Sandegaard Skyt, Jørgen Breede Baltzer Petersen and Peter Ballingu** Aarhus University, Denmark	Investigating the linear-energy-transfer and dose rate signal quenching of radiochromic silicone-based dosimeters for different compositions and curing times
12h00-13h00	**Lunch**		
13h00-15h15	**RISING STAR COMPETITION** Chairs : Geoff Ibbott and Sam Beddar		
13h00 -13h15	CT	**Marco Caprioli, L. Delombaerde M. De Saint-Huber, L. de Freitas Nascimento, S. S. Galvez Febles, K. Himschoot, B. van der Heyden, D. Vandenbroucke, R. De Roover, Leblans, W. Crijns** KU Leuven, Belgium	Fading correction for calibration of a novel 2D OSL-film dosimeters

13h15-13h30	CT	**Mads L. Jensen, Rosana M. Turtos, Ludvig P. Muren, Brian Julsgaard, and Peter Balling** Aarhus University, Denmark	3D optically-stimulated-luminescence-based dosimetry using LYSO:Ce scintillators
13h30-13h45	CT	**Mélodie Cyr, Victor D. Martinez, Devic Slobodan, Nada Tomic, David F. Lewis, Shirin A. Enger** McGill Uuiversity, Canada	Investigation of the Dosimetry Characteristics of the GAFCHROMIC® EBT3 Film Response to Alpha Particle Irradiation
13h45-14h00	CT	**Elise Konradsson, M Blomstedt, C Jamtheim Gustafsson, S Å J Bäck, C Ceberg and S Ceberg** Skane University Hospital, Sweden	Dosimetric verification of conformal small field irradiation at a preclinical research platform – a polymer gel dosimeter feasibility study
14h00-14h15	CT	**Aline Rousseau, C Stien, S Dufreneix, G Boissonnat, JM Bordy, V Blideanu** CEA, France	End-to-end quality assurance for Volumetric Modulated Arc Therapy with Fricke-Xylenol orange-Gelatin gel dosimeters and dual-wavelength cone-beam optical CT scanner
14h15-14h30	CT	**Steve Collins, A Ogilvy, D Huang, W Hare, M Hilts and A Jirasek** UBC-O, Canada	Iterative Image Reconstruction Methodology in Optical CT Radiochromic Gel Dosimetry
14h30-14h45	CT	**Simon V Jensen, T B Erichsen, M B Jensen, E Worm, P R Poulsen, J B B Petersen, P Balling, D Q S Le and L P Muren** Aarhus University, Denmark	Developing a 3D-printing-based method to create anthropomorphic dosimeters for radiotherapy-delivery verification
15h00-15h20	**Health Break**		
15h20-16h00	**Refresher #2** Chair : Boyd McCurdy		
14:30–15:05	IS	**Yves De Deene** Liverpool hospital/Ingham institute, Australia	Gel dosimetry: MRI
16:30–18:00	**Meet the speakers / posters of the day** (Tentative : Beer included!) The four proffered papers and the seven rising star competition presenters will be available at the poster boards to discuss their works. Slides presented will be displayed on the poster boards.		

17h00-18h00: **IC3Ddose Virtual Session**
+12h for China;
+13h for Japan;
+14h for Australia.

17h00-18h00	Chairs : Simon Doran and Kevin Jordan		
	Each presentation will be of 10 min with an additional 2 min for live questions		
17h00 -17h12	CT	**Yves De Deene and Daniel Mason** Liverpool hospital/Ingham institute, Australia	Optimization of MRI pulse sequences and Gadobutrol doped polymer gel for real time 4D radiation dosimetry on the MRI-Linac
17h12-17h24	CT	**Tanya Kairn, L. Jessen, J. Bodnar, P. H. Charles, S.B. Crowe** Royal Brisbane and Women's Hospital, Australia	Lung radiotherapy verification with a 3D printed thorax phantom and an ionisation chamber arrayc
17h24-17h36	CT	**Morgan J Wheatley, and Y De Deene** Liverpool hospital/Ingham institute, Australia	Loss and reintroduction of the radical initiator into the FlexyDos3D silicone dosimeter for 3D printing
17h36-17h48	CT	**Rachael Wilks, S.B. Crowe, P. Chan, R. Cheuk, T. Kairn** Royal Brisbane and Women's Hospital, Australia	A 3D printed patient-specific vaginal mould for brachytherapy
17h48-18h00	CT	**Ryosuke Narita and Yoshinori Sakurai** Kyoto University, Japan	A characterization of dye gel dosimeters for boron neutron capture therapy - dose response to gamma rays
18h00-18h12	CT	**Libing Zhu, Xincheng Xiang, Xiangang Wang** Tsinghua University, Beijing, China	Preliminary research on the dose response of polymer gel dosimeter with carbon beam irradiation

18h15: **DAY 1 CLOSED – FREE TO EXPLORE THE CITY**

DAY 2 – JUNE 21st

8h40		**On Site Registration / Badge pickup**	
9h10-10h30		**Refresher #3** Chair: Luc Beaulieu	
9h10-9h50	IS	**Rongxiao Zhang** Dartmouth University	Cherenkov dosimetry
9h50-10h30	IS	**Sam Beddar** University of Texas, MD Anderson Cancer Center, USA	Scintillation dosimetry
10h30-10h50		**Health Break**	
10h50-12h00		**Proffered session #2 and Vendor's talk #1** Chairs : Shirin A. Enger and Rongxiao Zhang	
10h50-11h05	CT	**Camilla L Nielsen, Rosana M Turtos, Brian Julsgaard, Ludvig P Muren and Peter Balling** Aarhus University, Denmark	Optimizing the transparency of nano-LiF:Cu/silicone nanocomposites for 3D optically stimulated luminescence dosimetry
11h05-11h20	CT	**Daphnée Bernier-Marceau, É. Cloutier, M. Yeh, G. D Orebi Gann and L. Beaulieu k** Université Laval, Canada	Characterization of a water-based liquid scintillator for use in megavoltage radiotherapy beams
11h20-11h35	CT	**Jacob S. Nyemann, Mads L. Jensen, Lia B. Valdetaro, Morten B. Jensen, Rosana M. Turtos, Brian Julsgaard, Jørgen B. B. Petersen, Ludvig P. Muren and Peter Balling** Aarhus University, Denmark	Proof-of-concept photon and proton measurements for large-area 2D dosimeters based on optically stimulated luminescence
11h40-11h50	CT	**François Therriault-Proulx, PhD** CEO and co-founder MedScint Inc.	Vendor's talk
12h00-15h00		**LUNCH ON YOUR OWN – ENJOY A TERASSE!**	
15h20-17h00		**Refresher #4 and Proffered #3** Chair: Peter Balling	
15h20-16h00	IS	**Michelle Hilts** UBC-Okanagan, Canada	CT and CBCT imaging of gel dosimeters: a review of the state of the art
16h00-16h40	IS	**Boyd McCurdy** CancerCare Manitoba, Canada	EPID exit dosimetry and 3D dose reconstruction - New developments and applications

16h40-16h55	IS	**Olivia Moluchi, Matt Mulligan and Kevin Jordan** London Regional Cancer Center, Canada	Bare spherical gel dosimeter with optical computed tomography scanning
17h00-18h00	**Meet the speakers / posters of the day** (Tentative : Beer included!) The five proffered papers will be available at the poster boards to discuss their works. Slides presented will be displayed on the poster boards.		

18h00: **DAY 2 CLOSED - Tentative if budget allowed - Night out event!**

 2630 (2023) 011001 doi:10.1088/1742-6596/2630/1/011001

DAY 3 – JUNE 22nd

8h40	**On Site Registration / Badge pickup**		
9h10-10h30	**Refresher #5** Chair: Sofie Ceberg		
9h10-9h50	IS	**Magdalena Bazalova-Carter** University of Victoria, Canada	A review of small animal dosimetry techniques: image-guided and spatially-fractionated therapy
9h50-10h30	IS	**Shirin A. Enger** McGill UNiversity, Canada	Applied non-std dosimetry : hydrated electron, clinical calorimetry and others
10h30-10h50	**Health Break**		
10h50-12h00	**Proffered session #4** Chairs: Louis Archambault and Anna Karlsson		
10h50-11h05	CT	**Emmanouil. Terzidis, F. Nordström, J. Götstedt, A. Bäck** University of Gothenburg, Sweden	Different aspects of plan complexity in prostate VMAT plans
11h05-11h20	CT	**Tenzin Kunkyab, Michelle Hilts, Andrew Jirasek, Derek Hyde** UBC-O, Canada	Spatial and Dosimetric accuracy of 3D polymer gel with CBCT readout – Varian HyperArc® SRS implementation
11h20-11h40	CT	**Morten Bjoern Jensen, P Balling, H L Riis, A S Bertelsen, S V Jensen, S J Doran, J B B Petersen, L P Muren** Aarhus University, Denmark	Characterisation and application of three-dimensional silicone-based radiochromic dosimetry in 1.5 T magnetic resonance imaging-guided radiotherapy
11h40-11h55	CT	**Kalin I. Penev, Kibret Mequanint** Western Ontario, Canada	Rheology modifiers in Fricke - gelatin dosimeters
12h00-13h00	**Lunch**		
13h00-16h00	**DOSIMETRY UNDER EXTREME CONDITIONS** Chairs : John Schreiner and Magdalena Bazalova-Carter		
	*****COMP registered delegates are welcome to attend this session**		
13h00-13h40	IS	**Arman Sarfehnia** University of Toronto-Odette Cancer Centre, Canada	MR-Linac dosimetry: current approaches and challenges
13h40-14h20	IS	**Sofie Ceberg** Lund University, Sweden	FLASH radiotherapy and the associated dosimetric challenges
14h20-15h00	IS	**James Renaud** NRCC Ottawa, Canada	The role of standards laboratories in reducing uncertainty in clinical dosimetry
15h00-15h20	**Health Break**		
15h20-15h50	**Panel discussions with all five invited speakers of the day** Chair: Geoff Ibbott		

Journal of Physics: Conference Series **2630** (2023) 011001 doi:10.1088/1742-6596/2630/1/011001

15h50-16h50	**Meet the speakers / posters of the day** (Tentative : Beer included!) The four proffered papers will be available at the poster boards to discuss their works. Slides presented will be displayed on the poster boards.		
17h30	**Joint IC3Ddose – COMP session Welcome Remarks**		
18:00	**Joint IC3Ddose – COMP Keynote speaker** Chairs: Luc Beaulieu and Wayne Beckham		
	KL	**L. John Schreiner** President, Medical Physics for World Benefit, Emeritus professor, Queen's University	$3D^2$: Three Decades of 3D Dosimetry

*** Will be followed by a joint IC3Ddose – COMP welcome social event

DAY 3 CLOSED

Journal of Physics: Conference Series **2630** (2023) 011001 doi:10.1088/1742-6596/2630/1/011001

DAY 4 – JUNE 23rd

8h30-9h30	IT	**COMP Panel (in COMP meeting room)**	FLASH in Radiotherapy - Mechanisms,Technology, Dosimetry and Delivery
9h40-11h20	**Refresher #6 (Back in IC3Ddose room)** Chair: Louis Archambault		
9h40-10h20	IS	**Jerry Battista** Western University, Canada	Evolution of 3D X-Ray Dose Computation Algorithms
10h20-10h40	**Health Break**		
10h40-11h20	IS	**Geoffrey Ibbott** MD Anderson Cancer Center, USA	Proton Dosimetry
11h20-12h00	**Awards, futurs meetings and finale** **Chair: Luc Beaulieu** **Vendor's talks : MedScint** **Discussion on proposals for the 13th edition of the IC3Ddose (2024)** Final words and closing of the 12th IC3Ddose		
12h00-13h00	**Lunch**		

END OF DAY 4 AND IC3DDOSE MEETING CLOSED

12h50-:	**IC3Ddose participants are invited to attend Jack Cunnigham Young Investigator Symposium (COMP YIS).** **Access to COMP meeting rooms with IC3Ddose badge if not fully registered for COMP meeting.**

Journal of Physics: Conference Series **2630** (2023) 011001 doi:10.1088/1742-6596/2630/1/011001

A WARM THANK YOU TO OUR SPONSORS

Commercial Sponsors

Elekta

varian
A Siemens Healthineers Company

Start-up Sponsor

medscint

Association Support

COMP Canadian Organization of Medical Physicists
OCPM Organisation canadienne des physiciens médicaux

AQPMC Association québécoise des physicien(ne)s médicaux cliniques

CERCLE DES AMBASSADEURS DE QUÉBEC

Academic Sponsors

UNIVERSITÉ LAVAL
Département de physique, de génie physique et d'optique

CENTRE DE RECHERCHE SUR LE CANCER

CHU de Québec

AXE ONCOLOGIE

Journal of Physics: Conference Series **2630** (2023) 011002 doi:10.1088/1742-6596/2630/1/011002

Peer Review Statement

All papers published in this volume have been reviewed through processes administered by the Editors. Reviews were conducted by expert referees to the professional and scientific standards expected of a proceedings journal published by IOP Publishing.

- **Type of peer review:** Undefined
- **Conference submission management system:** Morressier
- **Number of submissions received:** 42
- **Number of submissions sent for review:** 42
- **Number of submissions accepted:** 38
- **Acceptance Rate (Submissions Accepted / Submissions Received × 100):** 90.5
- **Average number of reviews per paper:** 2
- **Total number of reviewers involved:** 18
- **Contact person for queries:**
  Name: Clive Baldock
  Email: c.baldock@westernsydney.edu.au
  Affiliation: Western Sydney University

Journal of Physics: Conference Series **2630** (2023) 012001 doi:10.1088/1742-6596/2630/1/012001

3D^2: Three Decades of Three-Dimensional Dosimetry

L J Schreiner
Departments of Oncology and Physics, Queen's University, Kingston, Ontario, Canada K7L 3N6

E-mail: ljs2@queensu.ca

Abstract. The development of three-dimensional (3D) dosimetry was motivated by its promise as an effective methodology for the validation of the complex dose distributions achieved by modern techniques such as Intensity Modulated and Volumetric Arc Radiation Therapy. 3D techniques were first proposed in the 1980s when clinics were just starting to move from two-dimensional contour plan based delivery to more conformal techniques. Advances in dosimeter materials, readout systems, and workflow and software systems for the registration and analysis of the volumetric dose data have made 3D dosimetry more attainable, yet to date it has not made major inroads into the clinic. This keynote address will highlight some 3D dosimetry developments over the years, many of which were first revealed through the past twelve International Conferences on 3D and Advanced Radiation Dosimetry (IC3Ddose). These conferences resulted in the publication of more than 130 didactic review articles and over 650 proffered research papers, the majority openly available on the internet (years before the current drive to publishing in open access journals). In this joint keynote address to the IC3Ddose community (at the end of its conference) and to attendees of the Canadian Organization of Medical Physicists annual scientific meeting (at its start), I will briefly review the radiation sensitive materials used for 3D dosimetry, the imaging systems required to read out the volumetric dose information, the workflows and systems required for efficient analysis, and the protocols required for reproducible dosimetry, and how the dosimetry has come into the clinic. The address includes some personal reflections of the motivational and practical changes in 3D dosimetry over time. And as we are all meeting in person for the first time in over two years, the address will end with some observations on the importance of conferences for the exchange of ideas and associated debate necessary for scientific advancement.

1. Introduction

This year's meeting of the IC3Ddose community is the 12th in a series of conferences on three dimensional and advanced clinical dosimetry [1–11]. Therefore, it is fitting in this keynote address to review three decades of three-dimensional (3D) radiation dosimetry through the lens of these conferences. The field effectively started with Gore's seminal paper in 1984 [12] and then developed into an active discipline as the medical physics community addressed the challenges in dose measurement associated with the implementation of conformal radiation delivery techniques. These techniques which better limit the irradiated volume to the tumour were foreseen as early as the 1950s [13] but came into wide clinical practice in the 1980s and 1990s with advances in imaging, treatment planning, and external beam techniques such as Intensity Modulated Radiation Therapy (IMRT). The move to conformal dose delivery and planning was challenging and initial clinical implementation of IMRT took considerable effort. The careful investigations of using multileaf collimators (MLCs)

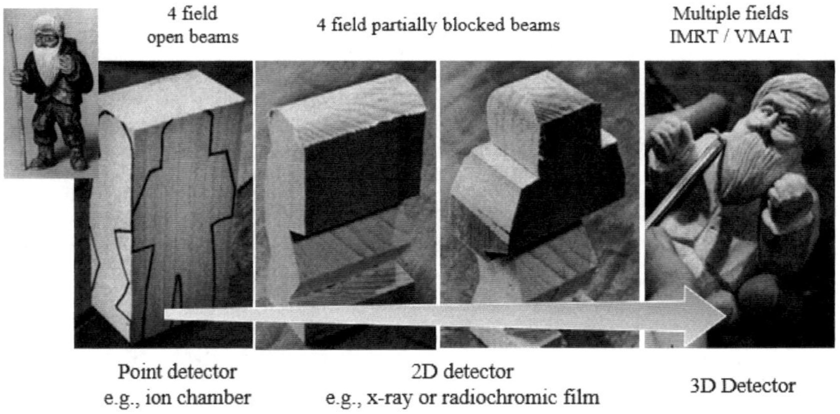

Figure 1. An illustration of the advancement of radiotherapy dose delivery over the last four decades to better treat a complex shaped target (upper left inset). The treatments have progressed from simple uniform doses delivered to the target and surrounding normal tissues as with a 4-field box, through treatments with individual ports conformally blocked to reduce normal tissue irradiation, to fully conformal dose delivery achievable with dynamic IMRT and VMAT deliveries. These dose delivery advances were anticipated from the 1980s onwards and it was thought that dose delivery validation would require commensurate advances in radiation dosimetry techniques which motivated the development of 3D dosimetry (as indicated in the text at the bottom of the figure).

rather than poured blocks for field delineation [14,15], of the impact of MLC leaf edges on treatment planning [16,17], and the introduction and validation of dose calculation algorithms able to predict dose from dynamic photon fluence patterns, etc., are mainly long forgotten. Within the next decade the use of conformal techniques increased in the clinic through Volumetric Modulated Arc Therapy (VMAT), Stereotactic Ablative and Fractionated Stereotactic Radiation Therapy (SABR and FSRT), and through improved procedures for high-dose rate brachytherapy. Of course, radiation therapy using these advanced delivery techniques would only give the intended benefit if resulting dose distributions correctly registered to the patient's anatomy. Thus, there was a heightened requirement for quality assurance (QA) both for technical components of the delivery (including the treatment planning systems, the treatment unit and its associated on-line imaging systems), and for the validation of the dose delivery planned for individual patients. The requirements for delivery validation motivated much of the initial interest in 3D dosimetry especially during the early stages of IMRT development when the main dosimetry was predominately with point detector ion chambers (see figure 1). The possibility of measuring actual IMRT delivered doses in 3D in homogeneous phantoms under standard irradiation conditions and in anthropomorphic phantom tests mimicking patient conditions was particularly attractive for IMRT implementation [18,19]. With time other tools were developed to provide dose delivery measurement and validation. Radiochromic film [20], 2D and 3D arrays of point detectors [21–23], electronic portal imaging detector (EPID) systems [23–26], scintillator dosimeters [27] and Cherenkov based dosimetry [28] became clinically very useful. But these dosimetries typically provide sparse 3D data and only surrogate validation of 3D dose delivery and are not considered full 3D dosimeters.

2. 3D Dosimetry

The features of ideal "true" 3D dosimeters were set by Oldham through his RTAP criteria [29]. Under RTAP a 3D dosimetry system (dosimeter and associated readout) would provide dose measurements throughout a 3D volume with 1 mm isotropic spatial resolution in less than one hour with 3% accuracy and a precision of 1%. Occasionally the criteria for the resolution, accuracy and precision have been relaxed in clinic practice, depending on the specific validation being performed (for example, in external dose delivery audits under IROC [30,31]). Nevertheless, the criterion for high resolution isotropic

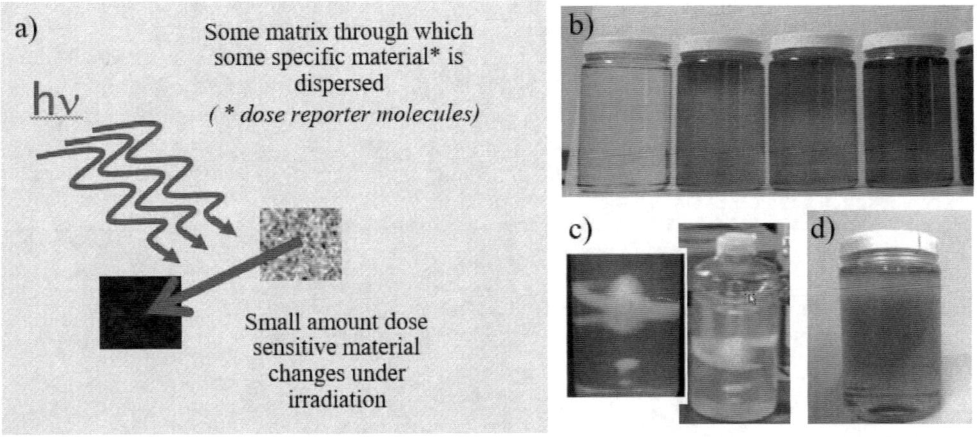

Figure 2. a) Three dimensional dosimeters are chemical systems with some specific molecular constituents that change under irradiation. The degree of change is dose dependent. Three different hydrogel dosimetry systems are shown in b) to d). b) Four different ferrous xylenol orange gel (FXG) dosimeters; the three dosimeters on the right have been irradiated by electron beams of increasing energy. The spatial information in these Fricke gels will disappear within a few hours because of diffusion of the indicator ferrous and ferric ions. c) Two views of a poly-acrylamide (PAG) polymer gel dosimeter after a Cobalt-60 beam irradiation. The photograph on the left was taken soon after the dose delivery in 1998, the image on the right was taken over two decades later in 2021. d) A radiochromic leucodye micelle gel dosimeter irradiated with a 6MV external beam delivery mimicking a prostate treatment. The photograph was taken a few weeks after the irradiation.

measurement limits full 3D radiation dosimetry to chemical radiation dosimetry based on quantifying the effects of radiation-induced chemical changes occurring within some volume of material [29,32,33]. The clinical applicability of 3D dosimetry has advanced considerably in the last decades through the development of improved dosimeters [33] (e.g., radiochromic plastics [34,35], radiochromic gel dosimeters [36,37] and normoxic and less toxic polymer gel systems [38,39]) and by improved readout protocols using optical computed tomography (CT) [40–45], x-ray CT [46,47], or magnetic resonance imaging[48,49]. These advances were often first reported in the proceedings of the DosGel and IC3Ddose International Conferences on 3D Radiation Dosimetry [1–11]. Moreover, the 135 invited review papers and 651 submitted proffered papers therein (including the material in these proceedings, see table 1) provide a detailed account of the evolution of three-dimensional dose measurement [50].

Each conference has included a strong didactic component with invited papers that provide the fundamental science and basic mechanisms that guided the development and response of various 3D dosimetry systems. The reviews extend also to various 3D readout techniques, describing the properties of the irradiated materials enabling dose measurement with the various imaging modalities, and providing details that inform readers of the imaging protocols they must adopt to achieve reproducible and accurate dose readout. Review and proffered papers also discuss the data analysis and data workflow integral in 3D dosimetry. Initially the focus was on data analysis concentrating on the scientific fundamentals of converting readout to dose, with emphasis on uncertainty and error analysis. Eventually the reports evolved more to description of the analytic tools required for fast efficient data analysis and for the comparison of 3D measurements to the dose distributions from treatment planning [51–53]. These reviews, and the proffered papers throughout the proceedings, provide multiple reports on practical issues in 3D dosimetry and offer a compendium of best practices for dosimeter preparation (including setting timescales for various steps in the preparation of the dosimeters, periods between sample preparation and irradiation, and then time to readout [33]) and include clear caveats and practices to avoid. The proceedings offer the medical physics community a valuable resource highlighting the importance of developing and using careful dosimetry protocols to ensure consistent

Table 1. A historical review of the twelve International Conferences on 3D and Advanced Radiation Dosimetry originally instigated as the DosGel conferences to provide a forum for researchers developing 3D dosimetry. The meetings were renamed in 2010 to acknowledge that a full complement of radiation dosimetry technologies is required in clinical dosimetry. Ten of the proceedings have been published in the refereed IOP open access Journal of Physics: Conference Series, the first two proceedings are available by request from the author. These proceedings offer a comprehensive summary of developments in the field, each proceedings offering didactic review articles on select topics and proffered papers presenting initial reports of many novel advances in 3D and other advanced dosimetry over the conferences' three decades.

Proceedings	Meeting	Location	Review papers	Proffered papers
Volume TBA **2023**	12th International Conference on 3D Radiation Dosimetry (IC3DDose), June 20-23, 2022.	Quebec City, Quebec, Canada	13	25
Volume 2167[11] **2022**	11th International Conference on 3D Radiation Dosimetry (IC3DDose), May 10-13, 2021.	Quebec City, Quebec, Canada (virtual)	-	34
Volume 1305[10] **2019**	10th International Conference on 3D Radiation Dosimetry (IC3DDose), September 16–19, 2018.	Kunshan, China	12	60
Volume 847 [9] **2017**	9th International Conference on 3D Radiation Dosimetry, November 7-10, 2016.	Galveston, Texas, USA	9	63
Volume 573 [8] **2015**	8th International Conference on 3D Radiation Dosimetry (IC3DDose), September 4-7, 2014.	Ystad, Sweden	12	79
Volume 444 [7] **2013**	7th International Conference on 3D Radiation Dosimetry (IC3DDose), December 4-8, 2012.	Sydney, Australia	17	93
Volume 250 [6] **2010**	IC3DDose: The 6th International Conference on 3D Radiation Dosimetry, August 22-26, 2010.	Hilton Head, South Carolina, USA	21	80
Volume 164[5] **2009**	DosGel: Fifth International Conference on Radiotherapy Gel Dosimetry, September 29 - October 3, 2008.	Hersonissos, Crete, Greece	10	55
Volume 56 [4] **2007**	DosGel: Fourth International Conference on Radiotherapy Gel Dosimetry, August 7–10, 2006.	Sherbrooke, Quebec, Canada	11	46
Volume 3 [3] **2004**	DosGel: Third International Conference on Radiotherapy Gel Dosimetry, September 13–16, 2004.	Gent, Belgium	12	39
2001 [2]	DosGel01: Second International Conference on Radiotherapy Gel Dosimetry, November 18-21, 2001.	Brisbane, Australia	9	41
1999 [1]	DosGel99: International Conference on Radiotherapy Gel Dosimetry, July 21-23,1999.	Lexington, Kentucky, USA	9	36
	Available in open access journal: *J.Phys.:Conf.Ser.*		117	574
		Total	135	651

and reproducible validation of clinical dose delivery. These advances enabled multiple groups to present examples of the use of 3D dosimetry in the technical and clinical implementation of new techniques and treatments within their clinics [50].

3. 3D dosimetry and more: lessons for COMP from DosGel and IC3Ddose

While the proceedings have from the onset [54] presented a strong motivation for the clinical use of true 3D dosimetry, they have signalled that the needs of the clinic [19, 55–58] extend the required dosimetry tools to beyond only 'true' 3D dosimeters [59,60]. Other dosimetry techniques are more efficient and effective in certain roles ([61] and Table 1 in ref.[50]). The meetings were renamed from '*DosGel'* to '*IC3Ddose'* to explicitly recognize that different dosimetry techniques are needed in the clinic. This extended the conference focus and encouraged attendance to a wider community with expertise in film, EPID based [23–26], scintillation [62,63], Cherenkov [28] and point array measurements [21,64]. This evolution enabled a more critical assessment of the role of the different systems and how they complement each other [64,65].

If the legacy of IC3Ddose was only in its multiple proceedings, then one might condense their impact to a simple technical description of the various 3D and other clinical tools available for dose delivery validation. A truer analysis would have to acknowledge that proceedings are only part of what a conference provides to a community. For example, multiple IC3Ddose presentations during the meetings have reported on the successful use of a particular dosimetry technique in some particular QA test or treatment verification, but questions and discussion after the presentations often deliberated whether the tests were appropriate, or if other techniques would have been better suited for the task. Hence, there was a scientific exchange during the meeting itself that benefited attendees and influenced further progress. This suggests some important closing points to consider in this keynote address as the current IC3Ddose meeting concludes and the Canadian Organization of Medical Physicists annual scientific meeting starts.

The first point is the technical challenge for the medical physicist that presents itself after nearly three decades of IC3Ddose experience: which particular dosimetry system should be used for a particular test in the clinic. The answer must take into consideration many components: including determining whether it is suitable for the specific radiation delivery; whether another dosimeter is simpler to use with measurements more easily analysed and interpreted; whether or not the software for a particular device (especially when commercially supplied) is well documented so that the output can be validated independently; whether the dosimetry system is readily commissioned and can be validated against other dosimetry systems already well-established in your clinic, etc. This leads to a first point (which has been called Schreiner's First Commandment [58]): '*know and understand your dosimetry system completely, including its limitations, before applying it to a particular validation task*' [61].

The second point is communal, recognizing the importance of regular interaction for scientific progress [66] and has been represented as Schreiner's Second Commandment [58]: 'engage in the clinical and scientific exchange of ideas and knowledge through publication in scientific journals, and, *perhaps more importantly, through regular communication, meetings and workshops with colleagues locally, nationally and internationally'* [61]. This commandment is perhaps especially relevant at our meetings in Quebec City as we have come from two years of scientific separation during COVID-19.

4. Conclusions

This keynote address to attendees of the IC3Ddose and COMP conferences is a personal review of the evolution of 3D dosimetry over nearly three decades. The review highlights the influence of the IC3Ddose meetings in this evolution. The work reported over four days at IC3Ddose 2022 in Quebec City and for four more days at the COMP AGM is important as it has considerable influence on the clinical success of radiation therapy [67,68].

5. Acknowledgments

The ideas presented in this address have been formed through research funded through Canadian CIHR project MOP-115101, CHRP projects CIHR CPG 151964 and NSERC CHRP 508528-17, and through many years of interaction with many colleagues in the IC3Ddose and COMP communities. I especially thank Kim McAuley, Tim Olding, Kevin Alexander, John Miller and Clive Baldock who could not make the meeting this year.

6. References

[1] 1999 *Proc. DosGel 1999: 1st Int. Workshop on Radiation Therapy Gel Dosimetry* Audet C and Schreiner LJ (eds) Edmonton, AB: COMP
[2] 2001 *Proc.DosGel 2001: 2nd Int. Conf. on Radiotherapy Gel Dosimetry* Baldock C and De Deene Y (eds) Brisbane, Australia: Queensland University of Technology
[3] 2004 *Proc. DosGel 2004: 3rd Int. Conf. on Radiotherapy Gel Dosimetry* C Baldock and Y De Deene (eds) *J. Phys.: Conf. Ser.* **3**
[4] 2006 *Proc. DosGel 2006: 4th Int. Conf. on Radiotherapy Gel Dosimetry* Lepage M (ed) *J. Phys.: Conf. Ser.* **56**
[5] 2009 *Proc. of DosGel 2008: 5th Int. Conf. on 3D Radiotherapy Dosimetry* Maris TG and Pappas E (eds) *J. Phys.: Conf. Ser.* **164**
[6] 2010 *Proc. of IC3DDose 2010: 6th Int. Conf. on 3D Radiation Dosimetry* Oldham M and Newton J (eds) *J. Phys.: Conf. Ser.* **250**
[7] 2013 *Proc. of IC3DDose 2012: 7th Int. Conf. on 3D Radiation Dosimetry* Thwaites D and Baldock C (eds) *J. Phys.: Conf. Ser.* **444**
[8] 2015 *Proc. of IC3DDose 2014: 8th Int. Conf. on 3D Radiation Dosimetry* Bäck S and Olsson L E (eds) *J. Phys.: Conf. Ser.* **573**
[9] 2017 *Proc. of IC3DDose 2016: 9th Int. Conf. on 3D Radiation Dosimetry* Ibbott G (ed) *J. Phys.: Conf. Ser.* **847**
[10] 2019 *Proc. of IC3DDose 2018: 10th Int. Conf. on 3D Radiation Dosimetry* Oldham M and Yin FF (eds) *J. Phys.: Conf. Ser.* **1305**
[11] 2022 *Proc. of IC3DDose 2021: 11th Int. Conf. on 3D Radiation Dosimetry* Schreiner LJ, Archambault L and Beaulieu L (eds) *J. Phys.: Conf. Ser.* **2167**
[12] Gore J C and Kang Y S 1984 *Phys. Med. Biol.* **29** 1189–97
[13] Wright KA *et al* 1959 *Radiology* **72** 101
[14] Cheng CW *et al* 2001 *Int. J. Cancer* **96** 385–95
[15] Huq MS *et al* 2002 *Phys. Med. Biol.* **47** N159-70
[16] Cadman P *et al* 2005 *J. Appl. Clin. Med. Phys.* **6** 74–86
[17] Cadman P *et al* 2002 *Phys. Med. Biol.* **47** 3001–110
[18] Schreiner LJ 1999 In *Proc.1st Int. Workshop of Radiation Therapy Gel Dosimetry* Audet C and Schreiner L J (eds) Edmonton, AB: COMP
[19] De Wagter C 2004 *J. Phys.: Conf. Ser.* **3** 4–8
[20] Devic S 2011 *Physica Medica* **27** 122–34
[21] Gutiérrez AN and Calvo O 2010 *J. Phys.: Conf. Ser.* **250** 12049
[22] Low DA 2015 *J. Phys.: Conf. Ser.* **573** 012009
[23] Bäck A 2015 *J. Phys.: Conf. Ser.* **573** 012012
[24] Mijnheer B *et al* 2013 *J. Phys.: Conf. Ser.* **444** 012011
[25] Mijnheer B *et al* 2010 *J. Phys.: Conf. Ser.* **250** 12020
[26] McCurdy BM 2013 *J. Phys.: Conf. Ser.* **444** 12002
[27] Beaulieu L *et al* 2013 *J. Phys.: Conf. Ser.* **444** 012013
[28] Pogue BW *et al* 2017 *J. Phys.: Conf. Ser.* **847** 012046
[29] Oldham M 2014 *Methods and Techniques for Comprehensive 3D Dosimetry* ed SKD and ABW Devon J. Godfrey (Madison, WI: Medical Physics Publishing)
[30] Ibbott GS and Thwaites D I 2015 *J. Phys.: Conf. Ser.* **573** 12002
[31] Ibbott GS 2010 *J. Phys.: Conf. Ser.* **250** 012001
[32] Schreiner LJ and Olding T 2009 in *AAPM Medical Physics Monograph: Clinical Dosimetry Measurements in Radiotherapy* **34**, Roger DW and Cygler JE (ed) (Madison, WI: Medical Physics Publishing) pp 979–1026
[33] Schreiner LJ 2015 *J. Phys.: Conf. Ser.* **573** 012003
[34] Adamovics JA *et al* 2006 *J. Phys.: Conf. Ser.* **56** 172–5
[35] Oldham M 2015 *J. Phys.: Conf. Ser.* **573** 12006

[36] Jordan KJ 2010 *J. Phys.: Conf. Ser.* **250** 12043
[37] Babic S *et al* 2009 *J. Phys.: Conf. Ser.* **164** 012044
[38] Fong DC *et al* 2001 In *Proc. 2nd Int. Workshop of Radiation Therapy Gel Dosimetry* Baldock C (ed) Brisbane, AU: Queensland University of Technology
[39] Senden RJ *et al* 2006 *J. Phys.: Conf. Ser.* **56** 156–9
[40] Jordan KJ 2004 *J. Phys.: Conf. Ser.* **3** 115–21
[41] Jordan KJ 1999 In *Proc.1st Int. Workshop of Radiation Therapy Gel Dosimetry* Audet C and Schreiner L J (eds) Edmonton, AB: COMP
[42] Oldham M 2004 *J. Phys.: Conf. Ser.* **3** 122–35
[43] Oldham M 2006 *J. Phys.: Conf. Ser.* **56** 58–71
[44] Doran SJ and Krstajić N 2006 *J. Phys.: Conf. Ser.* **56** 45–57
[45] Doran SJ 2009 *J. Phys.: Conf. Ser.* **164** 12020
[46] Hilts M 2006 *J. Phys.: Conf. Ser.* **56** 95–107
[47] Jirasek A and Hilts M 2009 *J. Phys.: Conf. Ser.* **164** 12038
[48] de Deene Y 2001 In *Proc. 2nd Int. Workshop of Radiation Therapy Gel Dosimetry* Baldock C (ed) Brisbane, AU: Queensland University of Technology
[49] De Deene Y 2010 *J. Phys.: Conf. Ser.* **250** 12015
[50] Schreiner LJ 2017 *J. Phys.: Conf. Ser.* **847** 012001
[51] Kozicki M and Maras P 2015 *J. Phys.: Conf. Ser.* **573** 12070
[52] Kozicki M *et al* 2014 *Phys. Med. Biol.* **59** 4111–36
[53] Alexander KM *et al* 2015 *J. Phys.: Conf. Ser.* **573** 012042
[54] Schreiner LJ 1999 In *Proc.1st Int. Workshop of Radiation Therapy Gel Dosimetry* Audet C and Schreiner L J (eds) (Edmonton, AB: COMP) 1-10
[55] Kron T 2001 In *Proc. 2nd Int. Workshop of Radiation Therapy Gel Dosimetry* Baldock C (ed) (Brisbane, AU: Queensland University of Technology) 2-9
[56] Yin F-F *et al* 2010 *J. Phys.: Conf. Ser.* **250** 12002
[57] Low DA 2015 *J. Phys.: Conf Ser* **573** 012009
[58] Knöös T 2015 QA *J. Phys.: Conf. Ser.* **573** 12001
[59] Schreiner LJ 2006 *J. Phys.: Conf. Ser.* **56** 1–13
[60] Schreiner LJ 2009 *J. Phys.: Conf. Ser.* **164** 12001
[61] Schreiner LJ 2011 *J. Med. Phys.* **36** 189-91
[62] Beaulieu L *et al* 2013 *Phys.: Conf. Ser* **444** 012013
[63] Guillot M *et al* 2010 *J. Phys.: Conf. Ser* **250** 012006
[64] Ceberg S *et al* 2010 *J. Phys.: Conf. Ser.* **250** 12051
[65] Olding T *et al* 2013 *J. Phys.: Conf. Ser.* **444** 012052
[66] Schreiner LJ *et al* 2022 *Phys. Eng. Sci. Med.* **45** 413–7
[67] Peters LJ *et al* 2010 *J. Clin. Oncol.* **28** 2996–3001
[68] Weber DC *et al* 2011 *Radiother. Oncol.* **100** 150–6

Journal of Physics: Conference Series **2630** (2023) 012002 doi:10.1088/1742-6596/2630/1/012002

Charting the development of optical CT and allied methods for 3-D radiation dosimetry: 2022 update - mapping the global research community

Simon Doran
Division of Radiotherapy and Imaging, The Institute of Cancer Research, London, UK

E-mail: Simon.Doran@icr.ac.uk

Abstract. To recognise the many and varied contributions from groups around the world over more than 25 years of research into 3-D radiation dosimetry using optical imaging (primarily optical computed tomography), a systematic review has been conducted with the aim of charting the development of this field. Some 493 publications matched the criteria of the review, and these have been coded with information relating to the category of research, the type of dosimeter material used and the geographical location where the research was conducted. The results of the analysis demonstrate a steady growth in the number of academic groups involved, from 4 at the first DOSGEL meeting in 1999 to over 30 in 2016. The important role played by the DOSGEL and IC3Ddose conference series is demonstrated by the marked biennial trend in publications. The division of publications between categories was 36% materials development, 34% scanner technology, 24% applications, 5% review and 1% teaching. The most common application areas were brachytherapy, proton/particle therapy and IMRT/VMAT. Prior to the introduction of PRESAGE® in 2003, polymer gels were the most popular material, whilst radiochromic dosimeters have dominated in recent years. These and other trends are discussed on the basis of the data presented. A brief discussion of the future of the field, including some personal opinions, ends this brief review.

1. Introduction

With the series of conferences inaugurated by the First International Workshop on Gel Dosimetry (DOSGEL '99 held in Lexington, Kentucky) now well into its third decade, it seems appropriate to attempt to provide an overview of what has become a significant body of work in the area of optical computed tomography (CT). A previous review article [1], prepared for the DOSGEL conference in 2008 (Hersonissos, Crete) aimed to capture the state of a rapidly-developing field; in 2010, optical CT was described briefly as part of a broader review of polymer gel dosimetry [2]; and in 2017, Khezerloo *et al.* reviewed the use of the radiochromic dosimeter PRESAGE® [3, 4]. However, none of these publications were formal, systematic reviews of the literature.

The work presented here adopts the systematic approach in order to chart the growth of the field from an originally small number of pioneering laboratories to its current state, to acknowledge the many groups around the world who have played a role in its development, and to provide quantitative bibliometric indicators of the distribution of effort in the various sub-fields.

I closed my previous DOSGEL review in 2008 with the words "Optical CT has now arrived" and the work conducted here confirms that, in the 12 years since that article, this statement has been borne out.

2. Review methodology

2.1. Review protocol

Following the overall outline set down in Figure 1 of Xiao and Watson [5] - itself the product of a systematic review of best practice - the procedure below was adopted.

- Formulate the problem and define the research questions.
- Search the literature:
  - Start with five "seed" datasets.
  - Combine and de-duplicate the "seed" datasets.
  - Screen the combined dataset according to the inclusion criteria.
  - Use Google Scholar iteratively (see below) to discover further data, employing a stopping criterion.
- Assess quality: no formal assessment was performed, because the research aim was to discover the full extent of the research community, not to answer questions regarding the accuracy or efficacy of this type of dosimetry.
- Extract data: the following pipeline was used: Google Scholar → reference manager software (EndNote 20, Clarivate Analytics) → spreadsheet (Excel, Microsoft) → analysis software (R v4.0.5 https://www.r-project.org) .
- Analyse and synthesise: data were first coded in Excel according to the headings described below; the dataset was then analysed using RStudio v1.4.1106 (https://rstudio.com) to produce quantitative statistics, which were mapped spatially.

2.2. Definition of research question

The aim of the research was to map the development of the technique of optical CT and allied methods for radiation dosimetry, highlighting the unique role that the DOSGEL and IC3DDose series of biennial conferences has played in advancing the field. From the very start, the meeting organisers aimed to be inclusive, welcoming a diversity of contributions, and this has been reflected in broad inclusion criteria here. To prevent the review from becoming unmanageable in size, the subject matter was restricted to 3-D optical imaging methods applied to "integrating" dosimeters, and the development of the materials used for these. This meant that a number of other exciting and no less important optical methodologies, in adjacent areas such as scintillation imaging and Cerenkov imaging, were excluded from the review. Similarly, the large body of work in gel dosimetry based on Magnetic Resonance Imaging was excluded, even though many of its goals are the same as for optical CT.

2.3. Literature search

The five "seed" datasets were:

- a PubMed query conducted in March 2022 with the following search expression: (("three-dimensional dosimetry") OR ("3-D dosimetry") OR ("3D dosimetry") OR (gel dosimetry) OR ("radiation dosimetry") OR (radiochromic) OR (Fricke) OR (PRESAGE)) AND (("optical computed tomography") OR ("optical CT") OR (spectrophotometer)) **161 entries**
- the reference list of the 2008 DOSGEL review by Doran [6] **123 entries**
- the reference list of the optical CT section of the 2010 review by Baldock *et al.* [2] **29 entries**
- the reference list of the Khezerloo *et al.* review of 2017 [3] **92 entries**
- references matched manually from the combined DOSGEL and IC3Ddose abstract books 1999 – 2021 **149 entries**

The five seed datasets were combined in EndNote 20 (Clarivate Analytics), which removed obvious duplicates. The resultant datasets were screened for inclusion, rejecting articles that did not relate to the research question. In most cases, this was possible simply by inspecting the title, but sometimes reading the abstract or scanning the entire text was necessary to determine relevance. The above steps are easily replicable and would yield close to identical results for all users.

From this point, the study proceeded iteratively, with the aim of discovering the most comprehensive dataset possible within the time available for the project. Extensive use was made of Google Scholar in the following ways:

- keyword search using similar phrases to the PubMed search above — Google Scholar returns an extensive list of hits, of gradually diminishing relevance. For example, the search string "gel dosimetry optical CT" returns 11,000 "hits" (some duplicated) making it, to all intents and purposes, impossible to conduct an exhaustive search. Items were displayed in "screens" each containing 10 references and for each search, the stopping criterion chosen was (n = 50 screens) OR (m = 5 consecutive screens with no "hit" of interest found).
- "backward" searching based on scanning the reference lists of selected articles already discovered;
- "forward" searching using Google Scholar to list all articles citing selected articles already discovered;
- author searches — based on a knowledge of the field, exhaustive searches were performed for key players involved in optical CT research.

This iterative part of the process was more time-consuming but yielded approximately 40% of the final total of **493 references**, demonstrating the importance of the interactive part of the search. By definition, this step is not replicable, not least because the results returned by the Google search engine are tailored to the individual performing the search.

2.4. Data coding and processing
Each publication in the Excel spreadsheet was "coded" with the following additional information:

- the city name associated with the academic group from which the publication came, taken from the affiliation of the lead or senior author as appropriate;
- the longitude and latitude of the city, found using the `geocode` function of R package `ggmap`, which in turn makes use of the Google Cloud Platform Geocoding API;
- flags marking the reference as follows:
  - conference abstract (e.g., IC3Ddose), research article, review or teaching article;
  - category: materials development, scanner technology or applications (some publications fell into more than one category);
  - type of gel;
  - type of application

An R script was written to return the bibliometric results presented below. The `get_googlemap()` function from the `ggmap` library (in turn calling the Google Cloud Platform Static Map API) was used to obtain a world map and the locations of research groups were plotted.

3. Results and Discussion
Figure 1(a) shows the variation with publication year of the number of publications meeting the review criteria, starting in the mid-1970's with early work by Gupta and co-workers (e.g., [7]. Despite the existence of an earlier conference abstract from Tarte *et al.* [8], the first experimental demonstrations of optical CT scanners are generally credited to Gore *et al.* [9] (scanned laser) and Winfree *et al.* (CCD-based tomography system) [10]. From these beginnings, the number of publications rose rapidly, peaking at around 45 per year in 2010. There was a subsequent decline but still significant interest throughout 2010's, followed by a sharp fall in 2020 largely related to the COVID-19 pandemic, although it may also reflect a trend towards other methods of dose verification. 2022 has seen a significant rebound and, once the IC3Ddose contributions (not yet analysed) for 2022 are included, a healthy

resurgence of the field is likely to be found. The biennial trend in abstract publications is strongly influenced by the DOSGEL and IC3Ddose conference series, which have long been an important forum for presenting this type of research. A weakness in the research methodology of this review is that, whilst the *Journal of Physics Conference Series* (where the DOSGEL and IC3Ddose proceedings are published) and *Medical Physics* (where the American Association of Physicists in Medicine conference proceedings are published) are both listed on Google Scholar, other sources of abstract information may not be, and so some relevant material meeting the inclusion criteria may not have been captured, thus biasing the results. Nevertheless, the extended abstract format of DOSGEL/IC3Ddose often provides important information that is not found in full journal articles and so a focus on these sources is valuable. Notice that the peaks shown in green in Figure 1(a) are irregularly spaced, which reflects occasions where proceedings were published in the year following the conference.

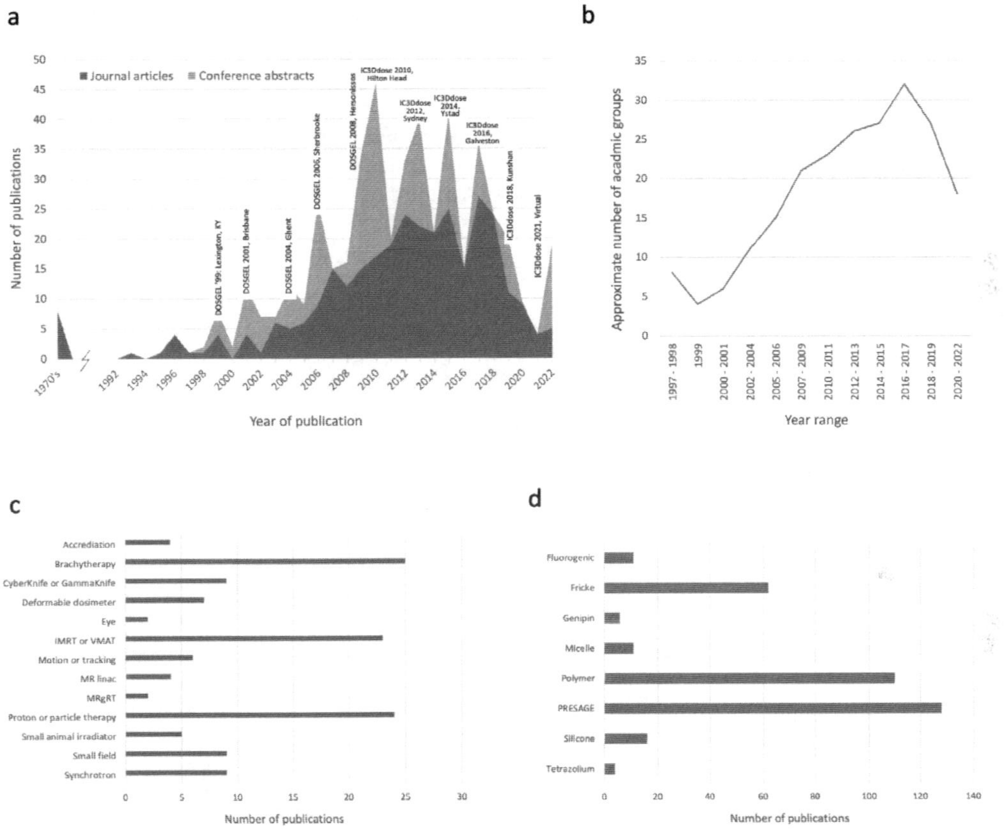

Figure 1: (a) Full journal publications and conference abstracts as a function of year of publication, with labels corresponding to the IC3Ddose conference series added. (b) Number of active research groups in the areas covered by the review vs year range. The left-hand bin represents relevant research prior to the first DOSGEL conference apart from the 1999 bin, all subsequent points include both the year of the conference itself and the year(s) between successive DOSGEL/IC3Ddose conferences. (c) Number of publications in each of the different application areas coded. (d) Number of publications for each of the types of dosimeter material coded.

Figure 1(b) shows the number of groups worldwide engaged in research in scope for the systematic review. This exhibited steady growth during the years from 1999 to 2017 and fell slightly in the 2018 – 2019 period. The subsequent larger fall during the 2020 – 2022 period again likely reflects the global impact on research due to COVID-19 rather than a genuine reduction in number of academic groups.

Year ranges for aggregation of results are not evenly spaced but reflect the timings of the DOSGEL/IC3Ddose conferences, as these have been major drivers of the global research effort. (A known methodological weakness here is that different academic groups in the same city are amalgamated and this will be addressed in future bibliographic work.)

Figure 1(c) presents an overview of the findings related to application area. Three applications dominate the uses described to date: brachytherapy, IMRT/VMAT and proton/particle therapy. Although the numbers involved overall are small (data not shown), in each case, the majority of publications occurred in the decade between 2005 and 2015 with only sporadic work performed since. Recent applications-related publications have primarily been in the areas of small-field treatments and MR linacs. These trends suggest that the radiotherapy community sees 3-D dosimetry as primarily of use in the early phases of introducing new technologies to the clinic and that, once this initial confidence-building phase is complete, other less unwieldy tools are used on a day-to-day basis.

Figure 1(d) shows results concerning the dosimeter type. Following Gupta's work [7] there was strong initial interest in Fricke dosimetry combining ferrous sulphate with xylenol orange and gelatin. For most of the 1990's and early 2000's, the materials most commonly employed were polymer gels, which were first introduced in the context of MRI-based dosimetry [11] and later characterized for optical measurements in [12]. However, since the invention of the PRESAGE® dosimeter in 2003[13], radiochromic dosimeters have once more predominated. Contrary to the situation described above for applications of 3-D dosimetry, research into new dosimeter materials remained strong until 2019, immediately before the pandemic. This suggests, on the one hand, a wealth of ideas and new chemical systems to investigate, but on the other, an acknowledgement that, despite more than 25 years of active research, no existing material fulfils all the criteria necessary for a tool that is easily usable in the standard clinical workflow. Arguably, the difficulty of obtaining either absolute dose quantification or results that are reproducible from batch to batch is a fundamental limitation of chemical methods of dosimetry.

Figure 2 maps the global distribution of research groups discovered by the systematic review for the entire period from 1974 to spring 2022, whilst Figure 3 presents the same information as a series of snapshots leading up to each of the DOSGEL and IC3Ddose conferences. The size of each marker symbol relates to the fraction of total world research output in the relevant time-window as conducted at each of the laboratories, and thus the symbol sizes do not represent the same absolute numbers of publications at different time points. The marker colours reflect the mix of categories (as defined on the inset Maxwell colour triangle) published by individual research groups.

This mode of presentation makes it easy to identify trends in research. For example, North America as a whole (with notable exceptions) has focused largely on scanner technology and applications, while Asia (again with exceptions) has concentrated on dosimeter development. The time series shows the interplay between categories of research. For the first two DOSGEL meetings, global research focused on the development of materials and early generations of optical CT scanners. During the period 2004 - 2008, scanning technology was the dominant theme of the literature, whilst from 2008 - 2014, the overall colour of plotted points changes from red to blue, indicating a readiness of the technology to tackle real-world radiotherapy applications. In more recent years, as limitations of the existing systems have become apparent, there has been a return to further rounds of materials development - with the emergence, for example of flexible dosimeters [14, 15] - novel scanner designs [16] and improved reconstruction techniques [17].

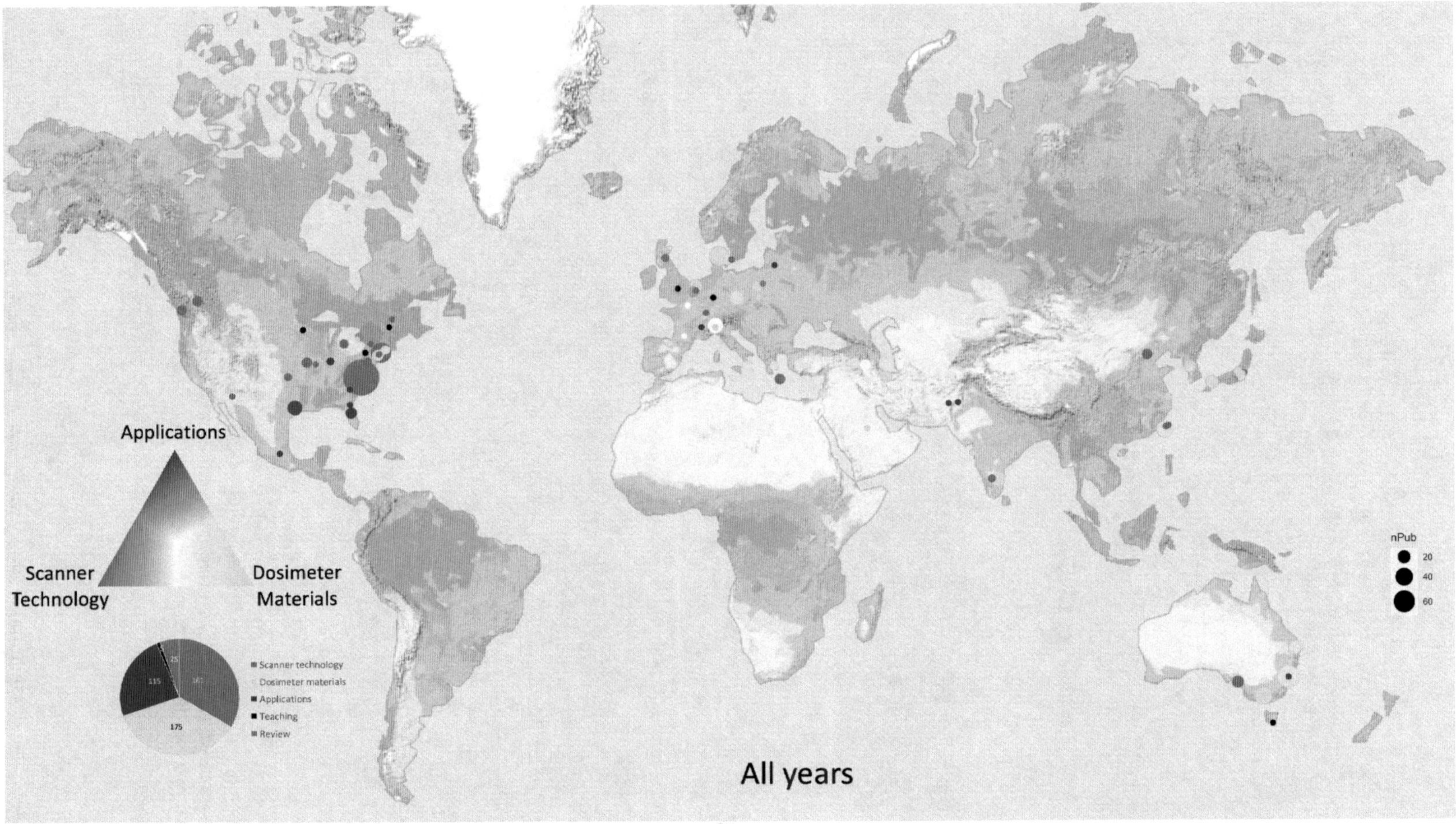

Figure 2: Location of research groups involved in the study of 3-D radiation dosimetry using the optical methods in-scope for this review (primarily optical CT). Each coloured dot has a size reflecting the publication output of the group. Dot colours reflect the mix of publication categories produced by the laboratory, as indicated by the colour triangle key. The inset pie chart shows the overall balance of research across the community between publication categories.

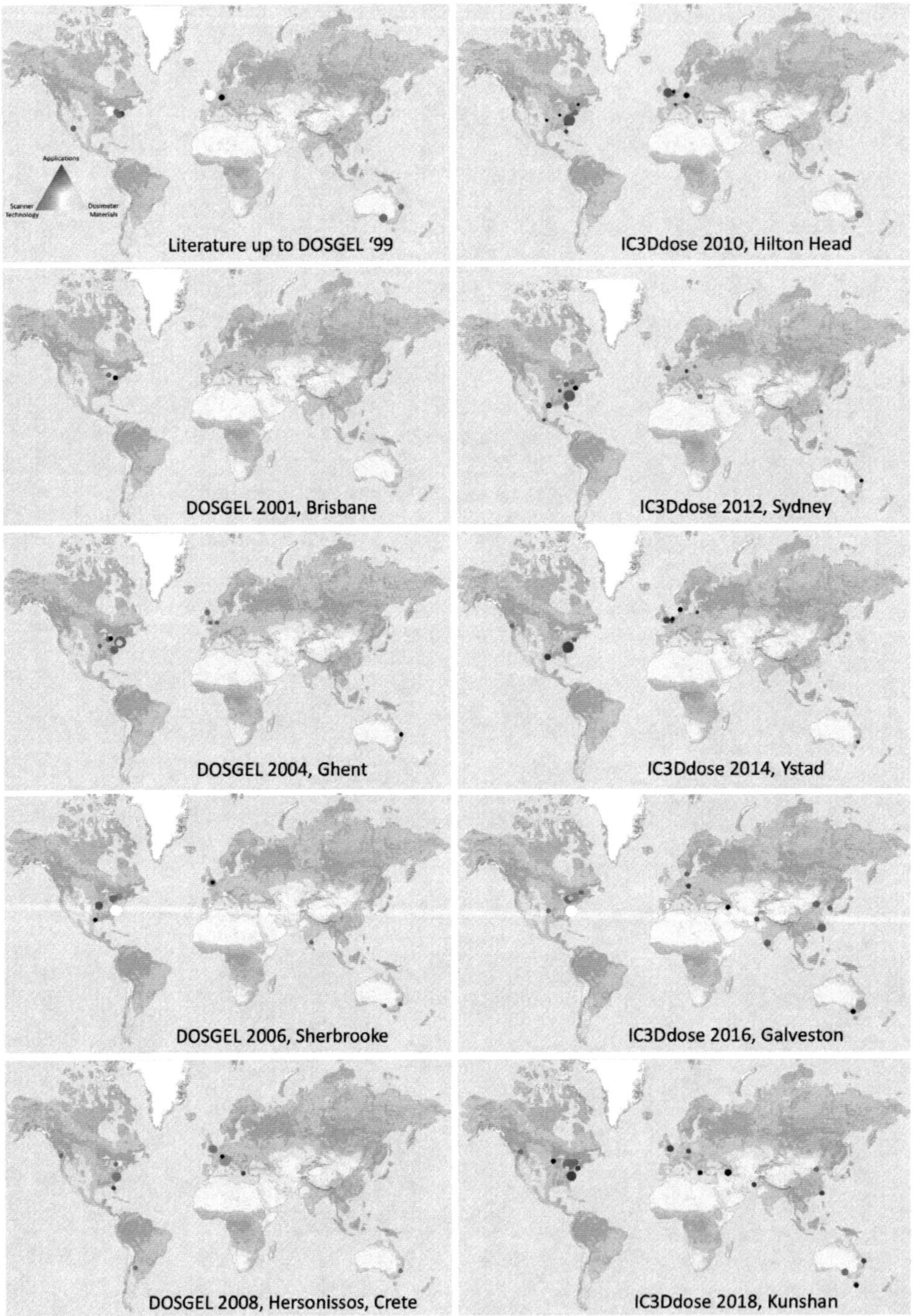

Figure 3: Tine-resolved snapshots of research activity in the field covered in this systematic review. Colour-coding of research categories is the same as for Figure 2. Note, however, that symbol sizes are relative to the total research activity across all centres *for the particular time period* and so a given symbol size reflects different absolute numbers of publications in each plot.

Predicting the future is, as ever, not easy. However, massive current interest in FLASH radiotherapy [18, 19] seems likely to translate into a need for dosimetry tools to monitor treatments that could be revolutionary but also have the potential to be dangerous if delivery errors occur. Previous research in synchrotron microbeam therapy [20] suggests that optical CT imaging of radiochromic plastics will have an important role to play in these high dose-rate scenarios, and I see this as a key application area.

In terms of equipment development, optical CT is still an inconvenient technique, relying as it does on refractive index matching via a liquid-filled tank. Over the years, a variety of studies have investigated the extent to which the index matching medium can be minimised [21] or dispensed with entirely [22, 23] but further progress is still needed towards a rapid-scanning tool usable in the clinic by regular (i.e., non research) medical physicists.

As discussed above, it is this author's belief that the non-reproducibility of dose-response calibrations between dosimeter samples (even from the same batch) will continue to be a major stumbling block in any bid to gain more widespread adoption of optical CT technology. To date, 3-D chemical dosimeters have been single-use devices. If each one continues to require lengthy calibration procedures, or is capable only of relative dosimetry, it will be difficult to incorporate this method into the clinical workflow. Attempts to create reusable chemical dosimeters (e.g., [24, 25]) have so far met with limited success, relying as they do solely on an unstable material response - itself a major problem, limiting flexibility in the timing of the device readout - that decays back to an equilibrium value over time. In this context, one exciting development on the horizon is the introduction of new imaging materials and methods based on optical stimulated luminescence (OSL) [26]. Distributing OSL particles within an elastomeric matrix holds out the tantalising prospect of a flexible and reusable 3-D dosimeter that is reliant on physical properties, rather than chemical reactions, and is erasable via an annealing process.

4. Conclusion

The last 25 years of research into optical CT imaging and related methods have seen a large and diverse literature develop with contributions from more than 30 different groups all around the world. It has become traditional to predict the demise of 3-D dosimetry, only to find that, at each successive IC3Ddose conference, new insights and discoveries revitalise the field, and it is my hope that this year will be no exception.

5. References

[1] Doran SJ 2009 *J. Phys.: Conf. Ser.* **164** 012020
[2] Baldock C *et al* 2010 *Phys. Med. Biol.* **55** R1
[3] Khezerloo D *et al* 2017 *Radiat. Phys. Chem.* **141** 88-97
[4] Khezerloo D *et al* 2017 *J. Cancer Res. Ther.* **13** 419-24
[5] Xiao Y *et al* 2019 *J. Plann. Edu. Res.* **39** 93-112
[6] Doran SJ 2009 *Appl. Radiat. Isot.* **67** 393-8
[7] Gupta B *et al* 1974 *Int. J. Appl. Radiat. Isot.* **25** 509-13
[8] Tarte B *et al* 1995 *Proc. 10th Conf. of Australian Optical Society*
[9] Gore JC *et al* 1996 *Phys. Med. Biol.* **41** 2695-704
[10] Winfree AT *et al* 1996 *Chaos* **6** 617-26
[11] Maryanski MJ *et al* 1993 *Magn. Reson. Imaging* **11** 253-258
[12] Maryanski MJ *et al* 1996 *Phys. Med. Biol.* **41** 2705-2717
[13] Adamovics J *et al* 2003 *Med. Phys.* **30** 1349
[14] Høye EM *et al* 2015 *J. Phys.: Conf. Ser.* **573** 012067
[15] De Deene Y *et al* 2015 *J. Phys.: Conf. Ser.* **573** 012025
[16] Ogilvy A *et al* 2020 *Phys. Med. Biol.* **65** 245012
[17] Dekker KH *et al* 2017 *Med. Phys.* **44** 6678-6689
[18] Bourhis J *et al* 2019 T *Radioth. Oncol.* **139** 18-22
[19] Wilson JD *et al* 2020 *Front. Oncol.* **17** 1563
[20] McErlean CM *et al* 2016 *Phys. Med. Biol.* **61** 320-37
[21] Chisholm K *et al* 2015 *Med. Phys.* **42** 2607-14
[22] Doran SJ *et al* 2012 E *Phys. Med. Biol.* **57** 665-83

Journal of Physics: Conference Series **2630** (2023) 012002 doi:10.1088/1742-6596/2630/1/012002

[23] Ramm D 2018 *Phys. Med. Biol.* **63** 045019
[24] Pierquet M *et al* 2010 *J. Phys.: Conf. Ser.* **250** 012047
[25] Colnot J *et al* 2018 *Physica Medica* **52** 154-64
[26] Nyemann JS *et al* 2022 *J. Phys.: Conf. Ser.* **2167** 012026

Gel dosimetry: MRI

Yves De Deene
[1]School of Science, University of Western Sydney, Sydney, Australia
[2]Fysikum, Stockholm University, Sweden
[3]Cancer Therapy Centre, South Western Sydney Local Health District, Liverpool Hospital, Australia

E-mail: Yves.DeDeene@health.nsw.gov.au

Abstract. While a vast amount of scientific literature is available on the topic of gel dosimetry with MRI readout, this wealth of information may seem at first overwhelming for medical physicists and newcomers in this rapidly evolving field of research. With this review, my mission is to streamline the wealth of information in the scientific literature and provide a quick guideline for those making their first steps in implementing gel dosimetry in a clinical environment, while still providing a lookout to new and emerging evolutions in the field. In a first section, the physical mechanisms behind the MRI contrast are briefly explained for both Fricke gels and polymer gels. In a subsequent section, an overview is given of the different MRI pulse sequences and pulse sequence optimization will be discussed. Emphasize is placed on the framework and formalism to calculate optimal parameters. The reliability of MRI-based polymer gel dosimetry will be discussed, and a quick beginner's guide is provided. Finally, a lookout to new and future developments of polymer gel dosimetry will be given.

1. Introduction

The paradigm of modern radiotherapy to deliver radiation to a target volume with high precision while sparing the surrounding healthy tissues, has been the practical motivation behind the rapidly emerging field of gel dosimetry. Gel dosimeters have never been aimed at replacing other radiation dosimeters, but because of their three-dimensional character they have a unique role to play in end-to-end verification of modern radiotherapy, especially where dose registration with other dosimeters is problematic because of steep dose gradients in all three dimensions [1].

While some early papers in the first half of the 20th century already mentioned the potential of using chemical dosimeters [2], gel systems [3-4] and polymers [5-7] for radiation dosimetry, the field of research into gel systems for 3D radiation dosimetry only took off when a research team at Yale university discovered that the radiation induced oxidation of ferrous (Fe^{2+}) ions in ferrous sulphate dosimeters, initially developed by Fricke and Morse [2], can be read out with NMR relaxometry [8]. A major obstacle for the use of Fricke gel dosimeters in radiotherapy hospitals is the diffusion of the ferrous (Fe^{2+}) and ferric (Fe^{3+}) ions in the gel matrix. Many studies have focussed on reducing the diffusivity of the ions by using different hydrogel materials and by use of chelators. While some methods have been able to reduce the diffusivity [9], none of the methods has been able to completely stop the diffusion. In 1993, the Yale group also suggested the use of polymer gel dosimeters with MRI readout [10]. Polymer gel dosimeters are hydrogels that contain vinyl monomers and often also a cross-linker monomer. When exposed to ionizing radiation, radiation induced water radicals will initiate a polymerization reaction which results in the formation of small polymer aggregates that become entangled with the hydrogel matrix. The drastic change in polymer mobility has an effect on the NMR transverse relaxation rate which can be exploited to acquire absorbed dose related quantitative R_2 maps. While the first dose maps acquired with Fricke gel dosimeters and polymer gel dosimeters demonstrated the potential of 3D dosimetry, gel dosimetry in the early days lacked robustness in many aspects. As the spatial integrity of polymer gel dosimeters appeared superior to the Fricke gel dosimeters, the interest in Fricke gel dosimeters declined over the years.

The challenge of increasing the accuracy and precision of 3D gel dosimetry has led to an active field of multi-disciplinary research which involved contributions from chemists, material scientists, medical physicists, engineers and MRI physicists. Scholars in the field gathered at biennial international conferences under the name "DOSGEL" (1999 – 2008) and "IC3DDose" (from 2010) to exchange ideas and collaborate. The conference proceedings are a valuable source of information on the topic and is comprised of proffered papers and elaborate review papers. Since 2004, the conference proceedings are published by the Institute of Physics [11-19]. A comprehensive review of polymer gel dosimetry has also been published by the journal Physics in Medicine and Biology [20] and in the form of book chapters [1, 9].

The accuracy and precision of polymer gel dosimetry has improved considerably over the years [21-23] as the physico-chemical mechanisms of gel systems become better understood [24-40], the radiation properties of gel dosimeters characterized [41-56], the NMR contrast mechanisms described [9,57-60], imaging artefacts in quantitative MRI compensated [61-71] and MRI pulse sequences optimized [72-74]. However, it is important to remain vigilant about all sources of uncertainty in order to obtain reliable dose maps [22, 75]. In addition to the scientific quest for a better understanding of gel dosimeters, several different gel dosimeters have been proposed and other readout techniques have been proposed, such as X-ray CT [76-78], optical CT [79-84] and ultrasonic imaging [85]. We also refer to the other review talks for in-depth discussions on these techniques. The focus of this paper is on quantitative MRI.

In order to conduct reliable polymer gel dosimetry, several experimental aspects need to be taken care of. The many parameters that need to be controlled in MRI-based polymer gel dosimetry often leads to the perception that polymer gel dosimetry is a time-intensive and labour-intensive technique and the comparison is often made with non-3D faster dosimetry techniques such as film dosimetry, TLD or OSL. While the author will not refute the claim that many aspects are involved in polymer gel dosimetry, the comparison with off-the-shelf dosimeters is not a fair comparison as the fabrication and calibration of other dosimeters is often taken out of the equation. Others also claimed that polymer gel dosimetry with MRI readout would be less accurate than optical CT and refer hereby to the many publications on MRI artefacts and studies of uncertainties in polymer gel dosimetry. A word of caution is here also pertinent as the large scale of studies on the uncertainty of a particular technique is not necessarily correlated with the uncertainty itself. It is the author's opinion that with enough expertise, MRI-based polymer gel dosimetry can be used as a well-established method in the hospital. At the same time, the author also advocates for a better understanding of the principles of MRI among medical radiation physicists.

2. Clinical applications – The Ghent University experience

Polymer gel dosimetry has been proven to have significant impact in radiotherapy treatment validation as is illustrated in figures 1-3 which shows some typical examples of clinical applications that have made significant impact in the Ghent University radiotherapy centre or changed the course of action of the treatment.

Figure 1 shows gel dosimetry in an anthropomorphic shaped phantom for the verification of an intensity modulated radiotherapy (IMRT) class solution for patients with nasopharyngeal carcinoma either as the result of loco-regional relapses or new primary tumours [86]. Gel dosimetry was still in its infancy at that stage (1998-1999) and the PAG gel was deoxygenated by bubbling the gel with nitrogen gas. In this experiment, a PVC head shaped cast was used and some penetration of oxygen through the plastic cast resulted in some inhibition near the cast. Despite, these physico-chemical artefacts, gel dosimetry experiments like the ones shown in figures 1-2 helped in establishing IMRT and gel dosimetry at the Ghent University Hospital and in benchmarking and creating confidence in the treatment planning. The treatment planning software was a beta-release of GRATIS® that was developed by George Sherouse [87] and was systematically complemented by in-house software [88-89] to accommodate IMRT treatment optimization. In the process of implementing IMRT in the centre, gel dosimetry was used next to a variety of other dosimeters such as TLD's and radiographic film in combination with the Rando phantom. Gel dosimetry was recognised for its full 3D capability that is perceived as crucial because of steep dose gradients in all three dimensions.

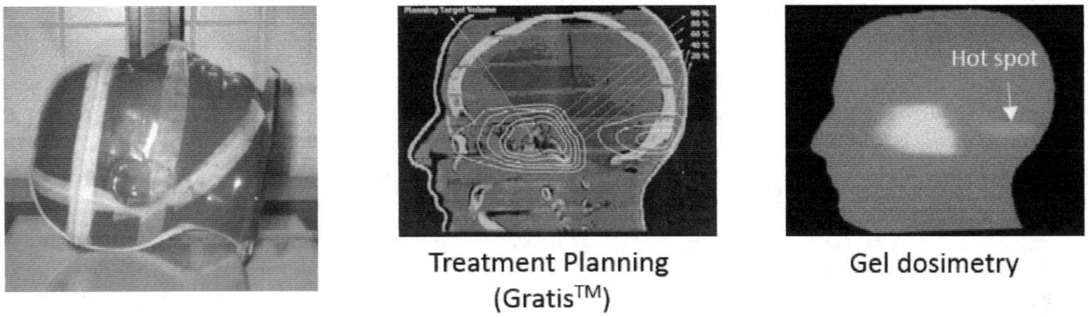

Figure 1. Nasopharyngeal tumor treated with IMRT. Initial treatment planning box was only covering the area near the target volume. A hot spot was detected on gel dosimetry. Increasing the treatment planning box confirmed the presence of the hot spot. The dosimeter gel was an anoxic polyacrylamide gelatin (PAG) gel. Figure adapted from De Neve et al 1999 [86] with permission of the publisher.

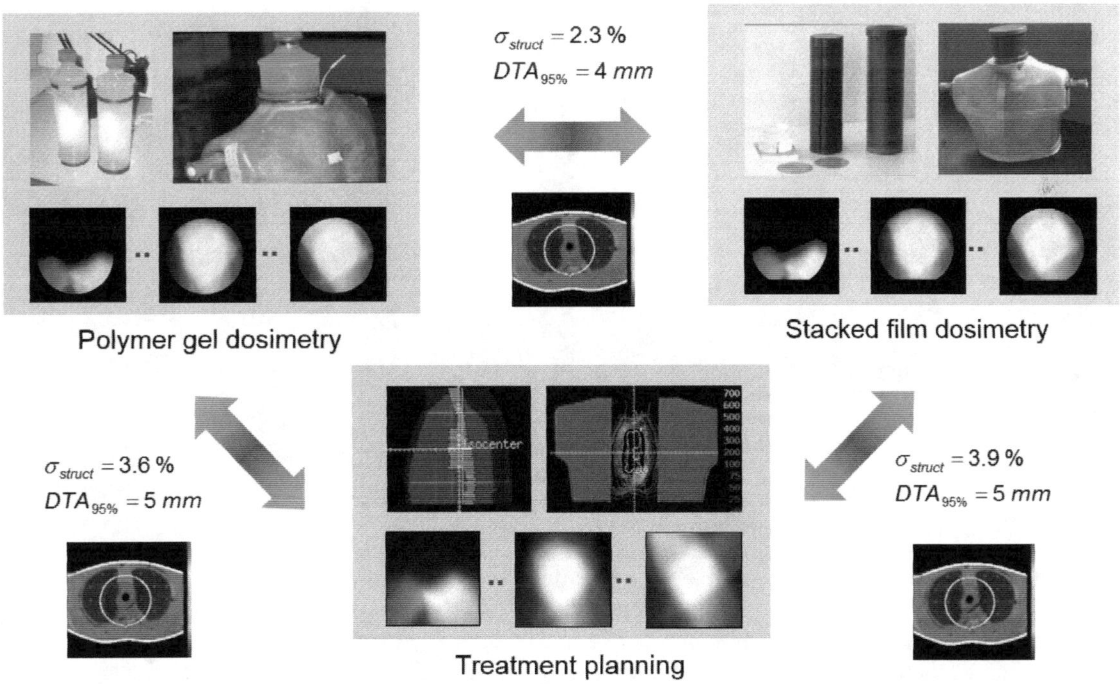

Figure 2. IMRT treatment of a mediastinal tumour consisting of 6 non-coplanar beams sparing the spinal cord. Comparison with stacked radiographic film dosimetry and treatment planning revealed a good correspondence which resulted in confidence about the whole treatment chain. The dosimeter gel was an anoxic polyacrylamide gelatin (PAG) gel that was cast in a cylindrical glass bottle that

was slid in a cylindrical cavity of a thoracic phantom. Figure adapted from De Deene et al 2000 [90] with permission of the publisher.

Figure 2 shows the use of polymer gel dosimetry in the verification of a non-coplanar IMRT treatment class solution of a mediastinal carcinoma [90] around the same period. Some improvements were made to the gel dosimetry approach in terms of MRI scanning and in avoiding oxygen permeation by use of a glass recipient. Here, a cylindrical gel dosimeter insert was used that could be easily replaced with a cylindrical insert containing a stack of 20 circular radiographic films. This way, we were able to compare the gel dosimetry dose maps with film dosimetry in a reproducible manner in full 3D. It also enabled us to study the effect of photon energy (6 MV and 25 MV) on the treatment outcome by inserting different bottles of polymer gel without moving the thoracic cast. In a later study, a similar phantom was used but including a lung cavity that could be filled with either air or water to study the effect of air cavities on the dose distribution [91].

It became generally accepted within the same centre that gel dosimetry fitted the purpose of validating class solutions of patients very well but was not cost-effective to validate individual patient plans. Since then, polymer gel dosimetry has been applied in a few cases were a new critical IMRT treatment was introduced such as the one shown in figure 3 which involved a sliding window IMRT treatment of a pituitary gland adenoma that was located very close to the optic chiasm [92]. Also in this case, a cylindrical insert was used and a comparison was made with both treatment planning and radiochromic micelle gel dosimetry using an in house built laser optical CT scanner [93]. Aside from some isolated spots in the gamma-maps, the gamma-index was below 1 in the entire 3D volume for both the MRI-acquired dose distribution and the optical CT-acquired dose distribution which provided sufficient confidence that the treatment plan matched with the actual dose distribution and gave 'green light' to start the treatment.

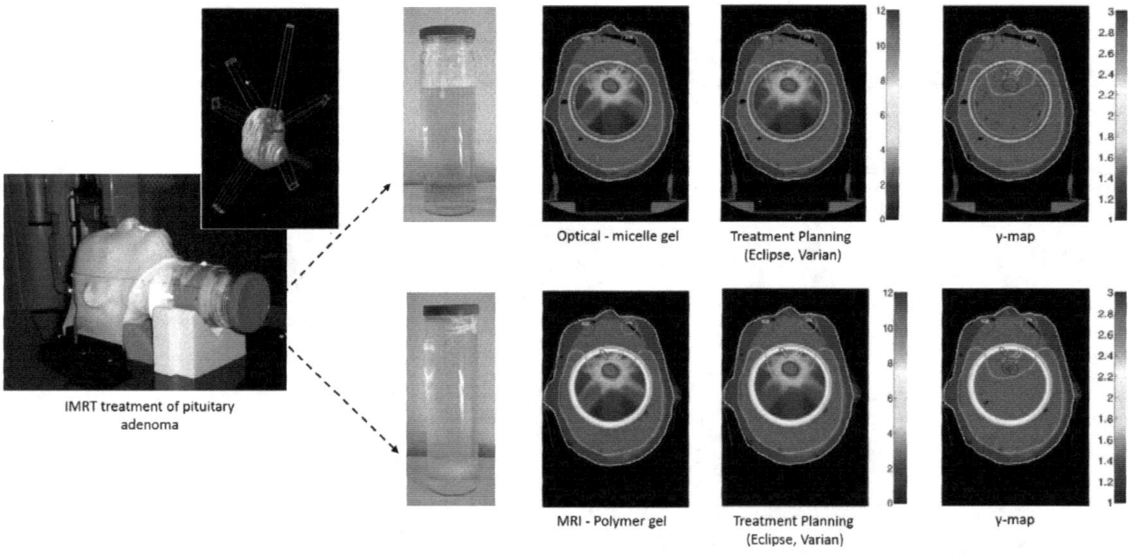

Figure 3. 3D dosimetry by both optical CT in combination with a micelle gel dosimeter (upper row) and MRI-based gel dosimetry using a normoxic PAGAT gel dosimeter (bottom row) provides confidence on the treatment plan of a pituitary gland adenoma using a sliding beam IMRT technique. Figure adapted from Vandecasteele et al 2013 [92] with permission of the publisher.

Not documented in the literature, but relevant to mention, is that in a particular case, a deviation between treatment plan and polymer gel dosimetry triggered further investigation into its cause. It was found that the deviation was caused by a misalignment error caused by a dislocation of one of the positioning lasers which was immediately corrected.

In the same centre, the use of MRI-based gel dosimetry has also played a similar role in the commissioning, validation and implementation of dynamic techniques such as intensity modulated arc therapy (IMAT) (Figures 4 and 5) and Tomotherapy® (Figure 6).

One of the advantages of MRI-based polymer gel dosimetry is that the dose distribution can be obtained in a large anthropomorphic volume as shown in figure 4 in the case of a whole abdominopelvic IMAT treatment of relapsed ovarian cancer [94]. To obtain reliable dose maps in such a large volume, compensation of B_1-field heterogeneity is required and compensation of temperature drift as a result of radiofrequency (RF) energy was compensated on the pulse sequence level [95].

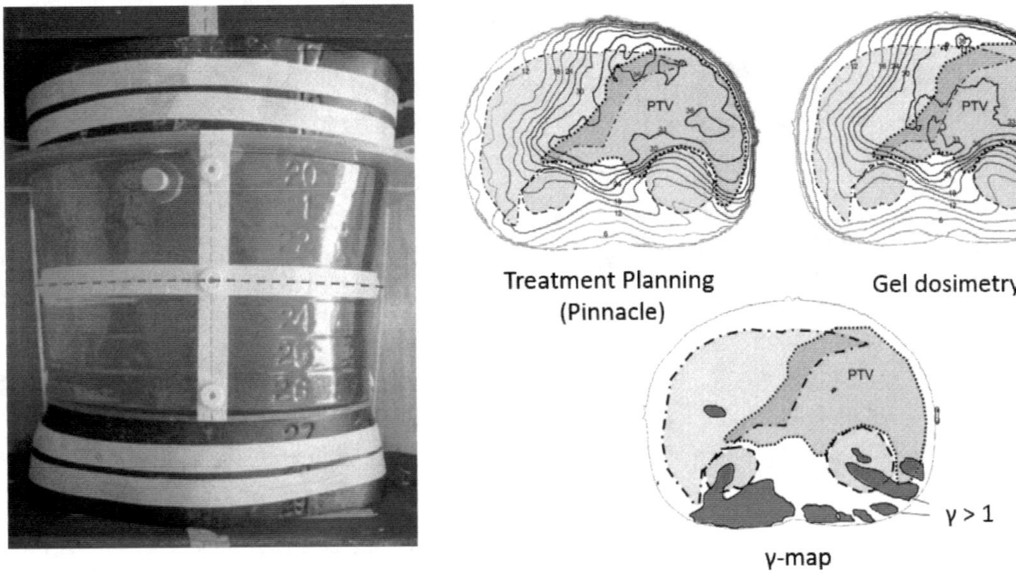

Figure 4. Whole abdominopelvic IMAT palliative treatment of patients with relapsed ovarian cancer. The abdominopelvic gel dosimeter phantom consists of a vacuum moulded Barex™ cast filled with a normoxic PAGAT gel and was surrounded by slaps of the Rando® phantom. The yellow shaded area corresponds to the liver, the green shaded areas to the kidneys and the pink shaded area corresponds to the planning target volume (PTV). The red regions in the gamma map are regions where gamma exceeds 1. Figure adapted from Vergote et al 2004 [92] with permission of the publisher.

In a European effort to consolidate the quality assurance of IMRT and IMAT (QUASIMODO-ESTRO), eight European institutions designed an IMRT plan for a horseshoe-shaped PTV surrounding a cylindrical organ-at-risk (OAR) in a simplified pelvic phantom [96]. Figure 5 shows a dose map obtained with polymer gel dosimetry and the corresponding treatment plan calculated using the GRATIS® treatment planning system that was complemented by in house software.

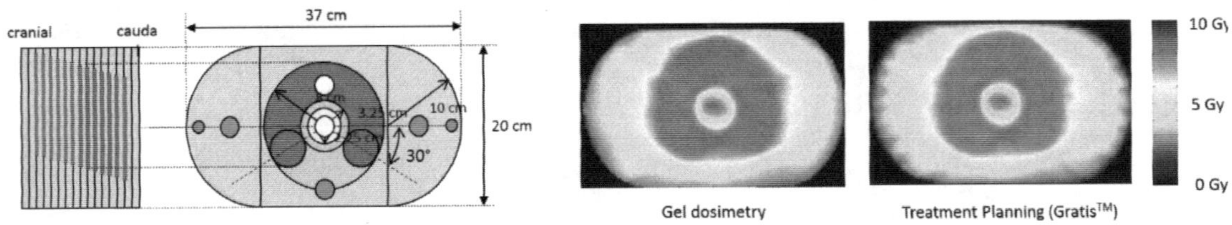

Figure 5. Pelvic phantom from the QUASIMODO-ESTRO audit with polymer gel dosimetry acquired dose map and treatment plan after IMAT. The first treatment plan showed some angular ripple which was not present in the gel acquired dose maps. Based on these results, it was concluded that the treatment plan had to be recalculated with a finer angular discretization. The polymer gel was a normoxic PAGAT gel that was poured in a vacuum moulded Barex™ cast.

Figure 6 shows polymer gel dosimetry of a Tomotherapy® treatment delivered at the Free University of Brussels. Because we were not satisfied with the absolute calibration, the data was not published but the dose distribution corresponded at least qualitatively well with the treatment plan.

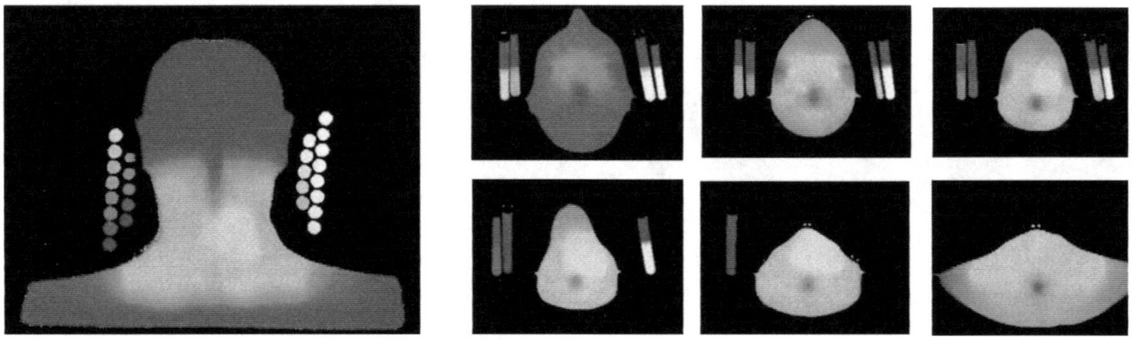

Figure 6. Polymer gel dosimetry of a Tomotherapy® Head & Neck treatment at the Free University of Brussels (VUB). The polymer gel was a normoxic PAGAT gel dosimeter.

Gel dosimetry is also very well suited to acquire the dose distribution of treatment techniques that are difficult to evaluate with other dosimetry, such as in spatially fractionated grid radiotherapy. The three dimensional nature of the dose distribution of a GRID treatment can be appreciated in figure 7.

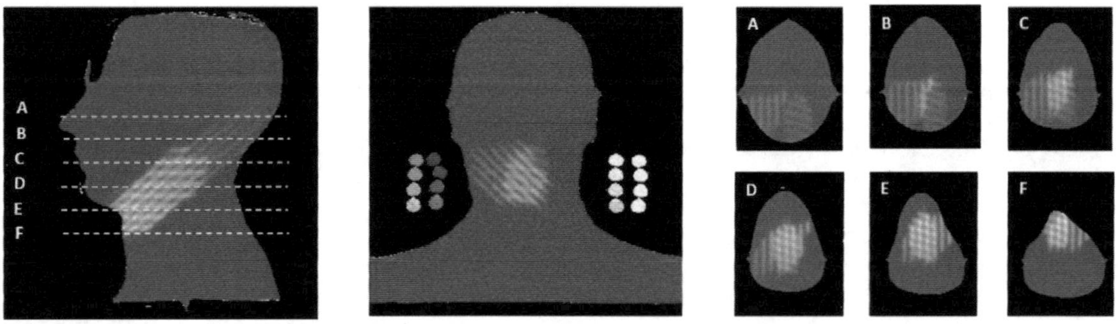

Figure 7. Polymer gel dosimetry of a GRID treatment. Normoxic PAGAT gel was used in this study.

Polymer gel dosimetry has also been useful in other spatially challenging dosimetry applications as illustrated in figures 8 and 9. Figure 8 shows a situation of electronic disequilibrium around a 25 mm diameter spherical air cavity that was inserted in a gel. The spherical hollow glass sphere was suspended in the gel by use of a glass rod.

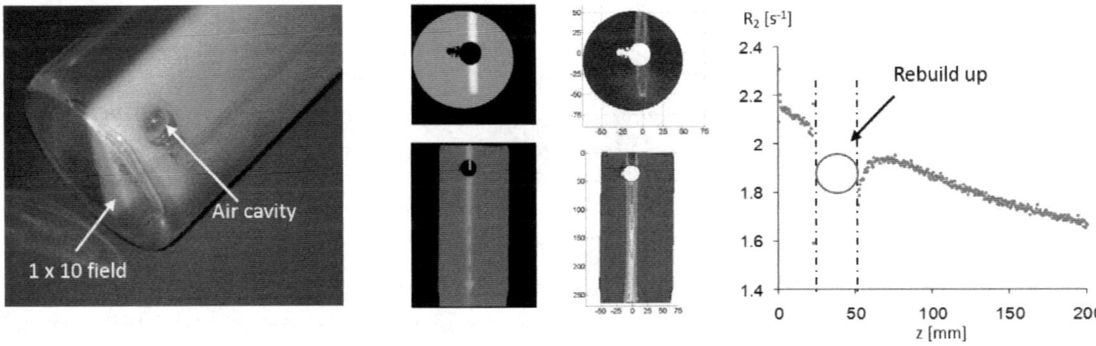

Figure 8. Polymer gel dosimetry in a situation of electronic disequilibrium resulting in a rebuild-up behind a spherical air cavity.

To investigate the dose disturbance caused by a tray containing lead markers for EPID spatial registration, a gel dosimetry experiment was performed which demonstrated a dose attenuation in the order of 10% (figure 9). From this experiment, it was concluded that the tray had to be removed before the start of the treatment beam delivery.

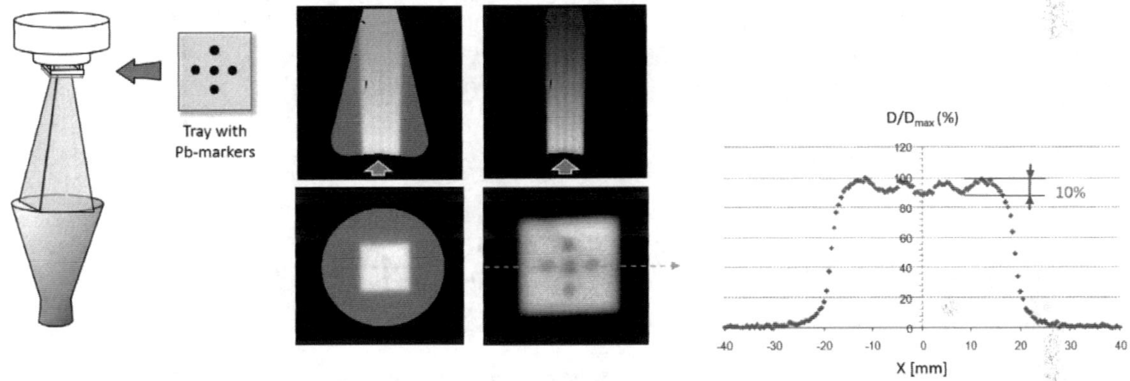

Figure 9. Tray with lead-markers producing a cross-reference location on electronic portal imaging. Polymer gel dosimetry was used to evaluate if the tray also created a dosimetric effect in treatment beams. As a dose attenuation of 10% was measured underneath the lead markers, it was concluded that the tray had to be removed before every treatment beam delivery. Figure adapted from De Deene et al 2002 with permission of the publisher.

It is recognised that polymer gel dosimetry with MRI has some limitations as compared to optical CT-based radiochromic dosimetry; First of all, some knowledge and expertise on quantitative MRI is required to assess the presence of imaging artefacts and to implement adequate compensation methods. In addition, access to an MRI scanner can be problematic in some centres. Secondly, the treatment of the gel is more delicate as temperature fluctuations during scanning may have a significant influence on the acquired R_2 values and thus on the dose maps. Thirdly, oxygen infiltration in the gel may result in a physico-chemical induced error in the acquired dose maps. In the next sections, we will elaborate more on these effects and provide some compensation strategies. Despite these difficulties, from the above-mentioned examples it will be clear that MRI-based polymer gel dosimetry (and X-ray CT based polymer gel dosimetry) has some inherent application-specific advantageous above optical CT based radiochromic dosimetry:

1. Polymer gel dosimetry can be performed in large anthropomorphic phantoms such as in the thorax and abdomen. The limitation is mostly from a practical perspective such as the

manufacturing of the gel. An example of 11 L of gel is provided in figure 4, which was only limited by the size of the recipients in which the gel was prepared. However, care is required in imaging large phantoms as gradient non-linearity may give rise to geometrical distortions and the B_1-field non-uniformity can lead to dose uncertainties. Both effects can be compensated.

2. With the move towards MR-only simulators in radiotherapy, the co-registration of the anthropomorphic gel dosimeter with a pseudo-CT becomes more straightforward and accessibility to MRI is increased.
3. Polymer gel dosimeters can be beaten into a foam to make lung-equivalent gel dosimeters [97] that can be readout with magnetization transfer (MT) MRI.
4. Recently, the potential of polymer gel dosimetry for real time dosimetry on an MRI-Linac has been demonstrated where MRI scans are taken during radiation delivery [98]. Figure 10 shows a series of MRI recorded dose maps during radiation on the experimental Australian MRI-Linac.

Whether MRI-based, optical-CT based or X-ray CT based gel dosimetry is the optimal choice depends on the application and is site dependent (i.e. accessibility to an MRI scanner or CT-scanner).

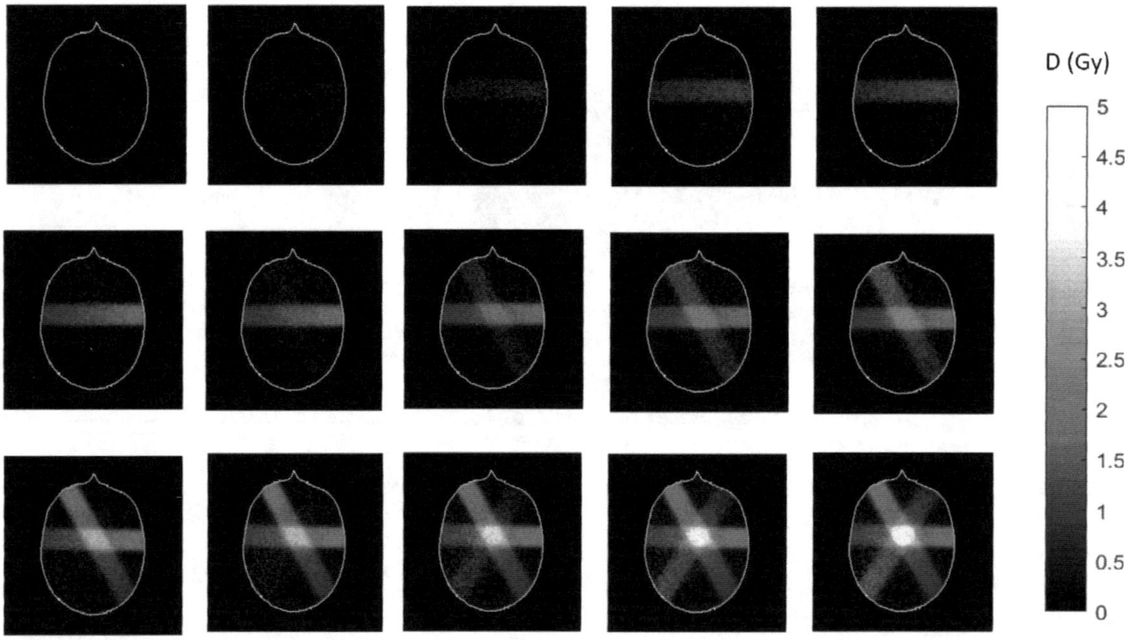

Figure 10. Dose maps of the central slice recorded dynamically on the MRI-Linac during radiation of a head phantom. The time between two adjacent images in a row is 44 seconds (every 4th recorded frame). Figure adapted from De Deene et al 2020 [98] with permission of the publisher.

3. Physical mechanisms of NMR relaxation and MRI contrast

A comprehensive overview of the physical mechanisms of NMR relaxation and MRI contrast of Fricke and polymer gel systems can be found elsewhere [9]. We here suffice with describing the main mechanisms with a focus on those aspects that affect the ultimate reliability.

3.1. Fricke gel dosimeters

Fricke gel dosimeters are based on the Fricke dosimeter [2] which consists of a ferrous (Fe^{2+}) sulphate in a 0.8 N sulfuric acid solution. Upon radiation, the Fe^{2+} ions are oxidized to Fe^{3+} ions, which results in a measurable physical change, either by electrometric titration or absorption spectrophotometry with a selective absorption peak of Fe^{3+} at a wavelength of 304 nm [99,100]. An interesting observation was that the amount of radiation induced oxidation was independent of the concentration of ferrous ions in the solution. Fricke dosimeters have played an important role in the development of the field of radiochemistry [101] and have been used for several decades as a dosimetry standard in many laboratories [102]. However, because of the strong dependence of radiochemical yield on the linear energy transfer (LET) of the radiation, the Fricke dosimeter fell into disuse [103] and has been replaced by other dosimetry standards such as calorimetry and ionization chambers as primary standards. For a more in-depth explanation of the radiation chemistry the reader is referred to textbooks [9, 104].

The radiation induced conversion of ferrous ions to ferric ions by ionising radiation also alters the magnetic moment and electron spin relaxation times of the metal ion. These changes result in a change in NMR relaxation rates. In 1984, it was demonstrated that the change in magnetic moment and electron spin moment resulted in a change of the transverse relaxation rate with radiation dose of $1.13\ 10^{-2}\ s^{-1}\ Gy^{-1}$ and a similar change in transverse relaxation rate of $1.21 \pm 0.27\ 10^{-2}\ s^{-1}\ Gy^{-1}$ [8]. It was also suggested that infusion of a hydrogel with the Fricke solution could result in a 3D dosimeter. Since then, several Fricke gel systems have been proposed. However, a major obstacle in Fricke gel dosimeters is the diffusion of Fe^{2+} and Fe^{3+} ions within the gel matrix and many studies have focussed on reducing the diffusivity of the ions through modification of the hydrogel matrix or the use of chelates.

The MRI dose sensitivity of a Fricke solution can be calculated by calculating the molar change with absorbed radiation dose and then by using the T_1 and T_2 relaxivity of Fe^{2+} and Fe^{3+} ions, the change in relaxation rate. The radiochemical yield (G) in the standard Fricke solution can be calculated from the chemical yields of the different water radical species [9,104] and amounts to about 15.6 Fe^{3+} ions per 100 eV of imparted radiation energy. The molar concentration of ferric ions created per unit of dose can then be obtained by multiplying the G factor in SI units (1.62 μmol/J) with the mass density (ρ):

$$\frac{\partial[\mathrm{Fe}^{2+}]}{\partial \mathrm{D}} = \mathrm{G} \cdot \rho = 1.62\ \mu\mathrm{M/Gy} \tag{1}$$

The theoretical calculation of the relaxivity of ions in solution involves a deeper understanding of the mechanisms of nuclear magnetic interactions which is beyond the scope of this paper but the reader is referred to other works for a full quantitative description of the different interaction terms [9, 105 - 108]. The NMR relaxation rate for an absorbed radiation dose D can eventually be written as:

$$\mathrm{R_k} = \mathrm{R_{k,0}} + \kappa[\mathrm{Fe}^{2+}]_0 r_{\mathrm{k,is}}^{\mathrm{Fe2+}} + \kappa\left(\frac{\partial[\mathrm{Fe}^{2+}]}{\partial \mathrm{D}}\right)\left(\mathrm{R_{k,is}^{Fe3+}} - \mathrm{R_{k,is}^{Fe2+}}\right) \cdot \mathrm{D} \tag{2}$$

where the index k refers to the longitudinal (k = 1) and transverse relaxation rate (k = 2) respectively. $R_{k,0}$ is the relaxation rate in free water, $r_k^{Fe^{2+}}$ is the relaxivity of ferrous ions and $[Fe^{2+}]_0$ is the initial concentration of ferrous ions (ferrous sulphate). The coefficient κ is a proportionality constant that is determined by the number of water molecules in the hydration sphere (i.e. the region where water is in immediate contact with the Fe^{2+} and Fe^{3+} ions. The relaxation dose sensitivity r_k^D is given by the factor preceding the radiation dose D in equation 2:

$$\mathrm{r_k^D} = \kappa\left(\frac{\partial[\mathrm{Fe}^{2+}]}{\partial \mathrm{D}}\right)\left(\mathrm{R_{k,is}^{Fe3+}} - \mathrm{R_{k,is}^{Fe2+}}\right) \tag{3}$$

After substitution of the theoretical values for κ, $\frac{\partial[Fe^{2+}]}{\partial D}$ (equation 1), $R_{k,is}^{Fe3+}$ and $R_{k,is}^{Fe2+}$ into equation 3, one finds for a field strength of 0.47 T, $r_1^D = 0.013\ s^{-1}Gy^{-1}$ and $r_2^D = 0.0178\ s^{-1}Gy^{-1}$. These values correspond very well with measured values by Gore et al: $r_1^D = 0.0113\, s^{-1}Gy^{-1}$ and $r_2^D = 0.0121 s^{-1}Gy^{-1}$ [8].

In Fricke gel dosimeters, the Fricke solution is dissolved in a hydrogel matrix. Both agarose, gelatine and polyvinyl alcohol (PVA) have been used as gelling agents. The requirement to obtain a solid gel matrix places constraints on the concentration of sulphuric acid [109]. Overall, a higher radiochemical yield is found in Fricke gel dosimeters as compared to Fricke solutions [110] with a 4-fold increase in the R_1-dose sensitivity in a 1% (w/w) agarose Fricke gel dosimeter and an increase in dose-sensitivity with a factor of 2.2 in a 4% (w/w) gelatine Fricke gel dosimeter [111, 112]. Additives such as benzoic acid [113], saccharides [116] and salt [117] have been added in an attempt to increase the sensitivity and/or to limit the diffusion of ions in the hydrogel after radiation exposure. Xylenol orange can be used as an optical indicator opening up the possibility for Fricke gel dosimeters to be scanned with optical CT [114] but comes at the cost of a decrease in the R_1-dose sensitivity [115].

In a Fricke gel dosimeter, the relaxation rate can be expressed by equation 4:

$$R_k = R_{k,0} + r_k^{Fe^{2+}}[Fe^{2+}]_0 + r_k^{gel}[gel] + r_k^D D \qquad (k = 1,2) \tag{4}$$

where $R_{k,0}$ is the relaxation in free water, r_k^{gel} is the relaxivity of the gel and r_k^D is the relaxation dose sensitivity. Note that as discussed above, the relaxation dose sensitivity is dependent on several chemical properties and thus also on the gel agent.

To compensate for the aforementioned drawback of ion diffusion, many attempts have been made to decrease the ion diffusion coefficient [9], either by adding a chelating agent or by changing the morphology of the gel [118, 119]. Strategies to further reduce the diffusion coefficient could focus on anionic hydrogels, the addition of viscosity increasing substances, the dispersion of ferrous ions in liposomes, the use of sorbent polymers and functionalization of the hydrogel backbone with complex forming groups. Challenges that come with these strategies are that the radiation induced oxidation reaction should not be compromised, that the gel dosimeter remains tissue-equivalent, that there is no significant loss in dose sensitivity, that the gel dosimeter is dose rate independent and that the resulting dosimeter remains affordable.

3.2. Polymer gel dosimeters

Polymer gel dosimeters consist of a gel matrix that is immersed with vinyl monomers (and a cross-linker monomer. When exposed to ionizing radiation, reactive water radical species will initiate a polymerization reaction that results in either the creation of highly cross-linked polymer aggregates that are entangled with the gel matrix or polymer chains that graft onto the gel matrix. For a more profound discussion on the radiation chemistry of polymer gel dosimeters, the reader is referred to other work [9, 33, and 34]. The spatial integrity of polymer gel dosimeters is superior to that of Fricke gel dosimeters. The change in molecular mobility of the hydrogen protons on the polymer and of bound water molecules is significantly reduced upon polymerization, which results in a significant increase of the relaxation rate, in particular of the transverse relaxation rate R_2. The fast exchange of hydrogen atoms between water molecules and polymer results in an average measured R_2 that carries the signature of the restricted mobility of polymer chains [9, 120]. Several monomers have been used as active components of the polymer gel dosimeter. The R_2-dose response follows typically a sigmoidal response (figure 11) which can be characterized by a half-value dose value ($D_{1/2}$), an R_2-dose sensitivity and a dynamic R_2-range depending on the monomer.

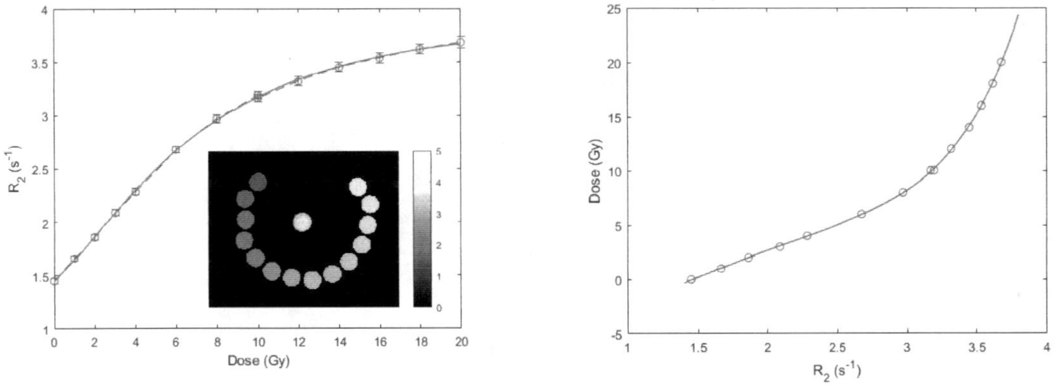

(a) **(b)**

Figure 11. Typical R_2-dose response curve (a) and corresponding dose-R_2 curve (b) of a PAGAT gel dosimeter, measured with a 32 echo multi-spin-echo pulse sequence on a 3T MRI scanner. Note the excellent fit.

The uncertainty of any dosimeter is to a large degree determined by its radiation properties such as the dependence of the dose response on the dose rate, energy, LET, the stability and environmental factors such as temperature and pressure. The radiation properties of several polymer gel dosimeters have been studied extensively.

4. MRI pulse sequences

An important aspect in setting up MRI-based polymer gel dosimetry is the choice of the MRI pulse sequence. This choice primarily depends on the kind of gel dosimeter (i.e Fricke or polymer gel) and is determined by the maximum achievable accuracy and precision. In 3D dosimetry, two kinds of accuracy need to be considered: spatial (geometrical) accuracy and dosimetric accuracy [9, 121]. In MRI, the spatial and dosimetric accuracy can be compromised by the presence of artefacts which can be scanner related such as inhomogeneity in static magnetic field (B_0), magnetic field gradients and radiofrequency (RF or B_1) field, or which can be object related such as magnetic susceptibility difference caused by magnetic field distortion and dielectric related standing waves. The dosimetric precision is determined by the interplay of dose response of the gel dosimeter and the sequence parameters as discussed in section 5.

Fricke gel dosimeters exhibit changes in both T_1 and T_2. Historically, Fricke gel dosimeters have been scanned with quantitative T_1, however it is equally possible to use quantitative T_2 sequences. In polymer gel dosimeters the radiation induced change in T_1 is at least an order of magnitude smaller than changes in T_2, hence for polymer gel dosimetry, quantitative T_2 sequences are preferred. In theory, it is possible to use any pulse sequence that generates images in which the signal intensity is uniquely correlated with the absorbed radiation dose (not necessarily linear). In theory, a T_1-weighted or T_2-weighted sequence can be used. However, in practice, as a result of B_0 and B_1-field heterogeneity, the uniqueness is not guaranteed which can severely compromise the accuracy. A better approach is to use quantitative T_1 or T_2 pulse sequences as the majority of these artefacts are canceled out. Below are some of the most common MRI pulse sequences that are applicable to acquire quantitative T_1 and T_2 maps.

In general, for every specific pulse sequence, the pixel signal intensity is determined on one hand by known pulse sequence specific parameters such as the flip angle of the applied RF pulses α, their phase φ and time intervals such as the echo time TE, the repetition time TR and inversion time TI and on the other hand by unknown gel (or tissue) specific NMR properties such as the proton density (ρ_H), the relaxation rates R_1, R_2, magnetization transfer parameters, magnetic susceptibility (χ_v) and chemical shifts ($\Delta\sigma$). The relationship between these parameters and the acquired pixel signal intensity is described by a signal equation $S_f(\alpha_k, TE_k, TR_k, \ldots, \boldsymbol{\rho_H}, \mathbf{R_1}, \mathbf{R_2}, \boldsymbol{\chi_v}, \boldsymbol{\Delta\sigma}, \ldots)$. For a comprehensive set of signal equations for different pulse sequences, the reader is referred to reference [9] and [122].

The unknown gel specific NMR parameters (in bold) can be determined by acquiring images for a set of k different sequence parameters and minimizing the χ^2 difference of the acquired signal intensity and the signal equation:

$$\min_{\alpha_k, TE_k, TR_k} \chi^2 = \min\left(\sum_{\forall k} \frac{\left(S_{image,k} - S_f(\alpha_k, TE_k, TR_k, \ldots, \boldsymbol{\rho_H}, \mathbf{R_1}, \mathbf{R_2}, \boldsymbol{\chi_v}, \boldsymbol{\Delta\sigma}, \ldots)\right)^2}{\sigma_S^2}\right) \tag{5}$$

where σ_S is the standard deviation (noise) of the measured signal.

Optimal NMR pulse sequences for T_1 or T_2 mapping are those for which equation 5 results in a set of equations that is well-determined and where for the sake of simplicity, the dimension of the phase space is small. Ideally, only one pulse sequence parameter needs to be varied to result in a set of unique signal relations. In addition, the pulse sequence should result in images with high spatial fidelity, i.e. are not geometrically distorted. Some commonly used pulse sequence are listed here.

4.1. Spin-echo, gradient echo and rapid acquisition with relaxation enhancement (RARE)

The most basic sequence to acquire R_1 or R_2 maps is a spin-echo sequence. The repetitive unit of the spin echo pulse sequence consists of an excitation RF pulse with flip angle 90° and a 180° refocussing pulse at half the echo time (TE/2). The time between two successive excitation RF pulses is the repetition time TR. The signal intensity in a spin-echo sequence is weighted by T_1 through the choice of the repetition time TR and by T_2 through the choice of the echo time TE. Thus, by acquiring several spin-echo images recorded with different repetition times, the longitudinal relaxation rate R_1 can be fitted on a pixel-by-pixel basis and by acquiring spin echo-images recorded with different echo times, the transverse relaxation rate R_2 can be determined. It is also convenient to apply an image threshold filter to filter out background voxels before the fitting. In addition to the parametric relaxation maps, parametric maps of fitting performance such as Pearson correlation coefficients can also be calculated. It is sometimes also advisable to treat the flip angle α of the excitation pulse as an unknown fitting variable to account for imperfect excitation (B_1^+-field).

Instead of acquiring a spin-echo, a gradient echo can also be acquired. The saturation recovery sequence is a gradient echo sequence with the flip angle of the excitation pulses equal to 90°. Here, all longitudinal magnetization that is rebuild during the repetition time TR is flipped in the transverse plane and gives way to measurable signal. Spoiler gradients are applied after each gradient echo readout to remove phase coherence in the transverse magnetization which would otherwise result in stimulated echoes. An increase in repetition time will result in an increase in signal intensity which also depends on the T_1 relaxation time. Spoiled gradient recalled echo imaging (SPGRE) in which the 90° pulses are replaced by smaller α pulses can be found in a technique called DESPOT in which base images are recorded with various flip angles [123] while keeping the echo time and repetition time fixed.

The spin echo sequence is a relatively slow sequence as the excitation/read out loop needs to be repeated for at least the number of phase encoding lines to acquire a single T_1/T_2 -weighted image. The read out can be accelerated by applying a series of 180° refocusing pulses after the first spin echo. This will result in a series of spin echoes that can be all phase encoded differently so that in a single pulse sequence repetition loop several phase encoding lines are acquired. This is the concept of a Rapid Acquisition with Relaxation Enhancement (RARE) sequence. The acceleration is determined by the number of echoes (or refocusing pulses) in the readout echo train (N_E). Different T_2 weighting of the different spin-echoes in the echo train will however lead to some broadening of the point spread function of each voxel.

4.2. The multi-spin-echo pulse sequence

The most common sequence that is used to acquire a set of T_2-weighted base images is the multi-spin-echo sequence [124, 125]. The pulse sequence scheme is very similar to the RARE sequence but instead of storing the different spin-echo signals in different k-space lines of the same k-space image, the different spin-echo signals are stored in corresponding k-space lines in separate k-space matrices (figure 12). A typical pulse sequence scheme of a multi-spin echo sequence is the Carr-Purcell Meiboom-Gill (CPMG) sequence where the 90-degree excitation RF-pulse has a 90-degree phase difference with respect to the successive 180-degree refocussing RF-pulses. By shifting the phase between the excitation and refocussing pulses, the sequence becomes more robust against B_1-field non-uniformity [126]. A CPMG multi-spin-echo sequence provides a series of differently T_2 weighted base images on which an exponential T_2-decay curve can be fitted on a pixel-by-pixel basis. Calibration vials are scanned together with the volumetric dosimetry phantom which can be used to convert the R_2 values in dose values.

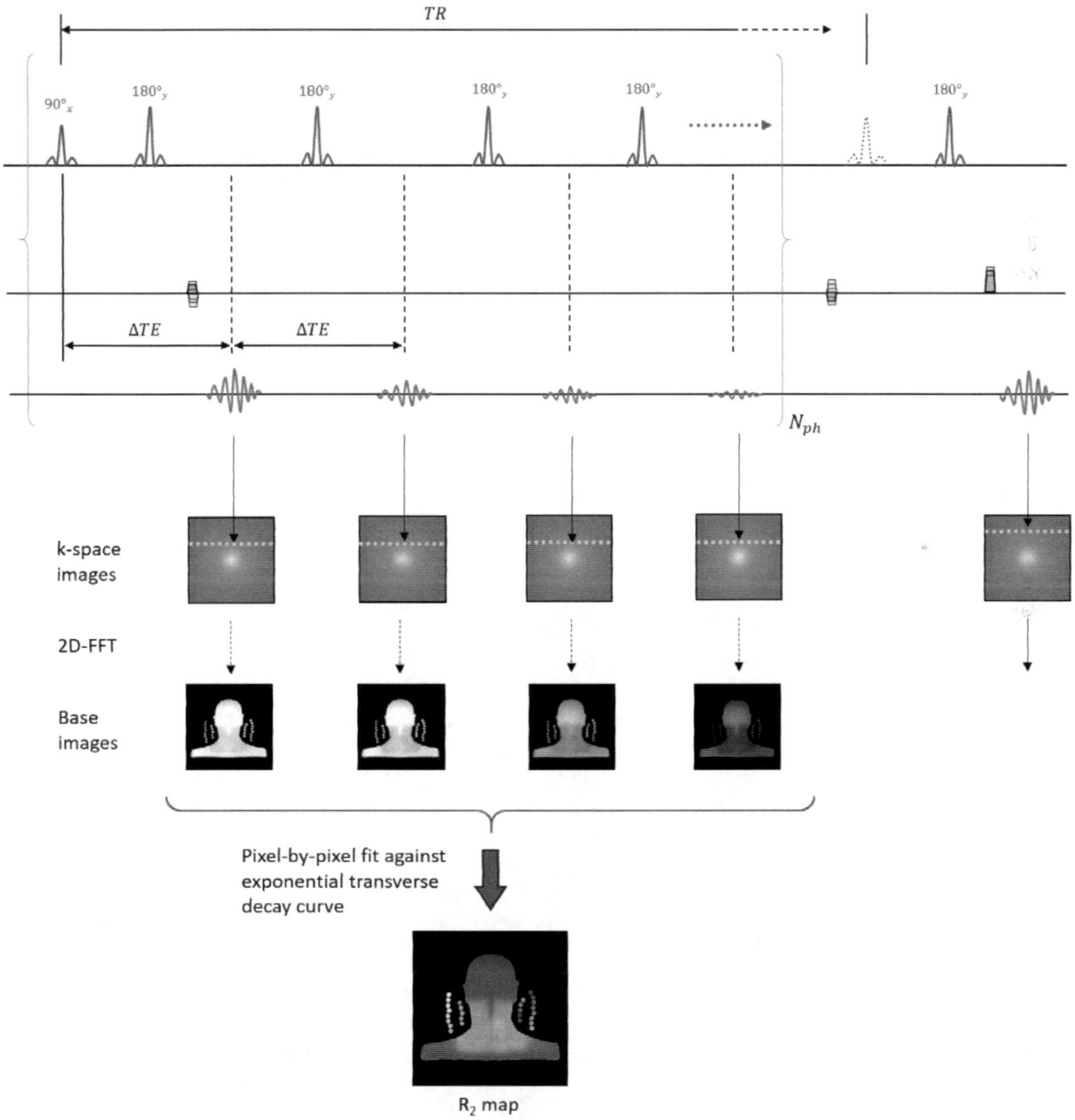

Figure 12. Quantitative R_2 maps can be obtained by using a multi-contrast CPMG spin echo sequence where several T_2-weighted signals are acquired in a single repetition time. The resulting k-space images are transformed in T_2-weighted base images on which a pixel-by-pixel exponential fit results in a quantitative R_2 map. Figure adapted from De Deene et al [9] with permission of the publisher.

4.3. Saturation and inversion recovery sequence

A second class of pulse sequences that can be employed to map R_1 are the saturation and inversion recovery sequences. The pulse scheme of a saturation recovery sequence and inversion recovery sequence are displayed in figures 13a and b respectively.

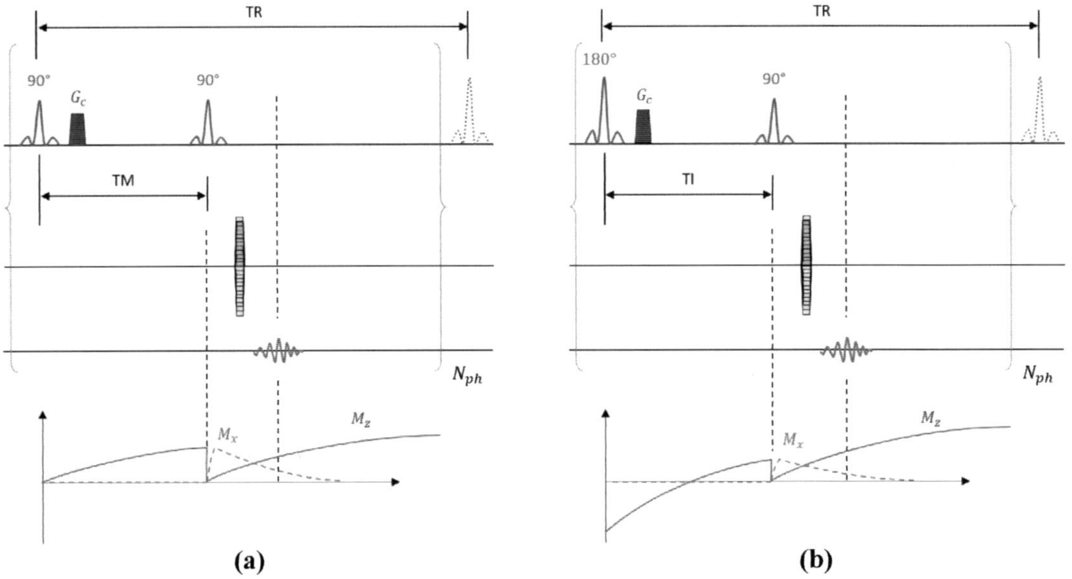

Figure 13. Saturation recovery (a) and inversion recovery sequence (b) scheme with the longitudinal magnetization (M_z) and transverse magnetization (M_x) indicated by a solid line and dotted line respectively. A crusher gradient (G_c) is added after the saturation or inversion pulse to spoil any remaining transverse magnetization. Read out and slice selection gradients are not shown. Figure adapted from De Deene et al [9] with permission of the publisher.

In the saturation recovery sequence, a 90° RF pulse followed by a crusher gradient will destroy the longitudinal and transverse magnetization. The longitudinal magnetization will start to recover immediately after the 90° pulse. After a recovery time TM, the longitudinal magnetization component is turned into a measurable gradient echo signal by use of a second 90° pulse. In an inversion recovery sequence (figure 13b), a 180° inversion pulse is applied. After an inversion recovery time TI, the longitudinal magnetization is turned into a measurable signal. In the case of an inversion recovery experiment, it is important to acquire the real component of the signal instead of the magnitude of the signal.

In principle, any kind of fast read out block can be applied after the 90° pulse, such as a spin echo, fast spin echo or a combination of a gradient echo train and spin echo train, also referred to as GRASE readout. For gel dosimetry, it is not recommended to use an Echo Planar Imaging (EPI) readout because of the significant spatial distortions related to the low bandwidth in the phase encoding direction in combination with magnetic field inhomogeneity as a result of magnetic susceptibility differences between air and gel.

4.4. Look-Locker sequences

A fast sequence to acquire R_1 maps is the Look-Locker sequence [127]. The Look-Locker sequence is similar to the inversion recovery gradient echo sequence, but a faster readout is achieved by replacing the 90° pulse with an RF pulse train with small flip angles α. Each α-pulse transfers some recovered longitudinal magnetization in the transverse plain which is readout as gradient echoes. Also, between each α-pulse, a small phase encoding gradient is applied so that each gradient echo results in a new k-space line. This means that for each inversion excitation part of k-space is sampled instead of only one k-space line in the case of the IRGE sequence. The relaxation weighting of the different k-space lines also results in broadening of the point-spread-function (PSF), thus compromising the intrinsic resolution. To avoid severe artefacts as a result of the relaxation between the different k-space lines, a k-space ordering scheme is often proposed. In this scheme the central k-space lines (close to $k_y = 0$) are from echoes recorded immediately after the first α-pulse.

A variation on the Look-Locker sequence is the TOMROP (T_1 by multiple readout pulses) sequence. In this sequence the gradient echoes are grouped and interleaved with some recovery time to allow T_1 relaxation to take place. Each group of gradient echoes is used in a separate T_1 weighted base image. The different T_1-weighted base images are then used to reconstruct an R_1 map. The Look-Locker sequence is particularly sensitive to RF pulse imperfections. Also changes in the flip angle distribution within the image (B_1-field non-uniformity) will affect the signal and B_1 inhomogeneity corrections may be necessary. In addition, more sophisticated numerical algorithms are required to solve for the relaxation rate R_1 [128].

4.5. Steady State Free Precession (SSFP) sequence

A sequence of RF pulse excitations separated by a time TR (without spoiler gradients) will bring the NMR signal in a steady state. In between two successive pulses the frequency encoding gradients can be placed in such a way that two echoes are obtained [122]. After recording images with various flip angles $\boldsymbol{\alpha}$ and repetition times TR, R_1 maps and R_2 maps can be calculated.

A special case of SSFP is when the net gradient area is zero at any of the three gradient axes during one TR interval (i.e. between two successive excitations). In this case, only one echo is obtained. This sequence is called balanced SSFP. Because both the spatial and dose accuracy are important in the 3D radiation dosimeters, careful validation of the quantitative MRI protocol is essential. Indeed, any imaging artefacts may result in a misinterpretation of the dose at a specific location or of the location where the radiation dose is administered. Benchmarking the accuracy of the dosimeter involves quality assurance experiments on blank (non-irradiated) phantoms and on phantoms with a well characterized golden standard phantom containing gels with well-known R_1 and R_2. Eventually, corrections for RF-field inhomogeneity can be applied [63,129].

4.6. Magnetization transfer pulse sequence

The MRI contrast mechanisms indicate a strong dependence of the MRI signal on magnetization transfer which has been experimentally demonstrated in a magnetization transfer experiment [130]. Magnetization transfer (MT) is preferred above R_2 as imaging property in polymer gel foams which find potential use as lung equivalent dosimeters. Not only is the dynamic range much higher for MT than for R_2, the measured MT ratio (MTR) is also less sensitive to variations in bubble size and density [97]. Magnetization transfer images can be acquired by use of an imaging pulse sequence that consists of an MT contrast preparation block and an image acquisition block. The MT preparation block has a series of off-resonance Gaussian shaped MT saturation pulses followed by spoiler gradients to get rid of residual transverse magnetization (figure 14). The MT saturation pulses will selectively destroy the magnetization on the polymer proton pool, which through chemical exchange affects the water proton pool (figure 14). For polymer gel foam dosimeters, the acquisition block requires the use of short echo times to limit diffusion attenuation of the signal. The resulting MT-weighted signal attenuation can be compared to the signal obtained with the same sequence but without playing out the MT saturation pulses. The ratio of the signal change in each voxel, with respect to pure water and the non-MT weighted signal can be mapped in an MT map. As the size of the polymer proton pool increases with radiation dose, the MT effect will increase. The MT contrast depends on the offset frequency of the MT saturation pulses. An optimum frequency offset can be determined and an MT versus dose response curve can be extracted. The MT map can be calibrated to dose by use of a set of gel foam calibration phantoms that have been irradiated with predefined doses.

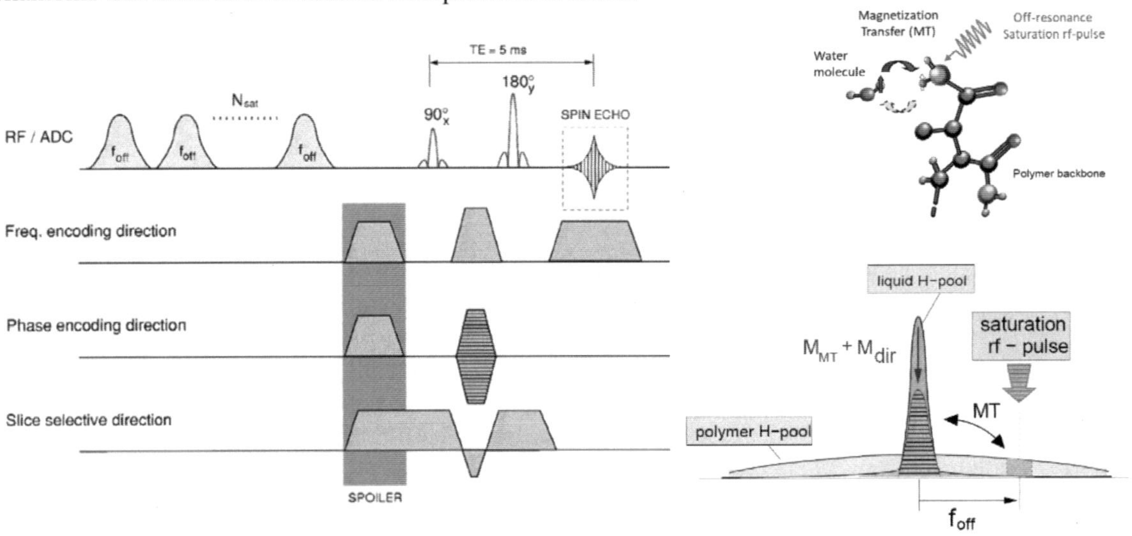

(a) (b)

Figure 14. Magnetization transfer imaging of polymer gel foam can be performed by use of a series of MT saturation pulses combined with a spin echo readout (a). The spin echo readout has a relatively short TE to avoid significant R_2 dispersion effects. The polymer gel foam proton pool has a relatively short T_2 and thus has a broadened spectral line shape (b). Off-resonance MT saturation pulses will selectively saturate the polymer gel proton pools which exchange with the water proton pools (b) resulting in an attenuation of the liquid hydrogen pool. Figure adapted from De Deene et al 2006 [97] with permission of the publisher.

5. MRI pulse sequence optimization

A more extensive treatise on the uncertainty in 3D radiation dosimetry can be found elsewhere [22, 23]. It is emphasized that dosimetric and spatial uncertainties are intermingled in the case of 3D gel dosimetry and it can be difficult to impossible to differentiate one from the other in a measured dose distribution. It is in this light that gamma-map comparisons have been used [131]. The overall uncertainty of 3D radiation dosimetry can be achieved through a reproducibility study of the complete dosimetry experiment from gel fabrication to dose distribution analysis (type A uncertainties) and by comparison against other non-3D dosimetry standards (type B uncertainties) [75]. In a clinical dose verification with gel dosimetry the uncertainty of the dose measured in each voxel comprises both the systematic inaccuracy and random errors, including stochastic image noise. To minimize systematic uncertainties, physico-chemical parameters (such as temperature and fabrication procedures) can be controlled carefully while MRI artefacts can be minimized by a careful choice of pulse sequences and artefact compensation techniques [62-64]. Stochastic image noise can be minimized by optimizing the imaging parameters such as repetition time, echo time and flip angle. The optimal imaging parameters are dependent on the type of gel (i.e. the T_1 and T_2 working range of the gel dosimeter).

5.1. Optimization for 3D radiation dosimetry

The optimization problem in 3D radiation gel dosimetry can be defined as "Find the imaging parameters that minimizes the stochastic noise in the dose maps for a predefined spatial resolution within a fixed measurement time and total imaging volume." A complication is that the total measurement time will be determined by the imaging parameters. Here it is important to keep in mind that taking more image averages (NEX) will result in a scaling of the standard deviation (noise) with $NEX^{-1/2}$. So in comparing different sets of imaging parameters, it is paramount to take the measurement time into account. On the other hand, in practice, the available measurement time can be constraint by the availability of the MRI scanner for the gel dosimetry readout. Note that this does not only apply to MRI but is also valid for X-ray CT and optical CT readout. Also, the total imaging volume and required spatial resolution are defined by the particular study (e.g. glioblastoma versus whole abdominopelvic radiotherapy).

Below are the general steps that can be followed to optimize the imaging parameters for a particular MRI pulse sequence and gel dosimeter (Figure 15):

1. Define the input parameters for the optimization problem:
   a. Select the MRI pulse sequence. Note that different pulse sequences can be selected and compared in terms of performance, both in terms of their susceptibility to artefacts and obtainable dose resolution (as to be determined).
   b. Define the parameter that is to be extracted from the MRI signal relation in the parameter space of one of the imaging parameters. A logical choice is R_2 for polymer gel dosimeters and R_1 or R_2 for Fricke gel dosimeters. For example, we define the signal in the parameter space of echo time $S_k = S(TE_k)$ for a polymer gel dosimeter, when we will acquire the signal S for different echo times TE_k for example by use of a multi-echo pulse sequence. However, in principle, it is not necessary to take the relaxation rates R_1 or R_2. A necessary condition is that the function preserves the uniqueness with respect to dose. Another logical choice could be for example:

$$P = \frac{S(par_1) - S(par_2)}{S(par_2)} \tag{6}$$

   where par_1 and par_2 are two different values of the imaging parameter 'par' (par = TE, TR , α, …). The motivation behind this functional form is that the subtraction in the nominator will remove any additive constant artefact, while the denominator will compensate (in first order) radiofrequency heterogeneity (B_1).
   c. Define the total volume-of-interest to be imaged and the imaging resolution. In multi-slice MRI, this will come down to the in-plane pixel resolution and the slice thickness. This will be translated in the matrix size (N_x, N_y) and the number of slices N_{sl} and voxel size $(\Delta x, \Delta y, \Delta z)$.
   d. Define the operating dose range and the available measurement time t_{tot} (i.e. the time in which the entire volume is to be scanned with the specified spatial resolution).
2. The signal intensity in the voxel with dimensions $(\Delta x, \Delta y, \Delta z)$ as a function of imaging parameters and radiation dose is determined by the type of sequence. See section 4 for some examples. In the signal equation, the relaxation rates are functions of dose ($R_k(D)$, k = 1,2), which depends on the type of gel dosimeter. In some cases, a linear relation can be assumed within a particular dose range. In other cases, a higher order functional relation (often sigmoidal) needs to be assumed. We have found that for many polymer gel types, a bi-exponential fit with offset fits the relation $R_2(D)$. While the inverse of such a bi-exponential function $D(R_2)$ is not trivial and requires numerical methods, we found that a bi-exponential function of the form described in equation 7 is applicable in practice [98].

$$D = a + be^{cR_2} - de^{-fR_2} \tag{7}$$

   where the fitting variables a, b, c, d and f are real positive.

3. The error propagation can now be either calculated or simulated. For a calculation of the dose error in the case of polymer gel dosimeters scanned with a single or multi spin-echo pulse sequence we refer to the literature [72, 74]. Alternatively, a numerical simulation can also be performed whereby Gaussian distributed noise is added to the signal S and propagated through the variable P(S) or $R_2(S)$ and further through the dose D by use of equation 7.
4. The noise propagation is repeated for different imaging parameters (TE, TR, α) until a minimum is found for the maximum standard deviation within the dose range $[0,D_{max}]$. The minimum can be found by either a brute force grid search over all imaging parameters or by use of a conjugate gradient minimization.

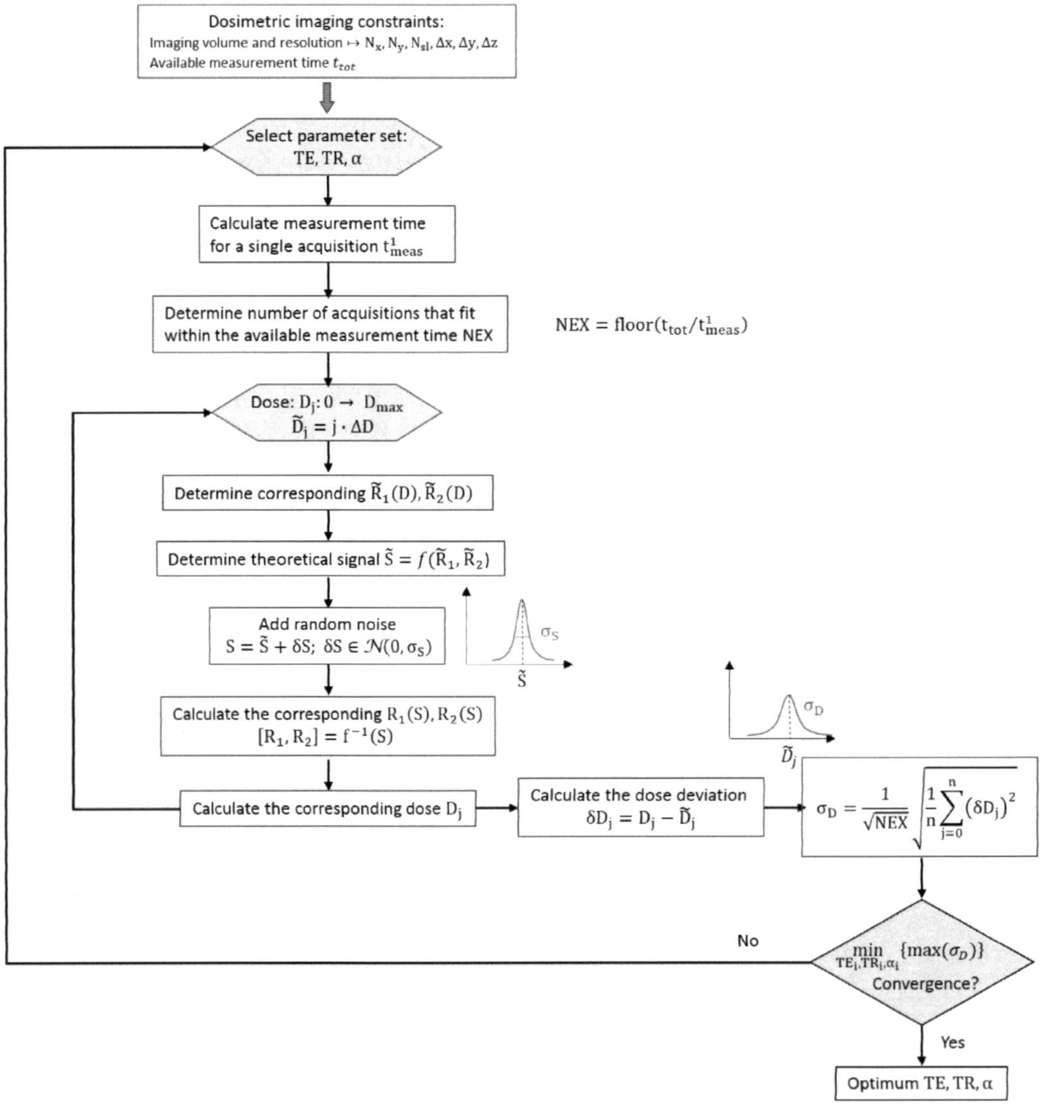

Figure 15. Flow chart of the optimization of sequence parameters TE, TR, α

5.2. Optimization for 4D radiation dosimetry with the MRI-Linac

The optimization of the parameters in a 4D measurement is similar to the optimization but here the aim is to obtain the smallest temporal uncertainty and the temporal resolution (number of scanned volumes within the measurement time) should match the temporal uncertainty as close as possible. Indeed, there is no advantage in having a higher scanning rate when the (stochastic) uncertainty in dose between different time frames is larger than the incremental change in dose. The temporal uncertainty (within a confidence level of p%) $\mathrm{TU_{p\%}}$ can be written as

$$\mathrm{TU_{p\%}} = \frac{\mathrm{D_\Delta^p}}{\dot{\mathrm{D}}} = \frac{\mathrm{k_p}\sqrt{2}\sigma_\mathrm{D}}{\dot{\mathrm{D}}} \tag{8}$$

where $\mathrm{D_\Delta^p}$ is the dose resolution and $\dot{\mathrm{D}}$ is the dose rate. For a confidence level of 95%, $k_p\sqrt{2} = 2.77$. Note that for a given dose rate, the temporal uncertainty is proportional to the standard deviation. The optimization problem can now be defined as "Find the minimum standard deviation within a scanning time frame that is as close as possible, but smaller than the temporal uncertainty $\mathrm{TU_{p\%}}$. Note that in contrast to the 3D optimization scheme described above, the available measurement time is flexible. Such an optimization scheme has been implemented in MATLAB for both the RARE sequence and a balanced SSFP readout, where in addition also the relaxation rates of the gel can be varied by adding an MRI contrast agent. The results are presented in another proceedings abstract [132]. We here describe briefly the approach and calculations for both sequences where the standard deviation on the dose is calculated (approximate) algebraically.

5.2.1. Rapid Acquisition with Relaxation Enhancement (RARE) sequence

The number of echoes that can fit within an effective echo time TE, which here also equals the echo train length (ETL) is given by $N_{ETL} = floor(TE/\Delta TE)$. The function $floor(\cdot)$ returns the highest integer value for which $N_{ETL} \cdot \Delta TE < TE$. The echo time spacing ΔTE is the minimum achievable echo time spacing in the sequence. This minimum value is to some degree also determined by the receive bandwidth. A larger receive bandwidth will typically result in a smaller achievable ΔTE, but this also results in a lower SNR.

The optimization approach starts with a choice of $\mathrm{N_{ETL}}$. The effective echo time then becomes $\mathrm{TE} = \mathrm{N_{ETL}\Delta TE} + \tau_\mathrm{D}$ where τ_D accounts for a dead time before another echo train can be switched. Another choice is the number of scanned slices within a single repetition time $\mathrm{N_{sl\,in\,TR}}$. The repetition time is then $\mathrm{TR} = \mathrm{N_{sl\,in\,TR}} \cdot (\mathrm{N_{ETL}\Delta TE} + \tau_\mathrm{D})$ and the minimum total time required to scan one k-space line for all ($\mathrm{N_{sl}}$) slices becomes:

$$\mathrm{T_m^1} = \mathrm{ceil}\left(\frac{\mathrm{N_{sl}}}{\mathrm{N_{sl\,in\,TR}}}\right)\mathrm{TR} \tag{9}$$

The measurement time for a single acquisition, is given by the product of $\mathrm{T_m^1}$ with the number of k-space line or phase encoding steps $\mathrm{N_{ph}}$, hence

$$\mathrm{T_m} = \mathrm{T_m^1} \cdot \mathrm{N_{ph}} \tag{10}$$

The total measurement for multiple acquisitions becomes

$$\mathrm{T_t} = \mathrm{T_m} \cdot \mathrm{NEX} = \mathrm{T_m^1} \cdot \mathrm{N_{ph}} \cdot \mathrm{NEX} \tag{11}$$

In the optimization problem, the number of acquisitions (experiments) is considered as an optimization variable.

Figure 16 illustrates the calculation of the total measurement time of a single scanning frame for the situation where $\mathrm{N_{ETL}} = 8$, $\mathrm{N_{sl\,in\,TR}} = 3$, $\mathrm{N_{sl}} = 5$, $\mathrm{TE} = 8$ ms, $\mathrm{N_{ph}} = 128$ and NEX = 3.

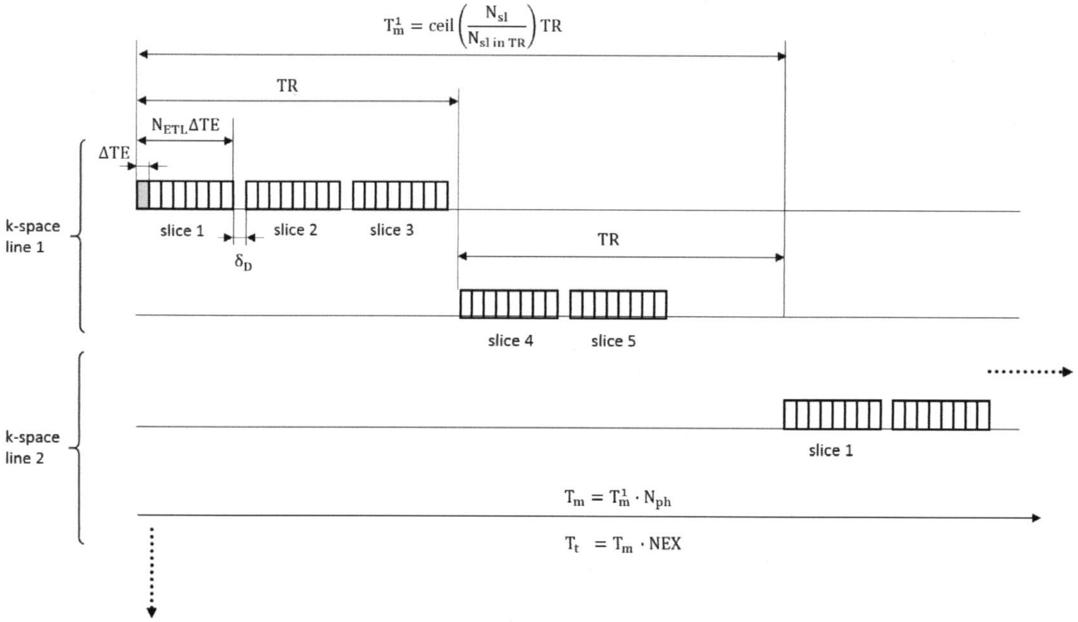

Figure 16. Sequence schematic for an example with $N_{ETL} = 8$, $N_{sl\ in\ TR} = 3$ and $N_{sl} = 5$. For the situation where $\Delta TE = 8$ ms, $N_{ph} = 128$ and NEX = 3, the minimum TR is 204 ms, the measurement time for one k-space line is $T_m^1 = 408$ ms, resulting in a measurement time for 1 acquisition $T_m = 52$ s and a total measurement time for 3 acquisitions (NEX = 3): $T_t = 2$ min 37 s

The relaxation rates can also be varied by adding a contrast agent such as gadobutrol [132].

The signal intensity can now be calculated for this set of imaging parameters and the concentration of contrast agent. For the RARE sequence, the signal intensity is given by:

$$S = S_0 e^{-R_2 TE}\left(1 - 2e^{-R_1\left(TR - \frac{TE}{2}\right)} + e^{-R_1 TR}\right) \tag{12}$$

In 4D radiation dosimetry, the absorbed dose D is estimated from the change in relaxation rate. Both the longitudinal relaxation rate $R_1(= 1/T_1)$ and transverse relaxation rate $R_2(= 1/T_2)$ depend on the dose, but R_1 is significantly less dependent on dose. As a first order approximation, one can assume a linear relationship between the relaxation rate and the dose.

$$R_1 = R_{1,0}^D + r_1^D D \approx R_{1,0}^D \tag{13}$$

$$R_2 = R_{2,0}^D + r_2^D D \tag{14}$$

where $R_{1,0}^D$ and $R_{2,0}^D$ are the intercept in the R_1-dose plot and R_2-dose plot respectively. r_2^D is the dose relaxivity, in other words, the change of transverse relaxation rate per unit of absorbed dose.
The R_2-dose response is only approximately linear in the interval [0 Gy, 2 Gy], while the R_1-dose response is independent of dose ($R_1 = R_{1,0}$). Note however that the non-linear R_2-dose response can also be considered in the calculations. When the polymer gel dosimeter is doped with a contrast agent, the intercepts $R_{1,0}^D$ and $R_{2,0}^D$ are also dependent on the concentration of contrast agent. There is also a small dependence of the transverse relaxivity r_2^D on the concentration of contrast agent but which in a first order approximation can be neglected [132].

$$R_{1,0}^{D} = R_{1,0} + r_1^{CA} D \tag{15}$$

$$R_{2,0}^{D} = R_{2,0} + r_2^{CA} D \tag{16}$$

The change in transverse relaxation rate ΔR_2 can be (in first order, neglecting R_1) obtained by taking the logarithm of the ratio of the signal intensity before radiation exposure and after radiation exposure.

$$\Delta R_2 = \frac{1}{TE} \ln\left(\frac{S(0)}{S(D)}\right) = r_2^D D \tag{17}$$

where S(D) is the signal intensity in the same voxel after absorbing an amount of radiation dose D and S(0) is the signal intensity in a voxel before radiation. Note the different meaning of S_0 and S(0).

The standard deviation on estimated dose can then be written as:

$$\sigma_D = \left|\left(\frac{\partial D}{\partial(\Delta R_2)}\right)\right| \sigma_{\Delta R_2} = \left|\frac{1}{r_2^D}\right| \cdot \sigma_{\Delta R_2} \tag{18}$$

and

$$\sigma_{\Delta R_2}^2 = \left(\frac{\partial(\Delta R_2)}{\partial(S(0))}\right)^2 \sigma_S^2 + \left(\frac{\partial(\Delta R_2)}{\partial(S(D))}\right)^2 \sigma_S^2 \tag{19}$$

which results in

$$\sigma_D = \frac{1}{r_2^D TE} \sqrt{1 + \left(\frac{S(0)}{S(D)}\right)^2} \left(\frac{\sigma_S}{S(0)}\right) = \frac{1}{r_2^D TE\ SNR} \sqrt{1 + e^{2 r_2^D D TE}} \tag{20}$$

In the derivation of equation 20, we ignored the T_1-effect and assumed that the R_2-dose response is linear. For a more exact approach, a numerical approach as outlined in section 5.1 can be followed.

A flow chart of the different steps in the algorithm to optimize the RARE sequence for inline MRI gel dosimetry on an MRI-Linac is shown in figure 17. First, the resolution and scanning volume are defined. This is characterized by the number of phase encoding lines (N_{ph}), number of points in the readout direction (N_{ro}), the number of slices (N_{sl}) and the voxel dimensions ($\Delta x, \Delta y, \Delta z$). The echo train length (N_{ETL}) is to be optimized. Also the repetition time is an optimization variable which is defined by the number of slices that are acquired within a repetition loop ($N_{sl\ in\ TR}$). The optimization algorithm will iterate and update these parameters until convergence is obtained. To account for the imaging time, it is allowed to change the number of acquisitions (NEX) that will contribute to the same image set. The three optimization variables (N_{ETL}, $N_{sl\ in\ TR}$ and NEX) will determine the total imaging time T_t (equations 9-11). For each set of optimization variables, the standard deviation in a predefined dose range [0, D_{max}] is calculated (equation 20) or simulated by adding normal distributed noise to the signal amplitude and propagating this to the estimated dose D. The standard deviation defines the temporal uncertainty. It is important to note that for each iteration step in the inner optimization loop, both the total imaging time T_t and temporal uncertainty $TU_{p\%}$ are updated. The optimization is continued until the total imaging time becomes larger than the temporal uncertainty, which defines the number of possible acquisitions as NEX-1 for this particular set of optimization variables. The maximum standard deviation on the dose within the dose range [0, D_{max}] is calculated and the corresponding temporal uncertainty is compared against the until-then obtained minimum value. The optimization of the variables N_{ETL} and $N_{sl\ in\ TR}$ is continued until convergence is obtained. The optimization can be conducted by use of a conjugate gradient minimization or by scanning in a two dimensional grid of the parameter space (N_{ETL},$N_{sl\ in\ TR}$). Both techniques can be combined to find the global minimum.

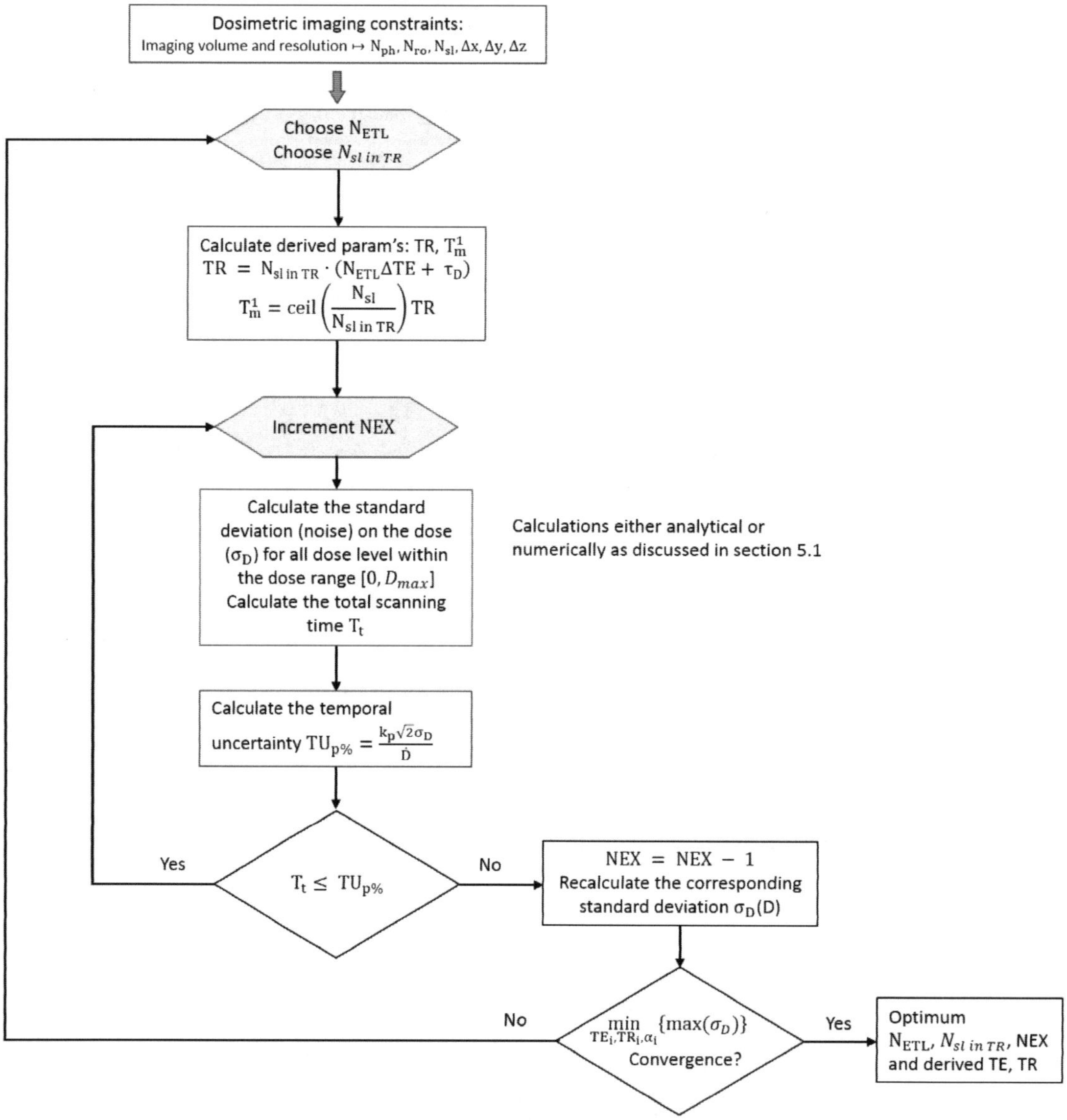

Figure 17. Flow chart showing the different steps in optimization of the RARE sequence: Two loops can be considered. The parameters N_{ETL} and $N_{sl\ in\ TR}$ are optimized in the outer loop.

5.2.2. Balanced Steady State Free Precession (bSSFP, trueFISP)

For the bSSFP pulse sequence the optimization is

$$\min_{\alpha \in \left[0, \frac{\pi}{2}\right]} \sigma_D \sqrt{T_m} \tag{21}$$

With bSSFP, it is a reasonable choice to consider the relative signal change $R = \frac{S(0) - S(D)}{S(0)}$ as the dose correlating metric.

The signal intensity in a voxel of a bSSFP image with TR $<<$ T_1, T_2 is given by:

$$S = S_0 \frac{\sin\alpha}{(1+\cos\alpha)+(1-\cos\alpha)(R_2/R_1)} e^{-R_2 TE} \quad (22)$$

A similar derivation leads to the expression for σ_D:

$$\sigma_D = -\left(\frac{\partial S(D)}{\partial D}\right)^{-1} \sqrt{1+(1-R)^2}\sigma_S \quad (23)$$

Minimization of the objective function σ_D towards concentrations of contrast agent and different flip angles can be performed numerically.

Note that with the bSSFP sequence, the scanning time is immediately defined by the minimum TR that is achievable in the sequence, the number of lines in the phase encoding direction and the number of slices.

6. How to start with gel dosimetry and MRI readout: Do's and don'ts

There are many different kinds of gel dosimeters and there exists a vast amount of scientific literature on the subject which may appear overwhelming at first and not all information is as valuable when starting with implementing gel dosimetry. The sequence below is not scientific but can be seen as a piece of advice that may help in bridging some obstacles in establishing gel dosimetry in a clinical setting.

1. ***Choose a gel dosimeter***: The choice of the gel dosimeter may depend on the application and practical considerations such as availability (e.g. commercial versus in-house fabrication). Paramount in the choice of a gel dosimeter are the physico-chemical and radiation properties, such as stability, temperature sensitivity and dose rate dependence. Some of the sensitivities such as temperature sensitivity can be compensated in the experimental design but requires a careful control during storage, irradiation and/or scanning. Not all the properties have been reported for all types of gel dosimeters (including commercial gel dosimeters) and it is important to be aware of possible related uncertainties.
   Gel dosimeters with the highest dose sensitivity are not always the best. Other properties of the gel dosimeter may result in higher dose uncertainties and stochastic uncertainties depend in a rather non-intuitive way on the interplay between dose response and MRI scanning parameters.
2. ***Choose an MR imaging sequence***: Several pulse sequences are available on standard clinical MRI scanners. However, most pulse sequences are designed for diagnostic imaging purposes where image uniformity and geometrical distortions are less of a concern. It is important to spend some time in optimizing the imaging protocol to acquire reliable quantitative MR images. For 3D polymer gel dosimetry, the most used pulse sequence is a multi-spin-echo (CPMG) sequence. To process the images in a quantitative MRI map, a pixel-wise fitting algorithm is required. Commercial and freeware software can be applied but it is important to keep in mind that the fitting algorithm also has an influence on stochastic dose uncertainties. For larger phantoms (such as whole abdominopelvic), correction for image non-uniformity may be required. In the case of gel dosimetry in regions with magnetic susceptibility differences such as air cavities or implants, B_0-field corrections may be required. For lung-equivalent gel dosimetry, a magnetization transfer imaging sequence may be required [97].
   Do not treat MRI as a black box. It is more efficient to obtain a good understanding of the inner workings of MRI, than trying to optimize the imaging design by trial-and-error.
3. ***Choose optimal sequence parameters:*** Optimization strategies are discussed in section 5 and the corresponding literature for both 3D [72, 74] and 4D [98] imaging. The importance of sequence parameter optimization should not be underestimated as the stochastic dose uncertainty can be decreased significantly by optimization.
4. ***Choose a recipient:*** Two kinds of approaches can be taken. If the dose distribution towards the skin of the patient is not important, one may choose to use a phantom with a cylindrical gel

insert and the gel dosimeter can be poured in a glass recipient. This is the approach that is mostly used with optical CT. However, if the entire dose distribution up to the skin is needed, a humanoid shaped cast is needed. Such a cast is preferably made from Barex®, glass or any other material with very low oxygen permeability.

5. ***Investigative testing:*** It is important to obtain a good sense of the expected uniformity and geometrical distortions. The recipient can be filled with a 'blank'-gel. The blank gel should have NMR relaxation rates close to the gel dosimetry phantom. Fiducial markers can be placed on the phantom. The phantom can then be scanned with MRI and/or CT and the image data set can be used as the template for the treatment planning (simulation). As the gel dosimeter will not have anatomical structures to use in the treatment planning (i.e. critical organs and CTV), anatomical structures can be virtually created by transferring them from a patient data set onto the planning template by use of a morphological affine transformation. The treatment should be planned on the basis of this data set and not on any other anatomical data set which may have a different geometry compared to the gel dosimeter phantom.
6. ***Fabricate the gel dosimeter:*** In fabricating the gel dosimeter make sure to control the temperature and storage conditions as much as possible. This will facilitate the reproducibility of the experiment. If a large volume of gel is used (e.g. > 10 L), it is important to make sure that temperature changes within the phantom occur relatively uniform. Avoid air pockets in the gel dosimeter. *Even normoxic gel dosimeters are still sensitive to oxygen* [21].
7. ***Perform the polymer gel dosimetry experiment:*** It may be necessary to scale the radiation output (MU) to optimize the response (relative dose resolution). However, such approach may not be possible in applications where the absolute dose as delivered to the patient needs to be matched, such as inline MRI-Linac treatments. In these cases, a highly dose sensitive gel dosimeter is required. The maximum delivered radiation dose may be limited as a result of loss of spatial integrity [45]. Some temperature stabilization and/or control may be required during the radiation and scanning. When scanning the gel dosimeter, it is advisable to scan both the phantom and calibration vials together to minimize any uncertainties as a result of instability. Calibration vials can be taped onto the phantom during scanning.
8. ***Compare the gel dosimetry acquired dose maps with the treatment plan:*** A gamma-evaluation can be used to assess the correspondence between the intended dose distribution and the delivered dose distribution. Use gamma pass rates with great care! The main purpose of a gamma evaluation is to highlight regions where the difference between both dose distributions exceeds an acceptable level. The spatial information on where gamma > 1 is important. In some regions one can tolerate a larger gamma-deviation than in others. Gamma pass rates depend strongly on the volume in which gamma is evaluated. For inter-comparisons of treatment approaches within the same centre and same treatment site (brain, H&N, abdominopelvic), a valid approach may be to extract gamma pass rates in a region defined by a lower isodose boundary. However, *much care is needed to use gamma pass rates as absolute instruments or for the comparison between different treatment sites as the location and size of the target may have a significant influence on the expected pass rate.*

7. Conclusion - An outlook to the future

Since the early developments of polymer gel dosimetry, scholars expected a broader application of gel dosimetry in clinical practice. There are some reports where gel dosimetry has been applied successfully in clinical validation of high precision treatments and where findings had an impact on patient treatment. The limited dissemination of gel dosimetry in the hospital is multi-faceted. There has been a continuous search for more sensitive, less toxic and more reliable gel dosimeter systems and readout technologies and physico-chemical and radiation parameters have been documented for several gel dosimeters. Optical CT scanning with radiochromic plastic dosimeters has been presented as an alternative method for 3D gel dosimetry as a more accessible and easier-to-use technology. However, until now, applications with radiochromic plastics have been limited to relatively small and canonical volumes (cylindrical). While some researchers have commented on the radiation properties of some radiation sensitive plastics, there are some reports that mention that materials such as PRESAGE® are not entirely immune to variations in dose response or physico-chemical induced uncertainties. The question on which dosimeter system to use depends on the application. For relatively small volumes that are confined to cylindrical shapes including inserts, radiochromic dosimeters with optical readout may be a preferred choice. However, for dosimetry in large anthropomorphic shaped regions and some emerging applications such as inline MRI-Linac with MRgRT tracking, lung dosimetry, dosimetry around deformable organs, polymer gel dosimetry with MRI readout is currently our preferred solution. However, improvements and further developments of MRI-based polymer gel dosimetry for these applications is still required. FLASH-RT poses another challenge for the entire dosimetry community. Both polymer gel dosimeters and radiochromic plastics are based on radical-induced reactions, where the equilibrium between initiation and termination reactions is responsible for dose-rate dependence. Dosimetry of FLASH-RT may be very challenging for such systems. However, gel dosimeters can be helpful in studying the mechanisms behind FLASH-RT.

8. References

[1] Mijnheer B 2018, *Clinical 3D Dosimetry in Modern Radiation Therapy*, CRC Press.
[2] Fricke H and Morse S 1927 *Am. J. Roentgenol. Radium Therapy* **18** 430–2.
[3] Day MJ and Stein G 1950 *Nature* **166** 146-7.
[4] Andrews HL, Murphy RE and LeBrun EJ 1957 *Rev. Sci. Instrum.* **28** 329-32.
[5] Alexander P, Charlesby A and Ross M 1954 *Proc. R. Soc. A*, **223** 392.
[6] Hoecker FE and Watkins IW 1958 *Int. J. Appl. Radiat. Isot.* **3** 31-5.
[7] Boni AL 1961 *Radiat. Res.* **14** 374-80.
[8] Gore C, Yang YS and Schulz RJ 1984 *Phys. Med. Biol.* **29**(10) 1189-97.
[9] De Deene Y 2020, Gel-based Radiation Dosimetry using quantitative MRI, In: *NMR and MRI of Gels* (ed: De Deene Y), RSC 275-357.
[10] Maryanski MJ, Gore JC, Kennan RP and Schulz RJ 1993 *Magn. Reson. Imaging* **11** 253-8.
[11] De Deene Y and Baldock C 2004 *J. Phys.: Conf. Ser.* **3**, doi: 10.1088/1742-6596/3/1/E01
[12] Lepage M 2006 *J. Phys.: Conf. Ser.* **56**, doi: 10.1088/1742-6596/56/1
[13] Maris T and Pappas E 2009 *J. Phys.: Conf. Ser.* **164**, doi: 10.1088/1742-6596/164/1/011001
[14] Oldham M 2010 *J. Phys.: Conf. Ser.* **250**, doi: 10.1088/1742-6596/250/1/011001
[15] Thwaites D and Baldock C 2013 *J. Phys.: Conf. Ser.* **444**, doi: 10.1088/1742-6596/444/1/011001
[16] Olsson LE, Bäck S and Ceberg S 2015 *J. Phys.: Conf. Ser.* **573**, doi: 10.1088/1742-6596/573/1/011001
[17] Oldham M 2017 *J. Phys.: Conf. Ser.* **847**, doi: 10.1088/1742-6596/847/1/011001
[18] Oldham M and Yin F-F 2019 *J. Phys.: Conf. Ser.* **1305**, doi: 10.1088/1742-6596/1305/1/011001
[19] Schreiner LJ, Archambault L and Beaulieu L 2022 *J. Phys.: Conf. Ser.* **2167**, doi: 10.1088/1742-6596/2167/1/011001
[20] Baldock C, De Deene Y, Doran S *et al* 2010 *Phys. Med. Biol.* **55** R1-63
[21] Vandecasteele J and De Deene Y 2013 *Phys. Med. Biol.* **58** 43-61
[22] De Deene Y and Vandecasteele J 2013 *J. Phys. Conf. Ser.* **444** 012015
[23] De Deene Y and Jirasek A 2015 *J. Phys. Conf. Ser.* **573** 012008
[24] Baldock C, Rintoul L, Keevil SF et al 1998 *Phys. Med. Biol.* **47** 3617-27
[25] Hepworth SJ, Leach MO and Doran SJ 1999 *Phys. Med. Biol.* **44** 1875-84
[26] De Deene Y, Hanselaer P, De Wagter C *et al* 2000 *Phys. Med. Biol.* **45** 859-78
[27] Jirasek AI and Duzenli C 2001 *Phys. Med. Biol.* **46** 1949-61
[28] Lepage M, Whittaker AK, Rintoul L *et al* 2001 *J. Appl. Polym. Sc.* **79** 1572-81
[29] Lepage M, Whittaker AK, Rintoul L *et al* 2001 *Phys. Med. Biol.* **46** 2827-39
[30] De Deene Y, Hurley C, Venning A et al 2002 *Phys. Med. Biol.* **47** 3441-63
[31] Salomons GJ, Park YS, McAuley KB *et al* 2002 *Phys. Med. Biol.* **47** 1435-48
[32] Jirasek AI and Duzenli C 2002 *Med. Phys.* **29** 569-77

[33] Fuxman AM, McAuley KB and Schreiner LJ 2003 *Macromol. Theory Simul.* **12** 647-62
[34] Fuxman AM, McAuley KB and Schreiner LJ 2005 *Chem. Engin. Sc.* **60** 1277-93
[35] Babic S and Schreiner LJ 2006 *Phys. Med. Biol.* **51** 4171-87
[36] Jirasek A, Hilts M, Shaw C *et al* 2010 *Phys. Med. Biol.* **51** 1891-906
[37] Kozicki M 2011 *Rad. Phys. Chem.* **80** 1419-36
[38] Jaszczak M, Wach R, Maras P 2018 *Phys. Med. Biol.* **63** 175010
[39] Jaszczak M, Kolesińska B, Wach R 2019 *Phys. Med. Biol.* **64** 035019
[40] Jaszczak M, Maras P and Kozicki M 2020 *Radiat. Phys. Chem.* **177** 109125
[41] De Deene Y, Vergote K, Claeys C *et al* 2006 *Phys. Med. Biol.* **51** 653-73
[42] De Deene Y, Pittomvils G and Visalatchi S 2007 *Phys. Med. Biol.* **52** 2719-28
[43] De Deene Y, Venning A, Hurley C 2002 *Phys. Med. Biol.* **47** 2459-70
[44] Taylor ML, Franich RD, Johnston PN *et al* 2007 *Phys. Med. Biol.* **52** 3991-4005
[45] Vergote K, De Deene Y, VandenBussche E *et al* 2004 *Phys. Med. Biol.* **49** 4507-522
[46] Hurley C, Venning A and Baldock C 2005 *Appl. Rad. Isot.* **63** 443-56
[47] Khan M, Hellemann G, Kuess P *et al* 2018 *Phys. Med. Biol.* **63** 06NT01
[48] Maryanski MJ, Audet C and Gore JC 1997 *Phys. Med. Biol.* **42** 303-11
[49] McJury M, Oldham M, Leach MO *et al* 1999 *Phys. Med. Biol.* **44** 1863-73
[50] Pappas E, Angelopoulos A, Kipouros P *et al* 2003 *Phys. Med. Biol.* **48** N65-N73
[51] Sedaghat M, Bujold R and Lepage M 2010 *Phys. Med. Biol.* **57** 5981-994
[52] Venning A, Healy B, Nitschke K *et al* 2005 *Nucl. Instrum. Meth. Phys. Res.* A **555** 396-402
[53] Venning A, Hill B, Brindha S *et al* 2005 *Phys. Med. Biol.* **50** 3875-88
[54] Brown S, Venning A, De Deene Y *et al* 2008 *Appl. Rad. Isot.* **66** 1970-4
[55] Trapp JV, Michael G, De Deene Y *et al* 2002 *Phys. Med. Biol.* **47** 4247
[56] Hill R, Holloway L and Baldock C 2005 *Phys. Med. Biol.* **50** N331-N344
[57] Lepage M, Whittaker AK, Rintoul L *et al* 2001 *Phys. Med. Biol.* **46** 1061-74
[58] Kennan RP, Richardson KA, Zhong J *et al* 1996 *J. Magn. Reson. B* **110** 267-77
[59] Spevacek V, Novotny J, Dvorak P *et al* 2001 *Med. Phys.* **28** 2370-8
[60] Quevedo A, Luo G, Halhardo E *et al* 2018 *Phys. Med. Biol.* **63** 2370-8
[61] Vandecasteele J and De Deene Y 2013 *Phys. Med. Biol.* **58** 63-85
[62] De Deene Y, De Wagter C, De Neve W 2000 *Phys. Med. Biol.* **45** 1807-23
[63] De Deene Y, De Wagter C, De Neve W 2000 *Phys. Med. Biol.* **45** 1825-39
[64] De Deene Y and De Wagter C 2001 *Phys. Med. Biol.* **46** 2697-711
[65] De Deene Y, Reynaert N and De Wagter C 2001 *Phys. Med. Biol.* **46** 2801-25
[66] De Deene Y 2004 *J. Phys. Conf. Ser.* **3** 34-57
[67] Lepage M, Tofts PS, Back SAJ *et al* 2001 *Magn. Reson. Med.* **46** 1123-29
[68] Hurley C, De Deene Y, Meder R *et al* 2003 *Phys. Med. Biol.* **48** 3043-58
[69] Gochberg DF, Fong PM and Gore JC 2001 *Phys. Med. Biol.* **46** 799-811
[70] Vergote K, De Deene Y, Duthoy W *et al* 2004 *Phys. Med. Biol.* **49** 287-305
[71] Watanabe Y, Perera GM and Mooij RB 2002 *Med. Phys.* **29** 797-802
[72] De Deene Y, Van de Walle R, Achten E *et al* 1998 *Sign. Proc.* **70** 85-101
[73] Baldock C, Murry P and Kron T 1999 *Phys. Med. Biol.* **44** N243-N246
[74] De Deene Y and Baldock C 2002 *Phys. Med. Biol.* **47** 3117-41
[75] Vandecasteele J and De Deene Y 2013 *Phys. Med. Biol.* **58** 19-42
[76] Hilts M, Audet C, Duzenli C 2000 *et al Phys. Med. Biol.* **45** 2559-71
[77] Hilts M 2006 *J. Phys. Conf. Ser.* **56** 95
[78] Jirasek A, Hilts M, McAuley KB 2010 *Phys. Med. Biol.* **55** 5269-81
[79] Gore JC, Ranade M, Maryanski MJ *et al* 1996 *Phys. Med. Biol.* **41** 2695-704.
[80] Kelly RG, Jordan KJ, Battista JJ 1998 *Med Phys* **25** 1741-50.
[81] Oldham M, Siewerdsen JH, Shetty A *et al* 2001 *Med. Phys.* **28**:1436-45.
[82] Doran SJ, Krstajić N 2006 *J. Phys. Conf. Ser.* **56** 45
[83] Doran SJ, Koerkamp KK, Bero MA *et al* 2001 *Phys. Med. Biol.* **46** 3191-213

[84] Campbell WG, Rudko DA, Braam NA *et al* 2013 *Med. Phys.* **40** 061712.
[85] Mather ML and Baldock C 2003 *Med. Phys.* **30** 2140-8
[86] De Neve W, De Gersem W, Derycke S E *et al* 1999 *Radiotherapy and Oncology* **50** 301-14
[87] Sherouse GW, Thorn J, Novins K *et al* 1989 *Med. Phys.* **16** 466
[88] De Gersem W, Claus F, De Wagter C 2001 *Int. J. Radiation Oncology Biol. Phys.* **51** 849-59
[89] De Gersem W, Claus F, De Wagter C 2001 *Int. J. Radiation Oncology Biol. Phys.* **51** 1371-88
[90] De Deene Y, De Wagter C, Van Duyse B *et al* 2000 *Magn. Reson. Med.* **43** 116-25
[91] Vergote K, De Deene Y, Claus F *et al* 2003 *Radiotherapy and Oncology* **67** 119-128
[92] Vandecasteele J and De Deene Y 2013 *Phys. Med. Biol.* **58** 6241-62
[93] Vandecasteele J, Ghysel S, Baete S *et al* 2003 *Phys. Med. Biol.* **56** 627
[94] Duthoy W, De Gersem W, Vergote K *et al* 2003 *Int. J. Radiation Oncology Biol. Phys.* **57** 1019-32
[95] Vergote K, De Deene Y, Duthoy W *et al* 2004 *Phys. Med. Biol.* **49** 287-305
[96] Bohsung J, Gillis S, Arrans R *et al* 2005 *Radiotherapy and Oncology* **76** 354-61
[97] De Deene Y, Vergote K, Claeys C 2006 *Med. Phys.* **33** 2586-97
[98] De Deene Y, Wheatley M, Dong B *et al* 2020 *Phys. Med. Biol.* **65** 225031
[99] Hardwick TJ 1952 *Canadian Journal of Chemistry* **30**(1) 17-22.
[100] Scharf K and Lee RM, *Radiation Research* **16** 115-24.
[101] Fricke H, Hart EJ and Smith HP 1938 *Journal of Chemical Physics* **6**(5) 229-40.
[102] Allen AO 1962 *Radiation Research* **17** 255-61.
[103] Schuler RH and Allen AO 1957 *Journal of the American Chemical Society*, **79**(7), 1565-72.
[104] Spinks JWT and Woods RJ 1990 *Introduction to Radiation Chemistry*, 3rd edition, Wiley-Interscience.
[105] Bloembergen N, Purcell EM and Pound RV 1948 *Physical Review* **73**(7), 679-715
[106] Slichter CP 1978 In: *Principles of Magnetic Resonance* Springer-Verlag Berlin Heidelberg
[107] Solomon I 1955 *Physical Review* **99**(2), 559-65.
[108] Connick RE and Fiat D 1966 *Journal of Chemical Physics* **44**(11), 4103-7.
[109] Olsson LE 1991 *PhD thesis* Malmö, Lund University, Sweden
[110] Appleby A, Christman EA and Leghrouz A 1987 *Med. Phys.* **14**(3) 382-4
[111] Olsson LE, Petersson S, Ahlgren L *et al* 1989 *Phys. Med. Biol.* **34**(1), 43-52
[112] Schulz RJ, deGuzman AF, Nguyen DB *et al* 1990 *Phys. Med. Biol.* **35**(12), 1611-22
[113] Duzenli C, Sloboda R and Robinson D 1994 *Phys. Med. Biol.* **39**, 1577-92
[114] Appleby A and Leghrouz A 1991 *Med. Phys.* 1**8**(2), 309-12
[115] Healy BJ, Zahmatkesh MH, Nitschke KN *et al* 2003 *Med. Phys.* **30**(9), 2282-91
[116] Hill R, Back SAJ, Lepage M *et al* 2002 *Phys. Med. Biol.* **47**, 4233-46
[117] Gambarini G, Arrigoni S, Cantone MC *et al* 1990 *Phys. Med. Biol.* **35**(12), 1611-22
[118] Kron T, Jonas D and Pope JM 1997 *Magn. Reson Imaging* **15**(2) 211-21
[119] Chu KC, Jordan KJ, Battista JJ *et al* 2000 *Phys. Med. Biol.* **45** 955-69
[120] Woessner DE 1962 *J. Chem. Phys.* **37** 647-54
[121] De Deene Y 2006 *J. Phys. Conf. Ser.* **56** 72-85
[122] Bernstein MA, King KF and Zhou XJ 2004 Handbook of MRI Pulse Sequences, Elsevier Academic Press, Burlington, Masachusetts, USA, 2004.
[123] Deoni SCL, Rutt BK and Peters TM 2003 *Magn. Reson. Med.* **49** 515-26
[124] De Deene Y 2009 *J. Phys. Conf. Ser.* **164** 012033, 42 pages
[125] De Deene Y 2010 *J. Phys. Conf. Ser.* **250** 012015, 10 pages
[126] Meiboom S and Purcell EM 1958 *Rev. Sci. Instrum.* **29** 688-91
[127] Look DC and Locker DR 1970 *Rev. Sci. Instrum.* **41** 250-1
[128] Brix G, Schad LR, Deimling M et al 1990 *Magn. Reson. Imaging* 351-6
[129] Magnusson P, Back SAJ and Olsson LE 2000 *Magn. Reson. Imaging* **17** 1357-70
[130] Lepage M, McMahon K, Galloway GJ et al 2002 *Phys. Med. Biol.* **47** 1881-90
[131] Low D A *Med. Phys.* **25** 656-61

Journal of Physics: Conference Series **2630** (2023) 012003 doi:10.1088/1742-6596/2630/1/012003

[132] De Deene Y and Mason D 2023 *J. Phys. Conf. Series* In press [Reference to the proceedings paper in the same volume]

CT and CBCT imaging of gel dosimeters: a review of the state of the art

A. Jirasek[1], M. Hilts[1,2]
[1]Dept. Physics, University of British Columbia – Okanagan campus, Kelowna BC, Canada
[2]Medical Physics Dept., BC Cancer – Kelowna, Kelowna BC, Canada

E-mail: andrew.jirasek@ubc.ca

Abstract. This review covers the main developments in x-ray CT-based polymer gel dosimetry (PGD) and more recently on-board cone-beam CT (CBCT) polymer gel dosimetry. Fundamental work as well as clinically-based applications are described and advantages and disadvantages of approaches highlighted. The body of work present within the literature points towards an overall feasibility of CT PGD and increasingly CBCT PGD as viable tools for the verification of highly complex, high dose radiotherapy treatments.

1. Introduction

Currently there exist three primary modalities for imaging polymer gel dosimeters (PGD): magnetic resonance imaging (MRI), optical computed tomography (optical CT) and x-ray CT [1, 3]. Additional modalities have been shown as proof-of-concept, including ultrasound imaging (US) and spectroscopic techniques [4, 5], however, these have not been widely pursued in recent years. Of the three main imaging modalities used to image PGD, x-ray CT (herein referred to simply as CT) was the last to be introduced [3] and perhaps remains the least investigated. A primary reason for the low uptake of CT PGD is the overall low dose resolution of the system [6, 7]. Nonetheless, CT PGD has numerous practical advantages, including but not limited to: the access to CT imagers in the form of CT simulators within radiotherapy clinics; ease of use; integration within clinical workflow; and minimal additional operating cost. With these advantages in mind, this article provides a high-level review of the main areas of concentration within CT PGD research. Some effort is made to concentrate the review on more recent work, e.g. development of a deformable CT PGD system, as reviews of earlier works can be found in previous overviews [8-10]. Recently, preliminary work has been undertaken to establish a novel PGD technique using on-board cone-beam CT (CBCT) for imaging and this review will also introduce this new area of PGD research. The review is organized as follows: summary of main technique developments in CT PGD (section 2); an introduction to CBCT PGD (section 3); applications of CT and CBCT PGD (section 4); followed by summary and conclusions (section 5).

2. CT PGD Technique Development

A primary focus of researchers in CT PGD has been on improving the dose resolution of the overall system. This effort has been multi-pronged and has covered areas of gel development (section 2.1), image acquisition optimization (section 2.2), and understanding the effects of CT dose on image quality and accuracy (section 2.3). Further investigations have involved characterizing the overall accuracy and precision of CT PGD (section 2.4) and the use of Monte Carlo techniques to model a full CT PGD system (section 2.5).

2.1.Gel optimization

Early work in CT PGD used acrylamide based PGDs, due to these systems being, at the time, the most widely studied in the literature [3, 11, 12]. However, the desire for increased dose sensitivity, simpler PDG manufacturing processes and the use of less toxic monomers lead to the investigation of alternate formulations of CT-based PGDs. Hill *et al* [13, 14] and Koeva *et al* [15] provided initial results on efforts to utilized normoxic formulations and to improve PGD dose sensitivity for CT-based PGD. Further work, both with and without using additional co-solvents, lead to a PGD based on (i) N-isopropyl-acylamide (NIPAM) as monomer and (ii) tetrakis-hydroxymethyl-phosphonium chloride as antioxidant [16-18]. This NIPAM-based PGD, formulation as given in table 1 [19], remains the standard formulation for CT PGD used by many researchers.

Table 1. PGD formulation for CT imaging.

	Concentration (by weight)
N-isopropylacrylamide (NIPAM)	15%
N,N'methylenebisacrylamide (Bis)	4.5%
Gelatin	5%
Water	75.5%
Tetrakis hydroxymethyl phosphonium chloride	5-10mM

While it has been shown that the formulation of table 1 does not provide a linear dose response [19], the system does produce an overall ~17-20H change in CT number over 20Gy (i.e. global average response of ~0.8-1H/Gy) [19]. It has been further shown that the formulation of table 1 does exhibit some dose rate dependence [20], although the severity of the variability is highly application dependent. Additional studies looking at alternate formulations have been conducted with the goal of improving dosimeter sensitivity, however, a more sensitive dosimeter than that described in table 1 has not, to the best of our knowledge, been reported nor widely adopted [21, 22].

2.2.Imaging optimization

2.2.1.*Parameter selection and background subtraction.* Early work in establishing the optimal scanning protocol concentrated on utilizing averages of multiple acquisitions of a single slice, with concomitant subtraction of a similarly-acquired background (blank, or unirradiated) gel [3]. Since this time, efforts have focused on optimization of CT image acquisition protocols, both for single and multi-slice CT scanners, for PGD [3, 23-27]. Cumulatively this work showed that tube voltage, current, slice thickness, acquisition time, and reconstruction filters all play an important role in establishing the overall image quality. The quantitative effects of imaging parameter on image quality are summarized in table 2. Johnston *et al* [26] illustrated that multi-slice CT scanners can be utilized for PGD. They achieved accurate results utilizing a slice-by-slice background subtraction technique to remove variability between simultaneously obtained slices. Given observed advantages over single slice CT of significant decreases in imaging time and tube load, multi-slice CT imaging is now recommended for x-ray CT PGD.

Table 2. Quantitative effects of imaging acquisition parameters on CT image noise [27].

Phantom diameter (d)	$\sigma_{NCT} \propto e^{d}$
Tube voltage (kV)	$\sigma_{NCT} \propto (kV)^{-1.4}$
Tube current (mA)	$\sigma_{NCT} \propto (mA)^{-0.5}$
Slice scan time (s)	$\sigma_{NCT} \propto s^{-0.5}$
Number of averages (NAX)	$\sigma_{NCT} \propto (NAX)^{-0.5}$
Pixel dimension (w)	$\sigma_{NCT} \propto e^{w}$ (or $w^{-0.65}$)
Slice thickness (h)	$\sigma_{NCT} \propto h^{-0.5}$

Additionally, some work has been done in establishing the effects of reconstruction technique on subsequent CT PGD image quality [28]. However, while promising, this avenue of investigation can require dedicated vendor agreements and support and has not been widely pursued.

2.2.2.*Image filtering.* Image filtering of averaged, background-subtracted CT images of PGDs has been widely investigated [22, 29-31]. In summary, image filtering has a profound effect on image signal to noise ratio (SNR) and has thus been established as in important tool in CT PGD due to the low contrast inherent in this system. While a multitude of filters can be employed [30], the adaptive mean (AM) filter has been shown to be effective at SNR image improvements (over unfiltered images) while maintaining edge preservation [29]. Figure 1 illustrates the potential efficacy of adaptive mean filtering on CT PGD images.

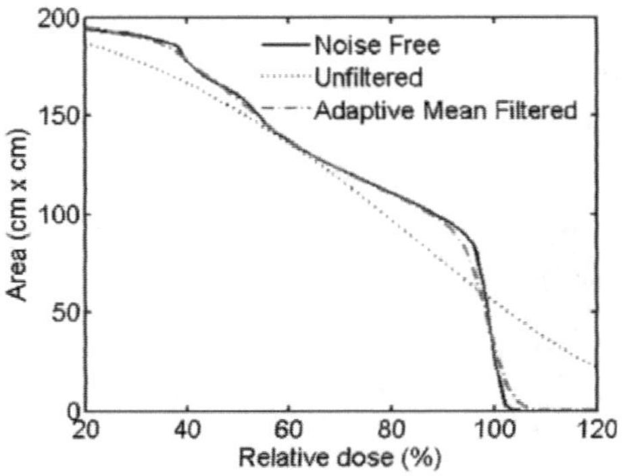

Figure 1. Representative dose-area histograms of noise-free (synthetic image), unfiltered (noise-added image), and adaptive mean filtered (window = 7, iterations = 2) PGD images [29].

In addition to basic statistical noise filtering, some work has been done to establish a more dedicated, remnant artefact removal technique for the removal of mild baseline inhomogeneities within processed PGD images [32]. The technique works via the application of a Savitsky-Golay filter of span *s* (width of filter), degree *d* (polynomial order) and number of iterations/applications *n*. The filter runs a first iteration that establishes the baseline signal. Artefacts above this baseline are removed, thus creating a new image to which the second iteration of the filter is applied. Figure 2 illustrates the efficacy of AM filtering and AM filtering + remnant artefact removal for irradiated PGD profiles. At present, AM filtering + remnant artefact removal is the standard for performing CT PGD.

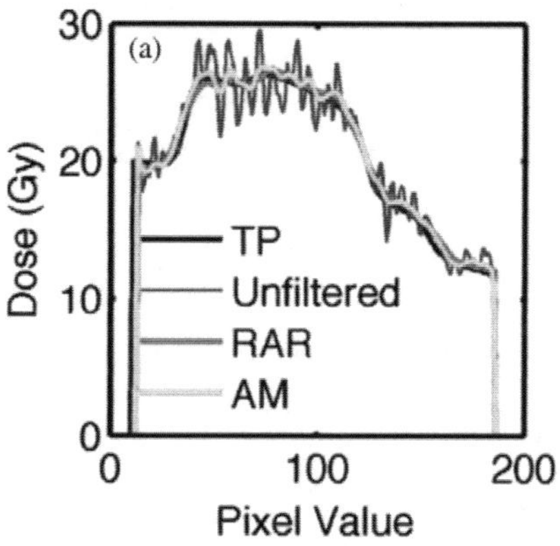

Figure 2. Profile through an irradiated PGD image. Shown are the treatment plan profile (black), unfiltered PGD profile (blue), adaptive mean filtered profile (green) and adaptive mean + remnant artefact removal (RAR, red) [32].

2.3.*CT Dose considerations.*

A central question from the beginning of CT PGD was what, if any, was the impact of CT radiation dose delivered due to the process of imaging the dosimeter? This question has been studied by Baxter *et al* [33] in some detail. In summary of their work, it was established that under optimized imaging protocols (table 2) with 16-25 image averages a ~<0.2 H change in CT number can be expected for acylamide based PGDs imaged while still radiosensitive. Given NIPAM-based PGDs (table 1) exhibit similar or lower sensitivity to the acrylamide formulation therein studied, we would expect a similar impact of CT dose on the NIPAM dosimeters. However, the full extent of this postulation has not been investigated.

Interestingly, Hill *et al* [14, 34] turned the question of CT dose around and used PGDs to *measure* CTDI within typical CT imaging applications and over a range of CT scanning parameters. Figure 3 illustrates CTDI measurements using PGDs and MRI imaging.

(a)

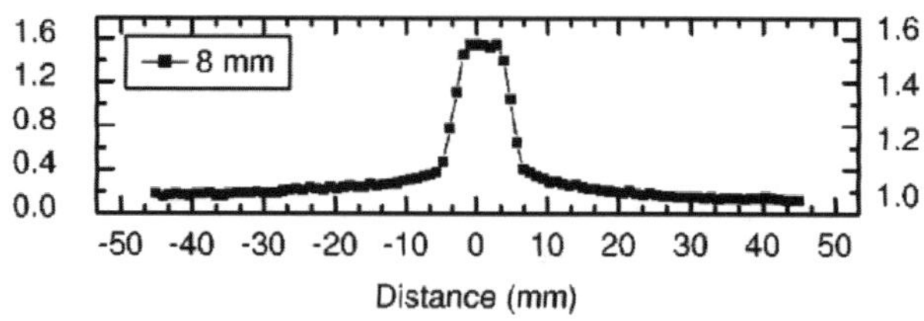

(b)

Figure 3. (a) demonstration of PGDs ability to measure CT dose within diagnostic CT scanners. (b) MRI profile through irradiated region of (a) for a nominal CT slice width of 8mm [14].

2.4.Accuracy and precision of x-ray CT PGD

Maynard *et al* [19] undertook an exhaustive evaluation of the accuracy and precision achievable with NIPAM-based (table 1) CT PGD. Results suggest that the average relative dose error between Monte Carlo dose calculations and CT-based gel measurements is, generally, <2%, depending on the calibration method employed, with poorly calibrated dosimeters more in line with <3% error, as shown in figure 4. Furthermore, 3%/3mm gamma pass-rates for 2D gel images as compared with Monte Carlo-calculated images were >97% for the cases shown in figure 4. It is clear that with careful experimental parameters, CT PGD can do well in accurately capturing radiotherapy-based radiation dose.

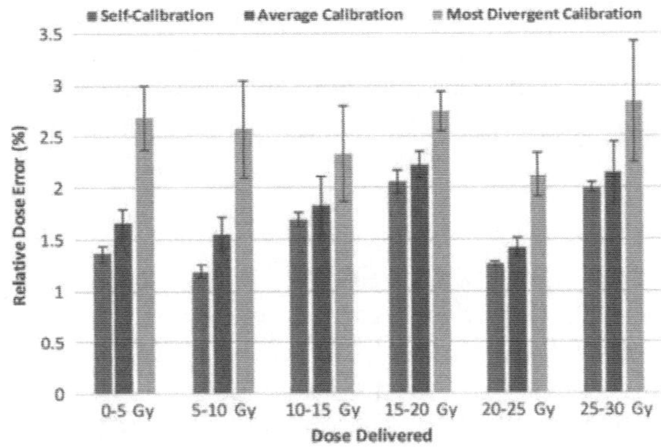

Figure 4. Relative dose error between Monte Carlo calculated 3-field dose delivery and CT PGD, as a function of delivered dose. Shown are several different calibration techniques, including self-calibration (blue), the average calibration curve generated over multiple gels (i.e. a "library" calibration file, red) and the largest error calibration curve (green) [19].

2.5.System modelling

Hayati *et al* [35] have undertaken Monte Carlo-based modelling of an x-ray CT PGD system, with the aim of creating a model that can be used for investigating and tuning CT and PGD parameters *in silico* for further optimization of the system. Figure 6 illustrates the ability of the Monte Carlo-based model in simulating PGD image reconstruction and dose response for a polyacrylamide-based PGD.

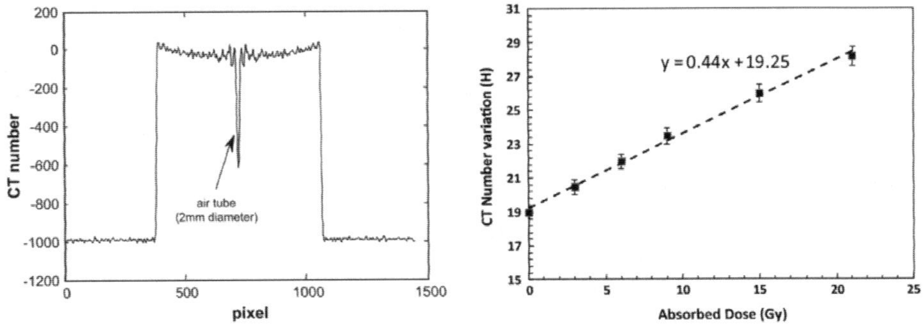

(a) (b)

Figure 5. (a) Monte Carlo-simulated reconstruction of a profile through a water phantom with a 2mm air tube in its centre. Clearly shown is the air tube at the centre of the water phantom. (b) Monte Carlo-simulated dose response of an irradiated polyacrylamide PGD imaged with a 140kVp scanner [35].

3. CBCT Gel Dosimetry

On-board imaging using CBCT has become a modern standard in linac based radiation therapy. The improved quality of CBCT images achievable with newer systems, along with the now relative ubiquity of these systems, has enabled the intriguing possibility of performing polymer gel dosimetry utilising CBCT as the imaging modality. CBCT image quality is reduced compared to traditional CT due to increased radiation scatter inherent in a cone-beam geometry, however there are unique and potentially highly advantageous features of CBCT imaging of gel dosimeters that make this pursuit interesting:

(i) The ability to irradiate and image in the same location. With CBCT integrated directly with linear accelerators, you can irradiate and image a gel without having to move the dosimeter. The key advantage here is the removal of re-positioning uncertainties incurred in moving the dosimeter from linear accelerator couch to CT or MRI bed for example in more standard gel imaging approaches.
(ii) The ability for close to real-time imaging of delivered dose distributions. With imaging integrated with the delivery system, the potential for near real-time imaging of dosimeters is apparent with CBCT.

Adamson *et al* [36] introduced CBCT PGD in their 2019 work where they outlined an on-line near real-time treatment verification workflow and demonstrated the technique's potential for characterizing high dose, steep dose-gradient deliveries. Dosimeter read-out by CBCT was performed directly following irradiation, 3 post-irradiation CBCTs were acquired for image averaging, and the total time between a pre-irradiation CBCT positioning scan and the final post-irradiation CBCT was 28min. This initial work was followed up by Jirasek *et al* [37] in 2020. In this work they showed that CBCT imaging 20-30 min post-irradiation allows for >90% polymer yield in NIPAM dosimeters and yields substantially improved image contrast when compared to imaging immediately following irradiation (see Figure 6). They employed a CBCT protocol optimized based on guiding principles in Hilts *et al* [27] which consisted of: 100kV, 1035mAs, head and DynamicGainFluoro kV modes, Titaniuum kV filter, full fan bowtie filter, 15 frames per second, smoothing filter and strong ring suppression. Up to 15 images averages were utilized to further improve image contrast to noise ratios.They achieved a 93% 3%,3mm Gamma pass rate for a 3 static field calibration measurement using 10 image averages and saw only marginal improvement at 15 averages. By comparison, using the standard CT PGD methodology achieved a 97% Gamma pass rate. As CBCT technology is still improving (e.g. inclusion of iterative reconstruction algorithms) the future may see image quality getting closer to CT and thus the potential for CBCT PGD an effective 3D dosimetry tool is very promising.

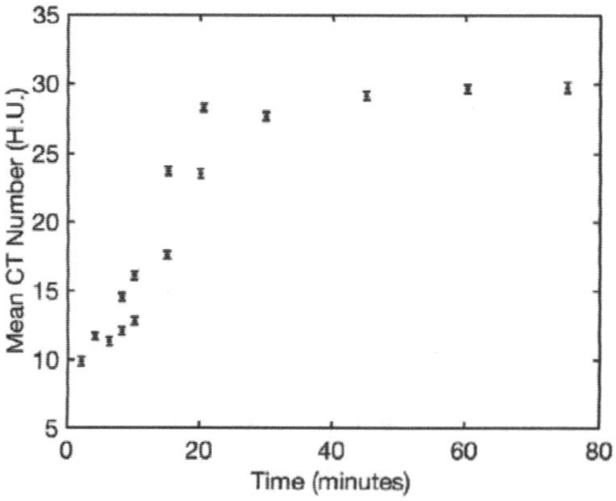

Figure 6. CBCT image contrast increases with time post-irradiation until full polymerization is achieved at approximately 45min. >90% polymerization is achieved within 20min [37].

Journal of Physics: Conference Series **2630** (2023) 012004 doi:10.1088/1742-6596/2630/1/012004

4. Applications

4.1. IMRT verification

Johnston *et al* investigated the application of NIPAM PGD using CT imaging for the verification of an IMRT treatment plan, as shown in figure 7 [38]. They employed an anthropomorphic head cast with a 1L gel insert to mimic a patient head and next treatment. Results were promising: 3%, 3mm gamma pass rate of 93.4% over the whole treated 3D volume and 99.3% in the single plane illustrated in figure 7, below. This work demonstrated the potential of CT PGD in providing quantitative dose information of high spatial resolution and signal to noise in clinically-based applications.

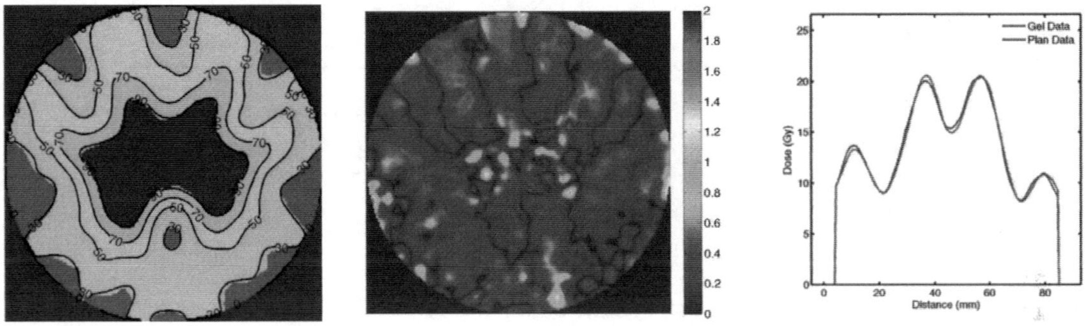

(a) (b) (c)

Figure 7. (a) IMRT treatment plan (contours) overlaid on top of IMRT-irradiated, dose-binned CT PGD image. (b) 3%, 3mm gamma map of PGD image and treatment plan (99.3% pass rate). (c) Representative profile through treatment plan (red trace) and PGD image (blue trace). The work demonstrated the excellent potential of CT PGD in characterizing complex radiotherapy treatment. From [38].

4.2. Deformable PGD

Building off earlier work of Liu *et al* in MRI-based PGD for deformable dose calculation verification [39], Maynard *etal* demonstrated the potential of CT PGD for deformable dose verification [40, 41], as shown in figure 8. Deformable PGD is complicated by the fact that a deformable container housing the PGD must remain oxygen impermeable in order to maintain the radiation sensitivity of the dosimeter. However, the work of Liu *et al* and Maynard *et al* highlight the fact that pliable (deformable), oxygen-impermeable containers can be used in concert with PGD to create a highly flexible system for measuring organ-like dose deformation in a radiotherapy context. Deformable PGD remains one of the only ways in which deformed doses can be experimentally measured in a clinical setting and, hence, an exciting area for further investigation.

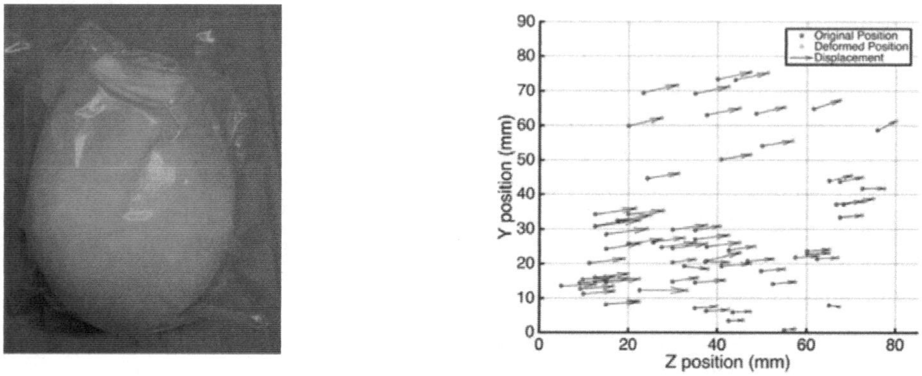

(a) (b)

Figure 8. a) a deformable gel dosimeter constructed in a latex balloon and sealed in low-density polyethylene. b) The measured displacement of wax-beads undergoing deformation within the gel dosimeter. Shown are the original position of the beads (red), their deformed position (green), and each bead's displacement vector (blue arrows) [40].

4.3.SRS Spatial Localization Accuracy

Verification of radiosurgery and other stereotactic treatment techniques offers excellent potential for CT based PGD given the high doses and steep dose-gradients that characterize these treatments. The first application of CT PGD was demonstrated in an early work that illustrated the potential of CT PGD to accurately localize a high dose irradiation achieved with linac, cone-based radiosurgery [11] Recently, Adamson *et al* [36] have illustrated this potential for CBCT PGD, as shown in figure 9. In this proof of principle study, CBCT PGD was able to verify the dose localization for a 6 target, multi-focal, 4-arc volumetric arc therapy (VMAT) stereotactic radiosurgery (SRS) treatment delivery to within 2mm of planned location The ability to measure and subsequently read-out the dose from a treatment delivery without moving the dosimeter, a feature inherent to CBCT, is demonstrated here and represents a clear advantage over other PGD techniques for cases such as this where an extremely high degree of spatial precision is required.

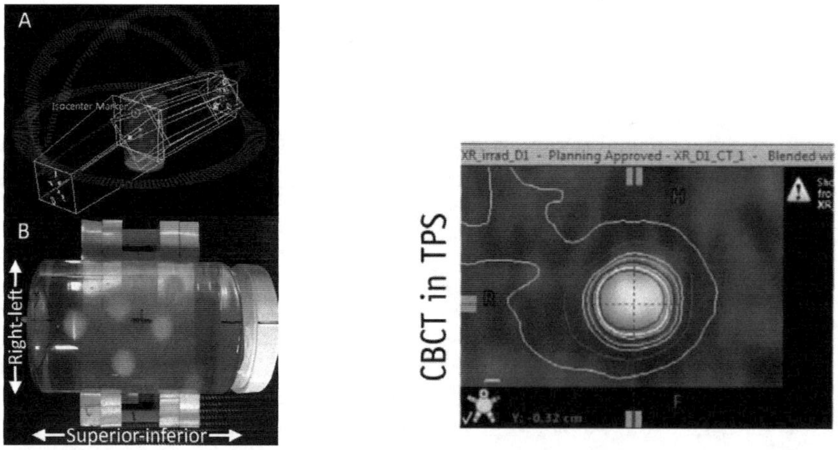

a) b)

Figure 9. a) 4 arc VMAT plan for a 6 multi-focal target SRS treatment was delivered to a PGD. The resulting polymerization in the 6 distinct target regions is clearly visible in the irradiated gel. b) Onboard CBCT (greyscale image) with treatment plan overlay (colour contours) for a single irradiated target within the multi-focal delivery [36].

Journal of Physics: Conference Series **2630** (2023) 012004 doi:10.1088/1742-6596/2630/1/012004

5. Summary and conclusions

It is clear that much work has been done in CT PGD since its first demonstration by Hilts *et al* [3]. Work on PGD formulation optimization, imaging parameter optimization, CT dose considerations, and Monte Carlo-based system modelling have all pushed CT PGD into the realm of being feasible as a truly 3D dosimetry method for, typically, high dose treatment verifications. The accuracy and precision of CT PGD is now sufficiently high such that many current problems involving high dose in 3D dose verification can be realistically tackled with the technique. In particular, the introduction of CBCT PGD offers new and exciting possibilities in onboard, close to real-time 3D dose verification using polymer gels.

6. References

[1] Gore JC *et al* 1984 *Phys. Med. Biol.* **29** 1189-97
[2] Maryanski MJ *et al* 1996 *Phys. Med. Biol.* **41** 2705-17
[3] Hilts M *et al* 2000 *Phys. Med. Biol.* **45** 2559-71
[4] Mather ML *et al* 2003 *Ultrasonics* **41** 551-9
[5] Rintoul L 2003 *Appl. Spec.* **57** 51-7
[6] Jirasek A *et al* 2014 *Phys. Med. Biol.* **59** 597
[7] Kakakhel MB *et al* 2011*Med. Phys.* **38** 5130-5
[8] Jordan K *et al* 2016 *J. Phys.: Conf. Ser.* **755** 011001
[9] Jirasek A 2013 *J. Phys.: Conf. Ser.* **444** 012005
[10] De Deene Y *et al* 2015 *J. Phys.: Conf. Ser.* **573** 012008
[11] Audet C *et al* 2002 *J. Appl. Clin. Med. Phys.* **3** 110-18
[12] Hilts M *et al* 2004 *Phys. Med. Biol.* **49** 2477-90
[13] Venning AJ *et al* 2005 *Phys. Med. Biol.* **50** 3875-88
[14] Hill B *et al* 2005 *Med. Phys.* **32** 1589
[15] Koeva VI *et al* 2009 *Phys. Med. Biol.* **54** 2779-90
[16] Chain JNMM *et al* 2011 *Phys. Med. Biol.* **56** 2091102
[17] Jirasek A *et al* 2010 *Phys. Med. Biol.* **55** 5269-81
[18] Jirasek A *et al* 2009 *Phys. Med. Biol.* **54** 907-18
[19] Maynard E *et al* 2017 *Med. Phys.* **44** 736-46
[20] Jirasek A *et al* 2015 *Phys. Med. Biol.* **60** 4399-411
[21] Sathiyaraj P *et al* 2018 *J. Can. Res. Therap.* **14** 287-91
[22] Javaheri N *et al* 2020 *Rep. Pract. Oncol. Radioth.* **25** 100-3
[23] Subramanian B *et al* 2006 *J. Med. Phys.* **31** 72-7
[24] Kawamura H *et al* 2013 *Appl. Radiat. Isotop.* **77** 94-102
[25] Trapp J *et al* 2001 *Phys. Med. Biol.* **46** 2939-51
[26] Johnston H *et al* 2015 *Med. Phys.* **42** 1666-77
[27] Hilts M *et al* 2005 *Phys. Med. Biol.* **50** 1727-45
[28] Hindmarsh J *et al* 2010 *J. Phys.: Conf. Ser.* **250** 012073
[29] Hilts M *et al* 2008 *Med. Phys.* **35** 344
[30] Hilts M *et al* 2004 *Med. Phys.* **31** 39-49
[31] Foist RB *et al* 2010 *Appl. Spectros.* **64** 1209-19
[32] Jirasek A *et al* 2012 *Phys. Med. Biol.* **57** 3137-53
[33] Baxter P *et al* 2007 *Med. Phy.* **34** 1934-43
[34] Hill B *et al* 2008 *Phys. Med.* **24** 149-58
[35] Hayati H *et al* 2016 *Radiol. Phys. Tech.* **9** 37-43
[36] Adamson J *et al* 2018 *Internat. J. Radiat. Oncol. Biol. Phys.* **103** 1271-9
[37] Jirasek A *et al* 2020 *Phys. Med. Biol.* **65** 225030

[38] Johnston H *et al* 2012 *Phys. Med. Biol.* **57** 3155-75
[39] Niu CJ *et al* 2012 *Med. Phys.* **39** 765-76
[40] Maynard E *et al* 2018 *Phys. Med. Biol.* **63** 075014
[41] Maynard E *et al* 2020 *Biomed. Phys. Eng. Express* **6** 035031

Journal of Physics: Conference Series **2630** (2023) 012005 doi:10.1088/1742-6596/2630/1/012005

MR-Linac Dosimetry: Current Approaches and Challenges

Arman Sarfehnia[1,2,3]
[1]Department of Radiation Oncology, Temerty Faculty of Medicine, University of Toronto, 149 College Street, Suite 504, M5T-1P5, Toronto, ON, Canada
[2]Department of Physics, Faculty of Science, Toronto Metropolitan (Formerly Ryerson) University, 350 Victoria Street, M5B-2K3, Toronto, ON, Canada
[3]Department of Oncology, Faculty of Medicine and Health Sciences, 5100 de Maisonneuve Blvd. West, Suite 720, Montreal, QC, H4A-3T2, Canada

Email: arman.sarfehnia@sunnybrook.ca

Abstract. The use of MR-integrated linacs (MR-linac) in clinical radiation oncology applications are becoming more prevalent. However, given the novel and radically different designs of these systems from conventional linacs, current radiation dosimetry protocols for high energy photons are no longer appropriate for use in modern MR-linac systems: First, given the presence of the magnetic field and the linac design, traditional reference conditions defined by previous high energy photon dosimetry protocols cannot be met; Second, the presence of the strong magnetic field can affect the performance of conventional equipment used for dosimetry. In this manuscript, we describe some of the challenges faced in radiation dosimetry in external MR-guided radiotherapy delivery systems, summarize some of the publications in this area, and finally discuss the early work by the American Association of physicists in Medicine (AAPM) Task Group 351 which is mandated with producing a protocol for reference dosimetry in MR-linac units.

1. Introduction

The phrase "adaptive radiation therapy" (ART) was described in an article by Di Yan *et al* [1] in 1997 as "a radiation treatment process where the treatment plan can be modified using a systematic feedback of measurements". ART was performed off-line in its early conception, by introducing changes to the treatment plan between treatment fractions. Magnetic Resonance image-guided radiation therapy (MRgRT) systems have allowed changes to a treatment plan to be made based on imaging immediately prior to treatment. An adapted plan created in this way not only accounts for changes in target size or position, but also changes to organs at risk in the vicinity of the target, and to the position of the patient on the treatment couch. On-line ART requires that high quality imaging of the patient be performed while the patient is in the treatment position, and that changes to the treatment plan be made quickly, before the patient's position alters. This is only possible when the imaging system is integrated with the treatment device.

However, the integration of the magnetic fields with a conventional high energy linac has been filled with challenges, not only from a technical and construction point of view, but also in other areas including radiation dosimetry and treatment planning. Magnetic fields can impact both the overall dose distribution in water as well as the radiation dosimeters we currently employ to measure this dose [2]. Furthermore, MR-linac delivery systems must be paired with accurate treatment planning systems that appropriately model the modified dose distribution due to the effects of Lorentz force on the high energy charged particles in various materials/media [3, 4]. This manuscript will only focus on radiation dosimetry and some of the work towards establishing a reference dosimetry protocols for MR-linac systems. Furthermore, of the current MR-linac systems under development/research (table 1), the discussions will be focused on the two commercially available systems: MRIdian system by Viewray Technologies Inc (Cleveland, OH, USA) [5], and Unity system by Elekta Solutions AB (Stockholm, Sweden) [6].

Table 1. Current MRI-linac systems/programs. Only first two systems are currently being widely used as commercial systems [7]. In the table's 'orientation' column, '⊥' refers to a setup with radiation being perpendicular to B field, whereas '//' refers to a parallel orientation.

System (Vendor/developer)	Radiation Energy	Magnetic Field Strength	Orientation (Radiation vs. B field)	Type of magnet
Unity (Elekta) [6]	7 MV FFF	1.5 T	⊥	Close, superconducting
MRIdian (Viewray) [5]	6 MV FFF	0.35 T	⊥	Split, superconducting
Australian MRI-linac [8]	4 & 6 MV	1.0 T	// (⊥ possible)	Open, Superconducting
Aurora-RT (MagnetTX) [9]	6 MV	0.5 T	// (⊥ possible)	Biplanar, high temp. super-conducting

2. Dose distribution in the presence of magnetic field

In the presence of strong magnetic fields, the Lorentz force will impact the trajectories of the charged particles. Given the perpendicular orientation of the magnetic field and radiation beam in both commercially available systems, the Lorentz force results in a change of some of the characteristics of the radiation dose distribution relative to what it would be in the absence of the magnetic field. This is evident in both the percent depth dose curve as well as profile shapes [10].

The Lorentz force will result in removal of contamination electrons, from air or the linac/cryostat itself, that would otherwise reach surface. These will spiral back due to feeling the Lorentz force of a perpendicularly oriented strong magnetic field to the direction of their travel [11, 12]. This would result in a lowering of the surface dose [13]. Inside the medium, the same effect results in charged particle created from photon interactions with the medium to spiral back and in general have a shorter range of travel. This means that there is a shift upstream of dose deposition by charged particles created in the medium, directly resulting in depth of maximum dose occurring closer to the surface [14]. Finally, as the charged particles exit the medium, they quickly spiral back in the lower density air (the so-called electron return effect, ERE) and contribute to additional dose inside the phantom and near the exit surface [14]. As a result of same phenomenon, the shape of an off-axis profile curve becomes asymmetric in the direction of the magnetic field, while remains symmetric in the other direction. The shape of percent depth dose curves for a

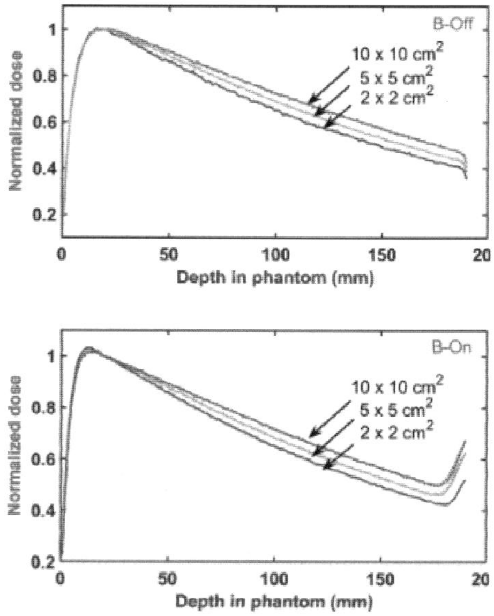

FIG 1: Normalized depth dose curves for various field sizes in a homogeneous water phantom. GEANT4 was used to simulate the figures using full Unity MR-linac spectrum. Top figure is with magnetic field turned off, and bottom figure is with it turned on. Images reproduced with permission from Ahmad *et al* [3].

Unity MR-linac spectrum simulated in presence and absence of the strong 1.5 T magnetic field for several field sizes are shown in figure 1. All the impacts described above are exaggerated at higher magnetic fields [2, 3, 14].

3. Reference Dose Detectors

3.1. Calorimetry

Calorimetry is one of the main radiation dose detection techniques used by radiation standard laboratories. Here, the radiation dose to medium D_{medium} is measured directly by measuring radiation-induced temperature rises $\Delta T_{\mathrm{medium}}$ through the relationship [15]:

$$D_{\mathrm{medium}} = c \cdot \Delta T_{\mathrm{medium}} \cdot \prod_{\mathrm{i}} k_{\mathrm{i}}\,, \qquad (1)$$

where c is the specific heat capacity of the medium, and k_{i} represents various often relatively small correction factors including heat transfer, heat defect, perturbation, etc. Of the many radiation detectors used in radiation dosimetry, only calorimetry can be categorized as a true primary absolute dosimeter. This is because in calorimetry, the conversion factor, specific heat capacity, relating the detector signal (i.e. temperature rise) to absorbed dose can be accurately determined in the absence of a known radiation beam, for example through electrical techniques. As seen from equation 1, calorimetry is theoretically insensitive to the presence of the magnetic field as the specific heat capacity c is independent of the magnetic field. This was later experimentally studied and verified by De Prez et al [16], D'Souza et al [17, 18], and Renaud et al [19] as calorimeters were shown to nearly display no sensitivity to the presence of the magnetic field.

There are two main types of calorimeters that are currently used in high energy photon radiotherapy [15]: (1) in *water calorimetry*, dose to water, is measured directly and absolutely at a point in water; (2) in *graphite calorimetry*, the medium of the detector is graphite, and hence the dose to graphite is measured directly and absolutely, and is converted to dose to water through simulated or measured conversion factors. Although, water calorimetry measures the quantity of interest in radiation dosimetry directly, its operation is challenging because of the low signal to noise ratio involved, and the need for operation of the entire device at 4° C to minimize heat transfer effects, among other factors. Relative to water calorimeters, graphite calorimeters produce higher signal to noise ratio for the same given dose, but require dose-to-graphite to dose-to-water conversion factors that were previously discussed and would in turn add to the overall dose uncertainty.

Water calorimeters [16, 18] and to a smaller extent a modified graphite calorimeter [19] have been used successfully in MR-linacs. Given the complexity of the detectors and high level of expertise required for their operation, calorimeters are not used clinically, but rather are used to characterize and calibrate other reference detectors used in clinics.

3.2. Ionometry

Ionization Detectors are the detectors of choice for radiation dosimetry in clinics. They are robust, easy to use, produce good signal to noise ratio, and their results are very reproducible. An ionization chamber is similar to a capacitor, where ionizations produced in the air volume of the cavity are separated and collected by the two electrodes defining the volume of the cavity and maintained at different polarity. Given the fact that Lorentz force changes the trajectory and affects the path length of the ions produced inside the chamber cavity, the measured signal by an ionization chamber is found to be dependent not only on the chamber type and radiation beam, but also on the orientation of the chamber with respect to both the radiation beam and the magnetic field involved [20, 21].

It is therefore of great importance that modern reference class ionization chambers be well characterized inside an MR-linac environment prior to their use. A few complications with ionization chamber reference dosimetry in the presence of strong magnetic field are noteworthy:

First, the electric field lines spanning the two electrodes are not perfectly homogeneous inside a cavity, and more importantly there may be so called dead volumes inside the chamber cavity in which the charge released is not collected. It has been shown that the change in trajectory of charges released, due to Lorentz force, may take them to areas where they are collected with different efficiencies, as such making the impact of dead volume on charge collection a function of chamber orientation with respect to the radiation and magnetic field [22-24].

Second, the effective point of the measurement (EPOM) of ionization chambers is shifted not only in the upstream-downstream direction but also laterally [25]. This change in EPOM is complex and depends on detailed analysis of radiation transport, and the impact of Lorentz force on charged particles.

Third, the presence of even a small air bubble adjacent to the outside wall of the ionization chamber could result in significant difference in charge collection given the impact of the ERE and the change in trajectory of the charged particle in this small air volume. This has been studied in detail following significant issues in measurement reproducibility of ionization chambers in solid phantom materials [26]. Although some have suggested the use of small amount of water or gel to fill the air gap that would have otherwise been present between the chamber and the solid water phantom cavity, the best approach is to perform dosimetry in water and visually ensure the absence of any air bubbles in close vicinity of the chamber volume to avoid dosimetric issues.

3.3. Chemical Dosimetry

Chemical dosimeters and their performance in the presence of the magnetic field have also been studied. The details of basic theory and operation of these chemical dosimeters can be found elsewhere [27]. However, in general they enable dose determination through a measurement of the concentration of chemical by-products of radiation-catalyzed reactions that are stable. Although rarely used for routine clinical dosimetry, both alanine and Fricke dosimeters are used by primary standard laboratories. Specifically, alanine can be used either as mailed audit dosimeter of choice or a transfer instrument for calibration standards. Studies towards their use in the same capacity for MR-linac have yielded promising results [28].

4. Reference Dosimetry Formalism

The American Association of Physicists in Medicine (AAPM) Task Group 51 [29] and its addendum [30] form the current primary reference dosimetry standard for high energy photon beams in North America. The protocol recommends the use of reference class ionization chambers with a traceable calibration factor back to a primary standard laboratory or equivalent. However, given the fact that modern MR-linac systems do not satisfy much of what is traditionally considered reference setup conditions required by the protocols, the AAPM created TG-351 in 2020 and tasked it with the development of a radiation dosimetry protocol in external MR-guided radiotherapy systems. The resulting protocol will follow the same format as the previous reference dosimetry guidelines, where it introduces the formalism, the reference condition setup, and practical recommendations and guidelines for best reference dosimetry practices in MRgRT environment, in addition to a sample in-depth uncertainty budget analysis.

The impending formalism will take similar form as that of the TG-51 and its addendum, and follows closely the structure introduced previously by several authors [12, 21, 23]. The general formalism determines dose to water at a point inside an external MRgRT system $D_{\mathrm{w}}^{\mathrm{Q,B}}$ from a corrected chamber signal reading M through the following relation:

$$D_{\mathrm{w}}^{\mathrm{Q,B}} = M k_{\mathrm{Q,B}} k_{\mathrm{Q}} N_{\mathrm{D,w}}^{\mathrm{Co-60}} \tag{2}$$

where $N_{\mathrm{D,w}}^{\mathrm{Co-60}}$ is the calibration factor of the chamber from a primary standard lab which is provided in a Co-60 beam under TG-51 defined reference conditions and in the absence of an magnetic field, and k_{Q} is the beam quality conversion factor that converts the chamber calibration factor from Co-60 to the beam quality of interest (i.e. that of the MR-linac beam) but in the absence of the magnetic field. These parameters, including the corrections necessary to convert a raw chamber signal to a corrected chamber signal *M*, have all been discussed in length in TG-51 and its addendum. The $k_{Q,B}$ quality conversion factor, however, will be introduced by the TG-351 and will convert the calibration factor of a reference class ion chamber from a zero magnetic field environment to one where the strong magnetic field is present. The $k_{Q,B}$ factors will be provided by the task group as a function of beam quality and the particular MRgRT system, the reference ion chamber type, and the orientation of the chamber with respect to the radiation beam and magnetic field. The data will be provided for a subset of all currently accepted reference class ion chambers that have been well studied and characterized either through repeated independent Monte Carlo studies and/or accurate experimental data. There have been a few articles as well that have summarized some of the available data for example by De Pooter et al [2], Iakovenko et al [32, 33] or work currently under review by Begg *et al* that provides local recommendations on MR-linac dosimetry in Australia.

The TG-351 protocol will also make further recommendations with respect to reference setup conditions and will build upon the recently published AAPM WGTG51 Report 374 on practical guidelines for accurate reference dosimetry [34]. Reference dosimetry must only be performed with waterproof chambers in a water phantom to avoid some of the issues discussed in section 3.3. Moreover, given that the magnetic field correction factors $k_{Q,B}$ have smallest magnitude when the ion chamber is oriented parallel with respect to the magnetic field, this chamber orientation is recommended for reference setup.

It is noteworthy that in strong deviation from its predecessors, TG-351 will recommend the ratio of the tissue phantom ratio at depths 20 cm and 10 cm in water, $TPR_{20,10}$, as the beam quality of interest in external MRgRT systems. Section 2 describes some of the changes to the shape of the PDD and its sensitivity to the presence of the magnetic field. $TPR_{20,10}$ acts as a more robust and reproducible beam quality specifier that is less sensitive to the effects of magnetic field.

Finally, the uncertainty budget described in the protocol can serve as a template for future users to perform their own uncertainty calculations by considering their own equipment and methodologies. The uncertainty budget will focus on both user dependent and independent criteria, including the effects of both linac as well as MR stability on final reference dose measurement

5. Conclusion

This manuscript discusses some of the existing external MR guided radiotherapy systems, as well as the basic physics behind the impact of a strong magnetic field on electron transport. It discusses several available radiation dosimeters, and their performance under the influence of a strong magnetic field. Finally, a brief introduction of the work done by the AAPM TG-351 tasked with producing a clinical protocol for external MRgRT reference dosimetry is presented.

6. Acknowledgements

The hard work of the entire AAPM TG-351 members including Drs. P. Alvarez, H. Bouchard, G. Ibbott, V. Malkov, O. Green, B. Muir, P. Yadav, L. de Prez, J. Dolan, B. Van Asselen, J. Wolthaus, and D.W.O. Rogers are acknowledged.

7. References

[1] Yan D *et al* 1997 *Phys. Med. Biol.* **42** 123-32
[2] De Pooter J *et al* 2021 *Phys. Med. Biol.* **66** 05TR02
[3] Ahmad SB *et al* 2016 *Med. Phys.* **43** 894-907
[4] Paudel MR *et al* 2016 *J. Appl. Clin. Med. Phys.* **17** 230-241
[5] Mutic S *et al* 2014 *Semin. Radiat. Oncol.* **24** 196-9
[6] Lagendijk J *et al* 2014 *Semin. Radiat. Oncol.* **24** 207-9
[7] Jelen U *et al* 2019 *J. Phys.: Conf. Series* **1305** 012010
[8] Keall PJ *et al* 2014 *Semin. Radiat. Oncol.* **24** 203-6

[9] Fallone G *et al* 2014 *Semin. Radiat. Oncol.* **24** 200-2
[10] Raaymakers BW *et al* 2004 *Phys. Med. Biol.* **49** 4109-18
[11] Hackett SL *et al* 2018 *Phys. Med. Biol.* **63** 095001
[12] Malkov V *et al* 2019 *Med. Phys.* **46** 5367-5370
[13] Keyvanloo A *et al* 2012 *Med. Phys.* **39** p. 6509-21
[14] Woodings S *et al* 2018 *Phys. Med. Biol.* **63** 085015
[15] Renaud J *et al* 2020 *Phys. Med. Biol.* **65** 05TR02
[16] De Prez L *et al* 2019 *Phys. Med. Biol.* **64** 035013
[17] D'Souza M *et al* 2019 *Int. Symp. on Standards, Applications and Quality Assurance in Medical Radiation Dosimetry* (IDOS 2019). IAEA
[18] D'Souza M *et al* 2020 *Med. Phys.* **47** 6458-6469
[19] Renaud J *et al* 2020 *Med. Phys.* **47** 1291-1304
[20] Meijsing I *et al* 2009 *Phys. Med. Biol.* **54** 2993-3002
[21] O'Brien DJ *et al* 2016 *Med. Phys.* **43** 4915
[22] Malkov V *et al* 2017 *Med. Phys.* **44** 4854-4858
[23] Malkov V *et al* 2018 *Med. Phys.* **45** 908-925
[24] Spindeldreier CK *et al* 2017 *Phys. Med. Biol.* 2017. **62**(16): p. 6708-6728
[25] O'Brien DJ *et al, Med. Phys.* 2018. **45**(2): p. 884-897
[26] O'Brien DJ *et al, Med. Phys.* 2017. **44**(7): p. 3830-3838
[27] McEwen M *et al* 2009 In *Clinical Dosimetry Measurements in Radiotherapy* 1027-1058. Madison, Wisconsin: Medical Physics Publishing
[28] Billas I *et al* 2020 *Phys. Med. Biol.* **65** 115001
[29] Almond PR *et al* 1999 *Med. Phys.* **26** 1847-70
[30] McEwen M *et al* 2014 *Med. Phys.* **41** 041501
[31] De Prez L *et al* 2019 *Phys. Med. Biol.* **64** 105025
[32] Iakovenko V *et al* 2020 *Med. Phys.* **47** 1940-1948
[33] Iakovenko V *et al* 2021 *Med. Phys.* **48** 2695-2697
[34] Muir B *et al* 2022 *Med. Phys.* **49** 6739-6764

3D dosimetry of proton beams and influences on response

G. Ibbott[1] and M. Carroll[2]
[1]Department of Radiation Physics, University of Texas MD Anderson Cancer Center, 1515 Holcombe Blvd., Houston, TX 77030 USA
[2]University of Wisconsin Health, 4547 Winnequah Rd, Monona, WI 53716 USA

E-mail: ibbott@me.com

Abstract. Proton beams offer the ability to conform dose distributions exquisitely to target volumes, allowing radiation oncologists to avoid exceeding tolerance doses to sensitive normal tissues. Several volumetric dosimeters have shown potential in conventional radiotherapy modalities, but suffer from dose under-responding, or signal quenching, in proton therapy. This reduced response has been seen with gels as well as with radiochromic plastics and may result from dependences on dose rate and linear energy transfer (LET). Investigations have shown that the under-response is affected by the chemical concentration of the active components of PRESAGE®, and changes to the composition have been shown to reduce quenching. It is not clear if similar changes can be made to gels. This report reviews studies into the impact of formulation changes on signal quenching, and the degree to which mathematical corrections might still be needed to achieve accurate dosimetry.

1. Frequency of errors in radiation therapy

Errors in radiation therapy occur relatively infrequently, but when they do occur, they can be disastrous. The likelihood and impact of errors in radiation therapy were highlighted in a series of articles in the New York Times, beginning in 2010 [1]. The articles indicated that the errors that make the headlines may be the most severe, but probably underestimate the total number of misadministrations as some may not be reported, and many may go undetected. The recognition that errors can occur at each stage of the treatment process: prescription, simulation, treatment planning, and delivery, as well as during quality assurance procedures and machine basic calibration, underscores the critical need for a comprehensive quality assurance program.

The frequency of errors in radiation therapy has been investigated by the Imaging and Radiation Oncology Core (IROC-Houston, formerly the Radiological Physics Center.) Participants in some clinical trials sponsored by the US National Cancer Institute are required to irradiate a phantom provided by IROC using the technique specified by the trial and achieve acceptable results. According to IROC, after 1,139 irradiations of a head-and-neck phantom between 2001 and 2011, performed at 763 institutions, the passing rate was 82% [2]. Passing was defined as delivering a dose within 7% and 4 mm distance-to-agreement, when compared with the institution's own treatment plan for the phantom. The irradiations were all delivered using IMRT techniques, with a variety of treatment planning and delivery equipment at the different institutions. The results indicate that, even when performed under ideal conditions, one out of 5 institutions failed to deliver the dose distribution prescribed by their own treatment plan, to within quite generous criteria. IROC determined that the causes for the failures included the use of incorrect data entered into the treatment planning system, inexact beam modeling, and software and hardware failures. These results suggest that improvements to institutional quality assurance programs might be warranted.

2. Benefits of 3D dosimetry in proton therapy

Proton therapy carries with it additional demands for quality assurance. Treatment with particle beams is attractive for some diseases because the particles penetrate tissue to a depth that is predicted by the energy at the entrance surface. Beyond that depth, the dose is much lower, and in the case of protons, is essentially zero. As shown in Figure 1, at the maximum depth of penetration, the *distal edge* of a proton beam falls rapidly from the dose in the peak to zero. The advantage of the rapid distal falloff is evident; tissues beyond that depth are spared while the tissue within the peak (known as the *Bragg peak*) receives a large dose. The goal in proton therapy is to position the Bragg peak within the target volume with the distal edge at the deepest margin of the target volume. This sometimes allows for the escalation of target dose while maintaining low normal tissue doses, resulting in improved therapeutic gain and better treatment outcomes. An example of the improvement achievable over photon treatment is shown in Figure 2.

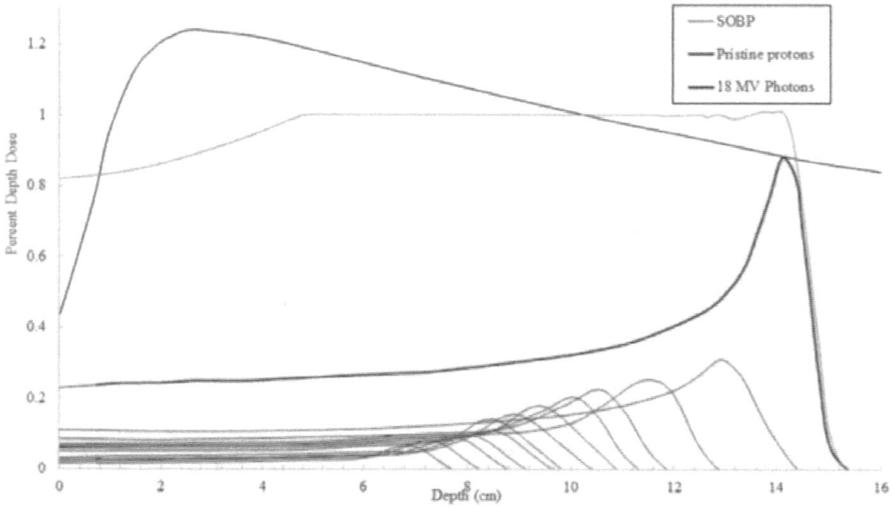

Figure 1. A pristine, or monoenergetic proton beam with a sharp Bragg peak is compared with a spread-out Bragg peak and an 18 MV x-ray beam [3].

A pristine proton beam as shown in Figure 1 is often a *pencil beam*, having transverse dimensions of 1 cm or less. Consequently, it must be broadened in the transverse direction to treat a typical target volume effectively. In addition, the width of the Bragg peak in the axial dimension is determined by the energy spectrum at that depth and generally is too narrow to be useful for treatment without modification. To broaden the peak in both the transverse and axial directions to a width comparable to the dimensions of a target volume, several options are available. The first, and simplest technique for transverse broadening is to insert a scattering foil into the beam. This is called *passive scattering*. The scattering foil broadens the beam considerably, and then collimation is required (comparable to a photon beam) to limit the transverse dimensions of the field as needed.

To broaden the beam in the axial direction requires that proton beams of several different energies be superimposed as shown in Figure 1. Creating such a *spread-out Bragg peak (SOBP)* has two effects: it broadens the high-dose region to the dimensions of the target volume in the axial direction, but it also raises the dose in the *plateau* upstream from the Bragg peak. As is clear from Figure 1, however, when compared to a photon beam, a proton beam provides a greater dose differential between the target volume and tissues both upstream and downstream.

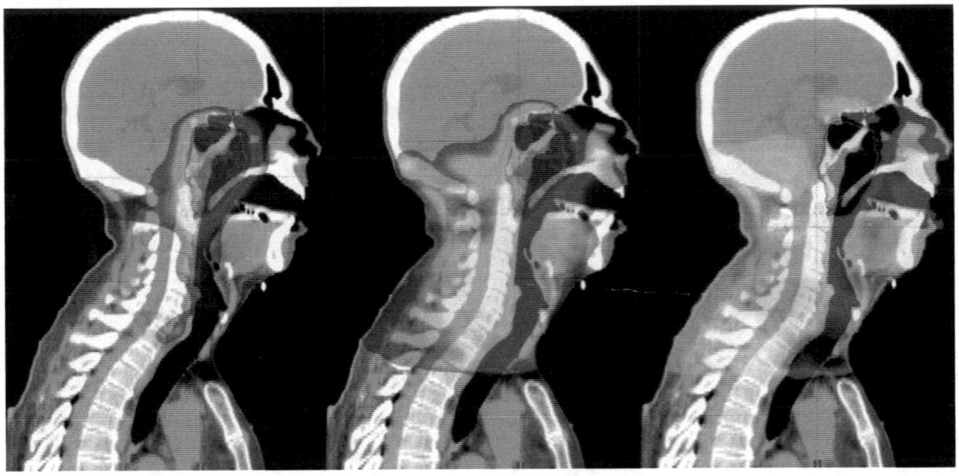

Figure 2. Treatment plans for a head-and-neck tumor are shown. The panel on the left is a calculated dose distribution prepared for intensity-modulated proton therapy; the center panel shows a treatment plan created for intensity-modulated photon beams; and the right-hand panel shows the difference, reflecting the additional dose delivered by photons to healthy tissues.

There are several techniques available to spread the Bragg peak in the axial direction. The first is typically employed when passive scattering is used and involves modulating the beam energy during treatment. This can be accomplished using filters of different thicknesses that can be inserted into the beam, with each thickness chosen to decrease the penetration of the beam in increments on the order of 1 cm. A *range modulator wheel* is generally used to rapidly modify the beam penetration during treatment.

A second method, which today is considered standard therapy, is to deliver the treatment through many pristine pencil beams, modulating the energy and position of each beam to build up the dose throughout the target volume. This technique, called *spot scanning*, allows the dose distribution to be tailored to the target volume in three dimensions. When spot scanning is used from multiple beam directions, and the dose deposited by each spot is adjusted to optimize the dose distribution, the technique is called *intensity-modulated proton therapy (IMPT).* As shown in Figure 2, calculated dose distributions using IMPT show substantially lower doses to normal tissues when compared with state-of-the-art photon therapy.

The range of the protons depends critically on not only the incident energy but also on the narrowness of the energy spectrum, and on the relative stopping powers of the tissues through which the beam passes. The incident energy spectrum can be controlled and accounted for in treatment planning, but the tissue stopping powers must be determined from CT scans of the patient. As shown in Figure 3, protons undergo several types of interactions with the atoms composing human tissue, including inelastic coulombic interactions with atomic electrons, elastic coulombic interactions with nuclei, and nuclear collisions. Proton interactions depend far more on the atomic composition of tissues than do photon interactions, and consequently there is not a direct relationship between Hounsfield Units and proton stopping powers as there is between HU and mass energy absorption coefficients. This leads to greater uncertainties in the effects of tissue characteristics on proton energy spectrum, depth of penetration, and lateral scattering. Furthermore, the treatment plans used in contemporary proton therapy are considerably more complex than suggested above, with complex optimization algorithms used to develop spot-scanned, intensity-modulated treatment plans using multiple beam angles. Assessing the quality of such complex plans using conventional point detectors or planar arrays is inadequate. Three-dimensional dosimetry systems such as gels and radiochromic plastics are ideal to evaluate the quality of such treatments.

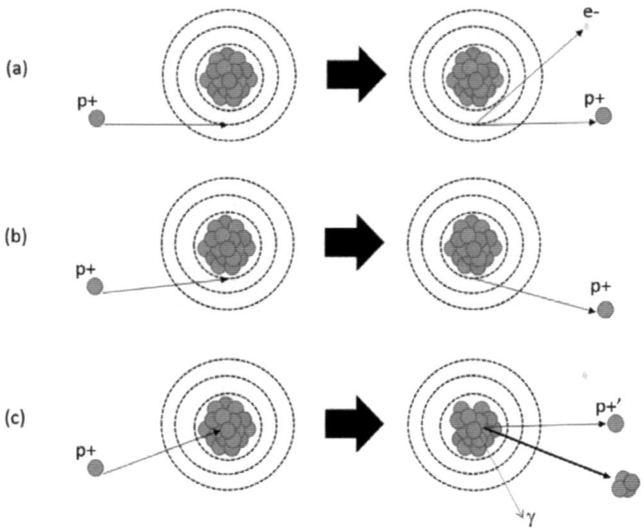

Figure 3. Protons interact with the atoms that make up human tissues through three general mechanisms: (a) inelastic coulombic interactions with electrons, (b) elastic coulombic interactions with nuclei, and (c) nuclear collisions [3].

3. Candidate 3D dosimeters for proton beams

As has been reported elsewhere, gel dosimeters offer the ability to measure and display 3D dose distributions [4-6]. Two broad classes of gel dosimeters are quite widely used: polymer gels and Fricke gels. Polymer gels record a dose distribution by converting local acrylic monomers to polymer clusters and chains. The polymers absorb and scatter light and a measurement of the optical density change can be related to the absorbed dose. Some polymer gels also change their paramagnetic characteristics and yield a signal on magnetic resonance imaging (MRI). The conversion of monomers to polymers raises the physical density of the gel matrix slightly, allowing a measurement by x-ray CT [7].

Fricke gels take advantage of the phenomenon first reported by Fricke and Morse and act by converting ferrous (Fe^{2+}) ions to ferric (Fe^{3+}) ions [8]. The ferrous sulfate is mixed with gelatin or another stabilizing agent, although some migration of the ferrous and ferric ions has been observed. The ferric ion concentration can be determined by optical methods, especially if a coloring agent such as xylenol orange is used, although MRI is often used [9]. A novel demonstration of the use of Fricke gels was shown by Lee *et al*, who used an MR-linac to image a gel as it was being irradiated, measuring the dose as it was being deposited [10].

Radiochromic plastics cannot be imaged with MRI and optical methods must be used instead. An example of a polyurethane matrix containing a radiochromic dose reporting system was described by Adamovics *et al* [11]. This dosimeter is available as a commercial product called Presage®. In this system, a dye, leuco malachite green, which is colorless, is converted to malachite green, which exhibits a green color. The converted dye absorbs light, particularly red light, and the resulting linear optical density coefficient (e.g., ΔOD/cm) exhibits a dose dependence. The mechanism requires an activator or radical initiator, for which chloroform or a similar compound can be used. Radiochromic plastics exhibit no oxygen sensitivity, as is the case with polymer gels, which simplifies their handling [12]. As the dose-reporting compounds are bound to the plastic, the signal is not reduced by diffusion, as is the case with ferric ions migrating through gelatin. Because the dosimeter is a solid, no container is required. And the dosimeter can be molded or machined into any desired shape or size. Disadvantages of radiochromic plastics include the complexity of manufacturing, which must be done in a fume hood and

Journal of Physics: Conference Series **2630** (2023) 012006 doi:10.1088/1742-6596/2630/1/012006

requires that the polyurethane be cured while under elevated atmospheric pressure. As mentioned above, optical imaging must be used for readout.

A fourth class of 3D dosimeter with the potential to demonstrate proton dose distributions has been described by Beddar *et al* [13]. This is a liquid scintillator solution that emits light when irradiated with a proton beam. The light is captured by video cameras that view a tank of the scintillator solution through mirrors, from different angles, and generate a 3D image from the multiple views. A quantitative relationship between the image and the absorbed dose has not yet been described.

4. Proton dosimetry with 3D dosimeters

Proton beams introduce several challenges not seen with photon and electron beams. First, there appear to be several mechanisms of signal loss that have to do with greater density of dose-deposition interactions. Jirasek and Duzenli evaluated a polyacrylamide gel system and showed through Fourier transform observations and track structure calculations that the high doses deposited close to the proton tracks saturated activation sites in the gel, reducing the sensitivity of the gel [14]. Gustavsson *et al* investigated a normoxic polymer gel and postulated that in regions of high linear energy transfer (LET), recombination of the radicals produced by ionization reduced the response of the gel [15].

Efforts have been undertaken to improve the agreement between 3D dosimeters and conventional dosimetry systems. Zeidan *et al* reported good agreement with treatment planning data in one of several polymer gels whose formulation had been modified to include larger or smaller amounts of the radical initiator [16]. Our own measurements in a formulation of Presage® intended for proton measurements demonstrated a broadening of the distal edge of a SOBP, suggesting both an under-response in the high dose, high LET region, and a small over-response in the low-dose, high-LET region [17]. This is shown in Figure 4.

The decrease in signal resulting from chemical physical processes is called *quenching* and is observed in nearly all chemical dosimeters [18]. In proton beams, it is attributed to the density of ionizations along the tracks of the protons. For Presage® dosimeters, and in regions of high LET, quenching can reduce the signal by 20 % or more.

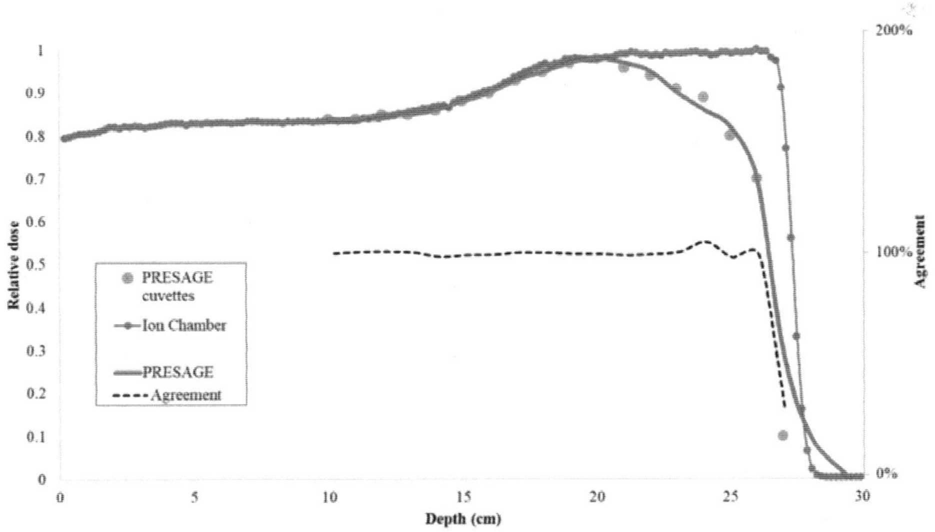

Figure 4. Relative depth dose comparison in a 225-MeV proton beam with 10 cm SOBP as measured by multi-layer ion chamber (blue), large volume Presage® formulated for proton beams (red), and cuvettes of Presage® (orange points). The agreement between large volume and cuvette measurements is shown (black line).

Journal of Physics: Conference Series **2630** (2023) 012006 doi:10.1088/1742-6596/2630/1/012006

5. Measurements in modified formulations of polyurethane radiochromic dosimeters

Polyurethane dosimeters of several different formulations were prepared and were irradiated with proton beams to selected doses to evaluate their sensitivity and linearity of response. The results are shown in Figure 5. The data indicate that greater concentrations of the radical initiators chloroform and bromoform both increase the sensitivity of the dosimeter, and that for comparable concentrations of the initiator, dosimeters containing bromoform are considerably more sensitive. All four preparations showed excellent linearity.

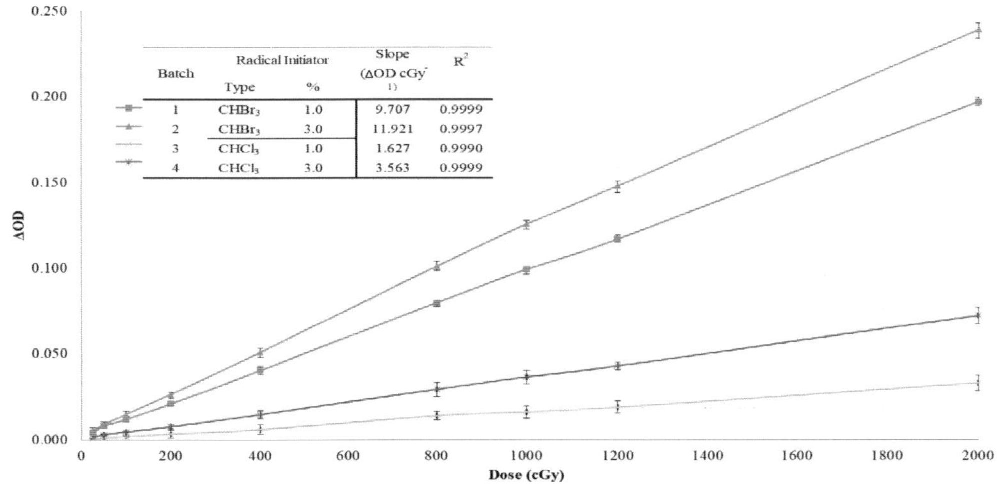

Batch	Radical Initiator Type	Radical Initiator %	Slope (ΔOD cGy⁻¹)	R^2
1	$CHBr_3$	1.0	9.707	0.9999
2	$CHBr_3$	3.0	11.921	0.9997
3	$CHCl_3$	1.0	1.627	0.9990
4	$CHCl_3$	3.0	3.563	0.9999

Figure 5. The sensitivities and linearities exhibited by four formulations of radiochromic polyurethane dosimeter.

On the basis of the data in Figure 5, the more sensitive bromoform-based formulation was selected for further modification to evaluate the effect on quenching. Figure 6 summarizes the results obtained when the bromoform concentration was varied between 1% and 27.5%, and the concentration of the dose reporting compound, leuco malachite green, was varied between 0.5% and 4%. The figure shows that bromoform concentrations between 10% and 15%, and LMG concentrations between 2.0% and 2.5% were successful in reducing the quenching observed in the high LET region to less than 5%.

The data shown in Figure 6 further indicate that quenching is most likely a result of saturation of the radical initiator. Quenching decreases as the concentration of the initiator is increased, suggesting that the availability of more initiators reduces the competition for interaction sites.

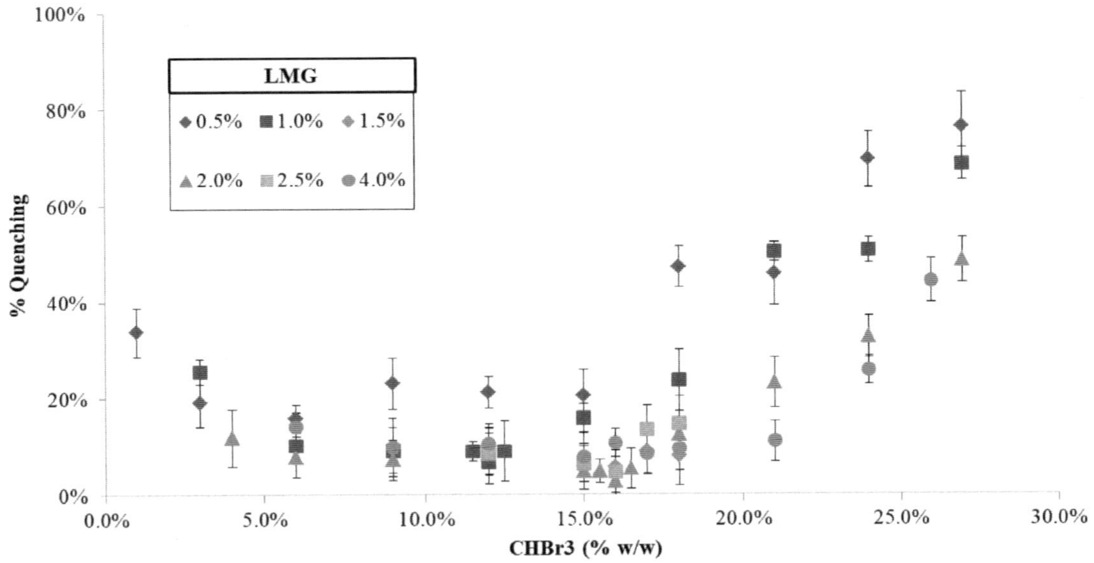

Figure 6. The percent of signal quenching measured at the distal-most point in the proton SOBP for several formulations using bromoform as the radical initiator. Error bars represent the standard deviation of three intrabatch cuvette comparisons.

6. Determination of empirical correction factor

To improve the accuracy of the selected formulation of polyurethane dosimeter, the data in Figure 6 were used to determine a Quenching Correction Factor. First, the correlation between polyurethane dosimeter response and the dose determined by ion chamber readings was determined along the depth-dose curve of a pristine proton beam:

$$r_P(d,E) = \frac{\varepsilon_{IC}^{P}(E) * M_P(d,E)}{M_{IC}^{norm}(d,E)} \qquad \text{Eq. 1}$$

where M_{IC}^{norm} is the normalized ionization chamber measurements, M_P is the polyurethane dosimeter signal, and ε_{IC}^{P} is the calibration factor of the dosimeter signal. The Quenching Correction Factor is therefore the reciprocal of the correlation above:

$$QCF(d,E) = \frac{1}{r_P(d,E)} \qquad \text{Eq. 2}$$

Because the QCF is a function of depth and incident energy, the application of a single model to another irradiation configuration becomes unrealistic. Consequently, an empirical correlation of QCF as a function of LET was made by fitting the calculated QCF from Eq. 2 with the voxel-averaged LETφ. The QCF-LET relationship was obtained and QCF as a function of LET (henceforth referred to as QCF(LETφ)) was derived allowing the application of the function to other irradiation configurations with LETφ similarly modeled using Monte Carlo. The derived QCF(LETφ) is shown in Figure 7 and was applied to subsequent measurements of proton depth dose obtained with the low-quenching formulation of polyurethane dosimeter.

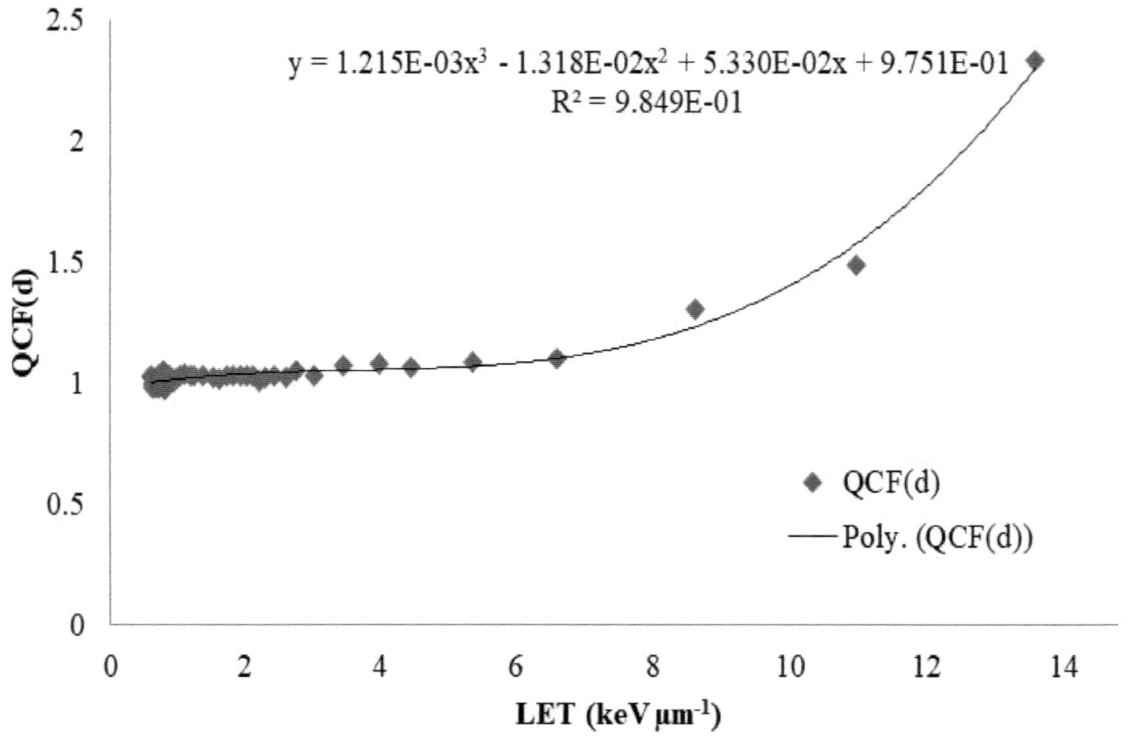

Figure 7. The Quenching Correction Factor as a function of LET. A polynomial fit to the data is provided to demonstrate the smooth variation of the function, and is the QCF(LETφ).

The low-quenching formulation of polyurethane dosimeter was employed to measure the depth dose of a passively-scattered 140 MeV proton beam with a 4 cm spread-out Bragg peak. The dosimeter results demonstrated disagreement with ion chamber measurements, as shown in Figure 8. However, after application of the QCF(LETφ) determined previously, much improved agreement with ion chamber measurements was seen. The corrected results are shown in Figure 8, and with expanded coordinates in Figure 9.

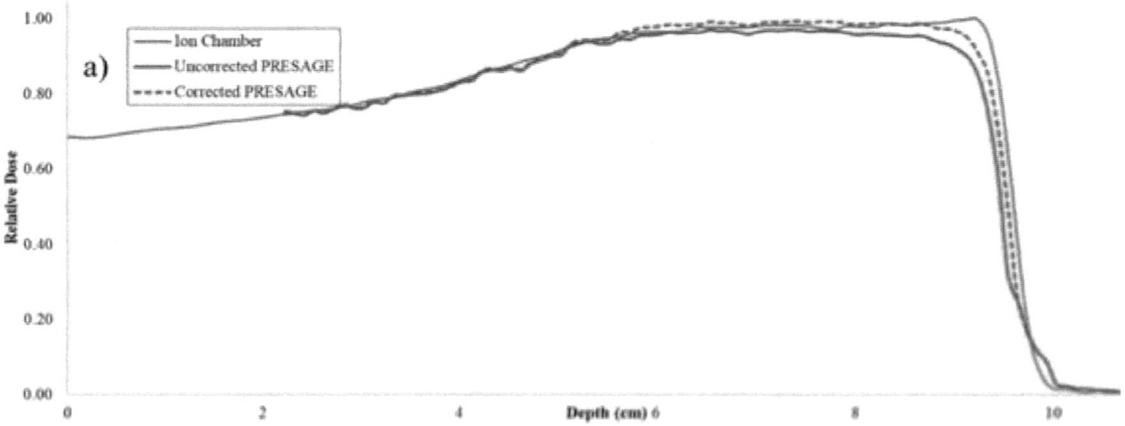

Figure 8. A comparison between ion chamber readings, the dose measured with the low-quenching formulation of polyurethane dosimeter, and the QCF(LETφ)-corrected dosimeter measurements.

Journal of Physics: Conference Series **2630** (2023) 012006 doi:10.1088/1742-6596/2630/1/012006

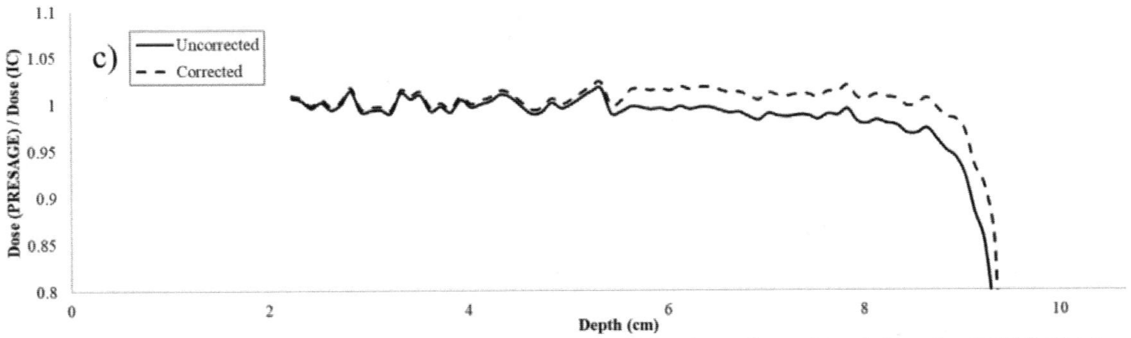

Figure 9. An expanded view of the data in Figure 7 showing the effect of applying the QCF(LETφ).

7. Conclusions

3D dosimeters offer many benefits beyond conventional dosimeters for evaluation of proton beams, both through remote audits and as a supplemental device in the clinic. These include more comprehensive dose analysis, and the potential to use time and resources more effectively. In addition to the demonstrated benefit with proton beams, 3D dosimeters have been shown to have applications in nearly all other facets of radiotherapy: IMRT, SRS, HDR Brachy, and MRgRT.

8. References

[1] Bodanich W. (2010) New York Times
[2] Molineu A *et al* 2013 *Med. Phys.* **40** 022101
[3] Carroll M *et al* 2017 *J. Phys.: Conf. Ser.* **847** 012027
[4] Ibbott GS *et al* 1997*Int. J. Radiat. Oncol.* **38** 1097-1103
[5] Ibbott G 2004 *J. Phys.: Conf. Ser.* **3** 58-77
[6] Baldock C *et al* 2010 *Phys. Med. Biol.* **55** R1-R63
[7] Jirasek A 2013 *J. Phys.: Conf. Ser.* **444** 012005
[8] Fricke H *et al* 1927 *Am. J. Roentgenol. Radium Ther. Nucl. Med.* **18** 430-432
[9] Lee HJ *et al* 2018 *Phys. Med. Biol.* **63** 025010
[10] Lee HJ *et al* 2018 *Phys. Med. Biol.* **63** 045021
[11] Adamovics J *et al* 2003 *Med. Phys.* **30** 1349
[12] Alqathami M *et al* 2016 *Phys. Med. Biol.* **61** 813-824
[13] Beddar AS 2019 *J. Phys.: Conf. Ser.* **1305** 012038
[14] Jirasek A *et al* 2002 *Med. Phys.* **29** 569-577
[15] Gustavsson *et al* 2004 *Phys. Med. Biol.* **49** 3847-3855
[16] Zeidan O *et al* 2010 *Med. Phy.* **37** 2145-52
[17] Grant R et al 2011 *Int. J. Radiat. Oncol.* **81**. S885
[18] Healy BJ *et al* 2003 *Med. Phys.* **30** 2282-91

Journal of Physics: Conference Series **2630** (2023) 012007 doi:10.1088/1742-6596/2630/1/012007

End-to-End QA in Radiation Therapy Quality Management

L J Schreiner
Departments of Oncology and Physics, Queen's University, Kingston, Ontario, Canada K7L 3N6

E-mail: ljs2@queensu.ca

Abstract. End-to-End (E2E) testing is a method originating from computer science that is designed to determine whether an application communicates as required with hardware, networks, databases, and other applications. This paper is to advocate that the quality management (QM) of modern radiation therapy (RT) would benefit from more regular use of E2E based quality assurance (QA) in the local clinic. The argument is that modern RT delivery is performed through some process linked by a chain of interdependent stages and actions mediated by complex interchanges during the patient's treatment. These actions along the chain are often modified due to decisions by clinical staff who are interpreting information acquired along the process. While physics QA can validate that each of these steps are technically achievable (e.g., through machine QA) such conventional QA does not guarantee that the overall process is being carried out as planned even when it has been described by a well-defined protocol and delivered by well-trained staff. The paper briefly reviews the changes in programmatic design as RT has become more complex, the associated changes in RT QM, and some past examples of E2E testing in RT clinics, usually performed during the implementation of some new RT technique or during external audits of the clinic's practice. The paper then makes the case for increased E2E QA based on the lessons learned from this experience and ends with some suggestions for implementing effective and sustainable E2E testing in a clinic's QM program.

1. Introduction

Since the initial IC3Ddose meeting in 1999, contemporary radiation therapy (RT) has moved fully to the conformal radiation therapy techniques foreseen in the 1960s [1]. Advances in treatment simulation imaging, computerised treatment planning, and radiation delivery enabled Intensity Modulated Radiation Therapy (IMRT) treatment tailored to each patient. The resulting spatially modulated delivered doses (typically called the 'dose distributions') fit more tightly around the specific target volumes so that tumour dose could be increased appropriately while dose to adjacent normal tissue was decreased. The progress in conformal RT came with a challenge: an increased potential of missing the target because of patient set-up errors, organ motion, or machine malfunction during treatment delivery. This risk motivated the development of on-line image guidance capabilities through the addition of imaging systems on the treatment unit [2–4] to enable patient set-up confirmation before dose delivery. Experience with image guidance over many patients' treatments also made the oncology community aware that the tumour region and surroundings could change between planning imaging and the first session at the treatment unit or afterwards during the course of the multiple fractions of radiotherapy [5–8]. Therefore, strategies were developed to account for and correct these changes, for example, by reassessing a patient's treatment plan and delivery at some dose level or fraction interval [9–11] to confirm that the initial plan is still appropriate or, alternatively, to adapt the further treatment to be given through replanning. More advanced Image Guided *Adaptive* Radiation Therapy (IGART) approaches envisage use of daily online imaging (cone beam computed tomography (CBCT) [2] or magnetic resonance imaging (MRI) [12,13] at the treatment unit) to monitor anatomy changes and provide for replanning without going back to the CT simulator [14]. Other sophisticated approaches for inter-fraction IGART extend the adaptation beyond corrections for only anatomical changes by incorporating dose delivery monitoring during each treatment (by electronic portal imaging devices (EPIDs) or some other in-vivo dosimetry) to enable dose distribution correction as fractions continue through the treatment course [14–16] with next treatment fractions using multileaf collimator (MLC) trajectories modified from the originally planned delivery.

All these IGART approaches involve a patient's treatment being modified during their radiation therapy based on new information gained during the treatment course in a process that involves data transfer between sophisticated advanced equipment, often with human evaluation [15,17,18] to assess and choose between different avenues of action at the multiple stages of the care (see figure 1). Clearly, well-executed IGART brings with it a commensurate challenge for the accurate implementation, and the associated validation, of very complex treatment activities. This challenge has been recognized and addressed by the RT community in recent decades (as outlined below). The objective of this paper is to review the evolution of practices to address these challenges by examining some pertinent literature and reflecting on the issues this literature raises. The author then proposes that the quality management of the systems required for modern RT would benefit from the inclusion of more regular testing to validate that the delivery provided by the cancer centre fulfils the intended treatment objectives, that is, by adopting end-to-end (E2E) testing in the clinic as an added tool [19–22] to ensure quality radiation delivery (the term Kron *et al.* use in their excellent 2022 review of radiotherapy quality management [17] to designate the RT that will deliver the intended patient treatment).

2. Evolution of Radiation Therapy Quality Management

The new technologies and the processes adopted during the uptake of IGART have improved the quality of RT and reduced errors. The modern treatment units are better controlled, have more stable radiation output [23], and provide greater functionality than the units in the 1980s and early 90s, and treatment planning systems and the imaging for treatment planning have improved considerably. Evidence has shown that the quality dose delivery enabled by IGART has provided new treatment options for patients and resulted in higher clinical control rates and reduced toxicity [7,24]. But IGART involves complex processes with many steps, some requiring human intervention, which introduces a potential for new error pathways [24–26]. The levels of complexity, integration, and implementation variability inherent to IGART processes with their image-based planning and computer-controlled delivery are such that device-centered quality assurance (QA) protocols alone can no longer ensure quality radiation delivery [27]. Therefore, modern radiation treatment techniques have compelled an associated reassessment of the quality management (QM) necessary for safe beneficial patient treatment: a rethinking of both the QM at a programmatic level of the complete radiation delivery service and the QA of the various components used within the IGART process. One should note that the characterization of radiotherapy as IGART should not be limited to external beam modalities alone; remote-afterloader high-dose-rate brachytherapy (HDR) techniques are increasingly image guided and involve processes with equivalent complexity, therefore they are considered to fall under the IGART umbrella for the purpose of this discussion.

2.1 Process Quality Management

While safe and appropriate RT delivery has always imposed strict requirements on a cancer centre's radiation program [17,27–31], the development of modern IGART has generated further guidelines from professional, national, and international bodies outlining how RT programs should function [32–36]. The various documents have many common elements dealing with the full scope of care, from the initial patient evaluation, through treatment and into subsequent follow-up [32–34,37,38]. Of particular interest for this IC3Ddose proceedings paper, these guidelines set standards for the programmatic environment for radiation treatment preparation (simulation and planning) and for delivery, with recommendations on the health care team staffing, on the roles and responsibilities of team members, on their training and on the maintenance of their qualifications. The guidance documents universally maintain that radiation services must develop and implement clear-cut protocols and procedures describing how patient care is to be delivered, and that they must also set up well-defined QM programs to ensure that the RT process is followed as required and that the radiation

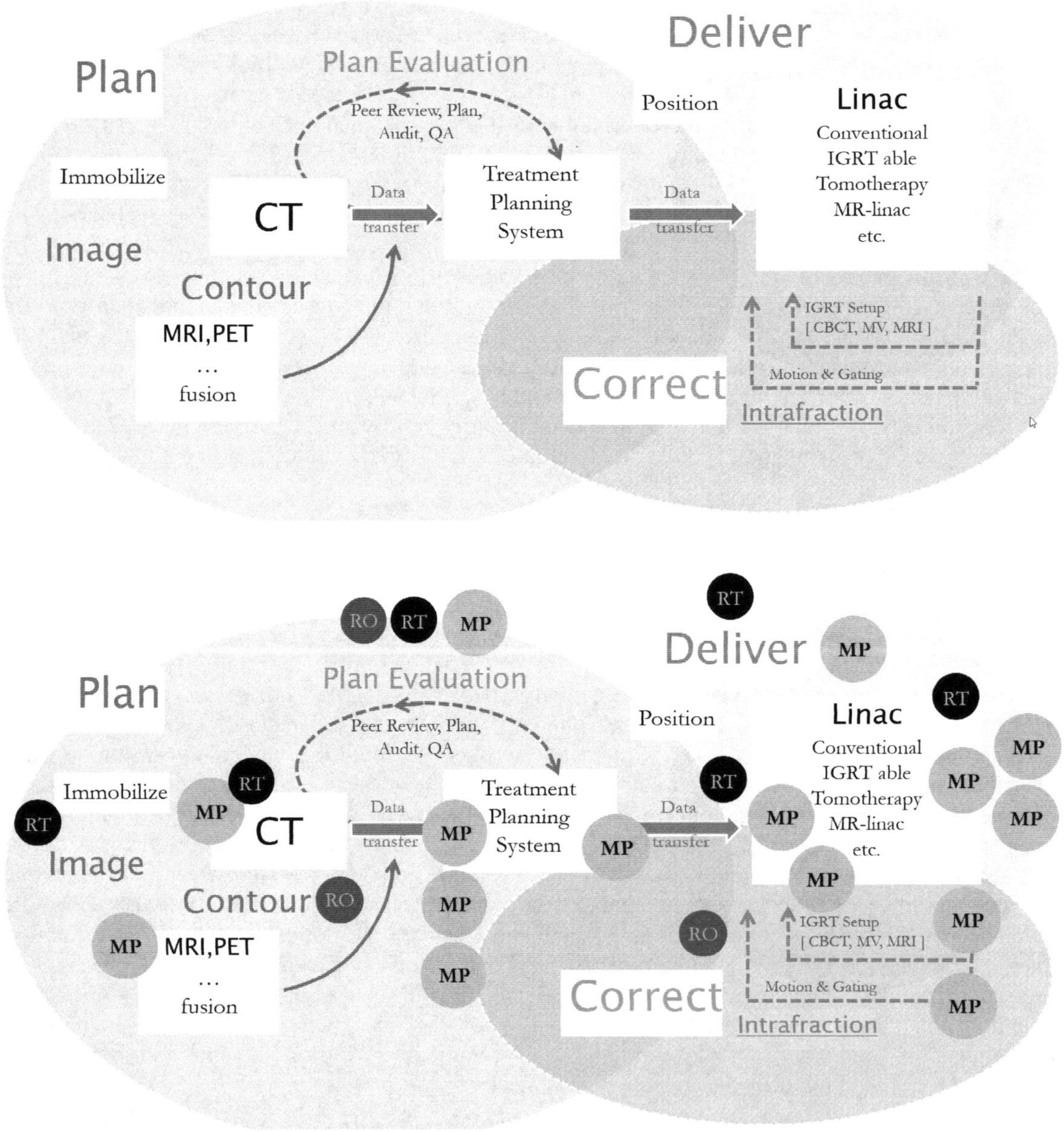

Figure 1. a) A sketch of the process steps for the simplest and most common implementation of image guided radiation therapy: patient set-up correction by on-line image registration to planning images and treatment plan followed by radiation delivery. Even this simple example involves multiple steps with data transfer, status assessment and decision making by various members of the treatment team. **b)** Multiple treatment protocols, task group reports, and technical quality control documents (see text)

specify QA measurements and tests that should be performed prior to and during technical implementation to assure the systems and equipment are able to perform to the accuracy and precision required for safe use. The coloured circles indicate some typical measurement points for the relevant QA performed by the medical physicists (MP), radiation oncologists (RO) and radiation therapists (RT). One could draw a similar figure to illustrate stages where clinical staff must make decisions to proceed with treatment based on observations and evaluations made at earlier points in the process. The figure would look similar to b) but with different players (for example, RTs rather than MPs) at different stages.
environment (equipment, hardware and software, and the radiation care team) is performing as needed. The evolving suite of guidance documents underscore that quality RT is not determined only by the equipment used but also by the complete environment in which the treatment process is provided. And RT programs have moved to programmatic analysis and design to improve this environment.

Two of the approaches to improve RT QM at the process level merit discussion at this point. They both specifically address the fact that the RT process is influenced by human factors which can affect treatment quality. The first approach has involved instituting systems to track and analyse incidents (defined as unwanted or unexpected deviations from the RT process that could potentially adversely affect patients, staff, or equipment). Specific local, national, and international incident reporting systems [26,39–41] have been created to report incidents to QM programs, to analyse the impact of these incidents, and then to report the results of investigations back to staff in order to raise awareness and improve process adherence. Incident reporting provides reactive tools to identify incidents that have occurred in the radiation treatment process, and analysis of incident reporting data can help guide the development of improved QM systems to avoid recurrence. The second tactic to improve QM has been to undertake analysis of how human factors affect RT processes [24,42] specifically addressing how human behaviour, abilities, and limitations influence the programmatic systems adopted under the RT program guidelines described in the last paragraph. Field observations and workflow analysis are then used to improve the design of the technologies, steps, and workflows of the work environment so to mitigate the treatment delivery variability introduced by these human factors. Results indicate that processes can be improved to account for human factors; however, some influences that weaken procedural compliance remain present [24,42]. This further motivates this article's promotion of more regular adoption of E2E testing to provide recurrent monitoring of select RT processes.

2.2 *Implication for Quality Assurance testing*

The technical advances in the imaging, planning and radiation delivery equipment and the networks connecting their interactions have also prompted an ongoing review of the quality assurance (QA) testing required in a well-developed QM program [27,29,43–48]. Select QA tests have come in and out of fashion during the transition from the simpler 2D treatments of the 1980s, through the implementation of IMRT techniques in the 1990s, to the comprehensive modern IGART. The QA evolution is because testing must reflect the particular technical environment in the clinic at a given time, and because professional opinions on requirements will change as this environment changes [23,46,49]. For example, the technical advances in the 1990s motivated the development of three-dimensional (3D) dosimetry to verify that the dose distributions from the novel IMRT techniques delivered the correct dose to the right location. At that time, linear accelerators were just being modified to include MLCs, computerized treatment planning systems (TPS) were just being improved to included CT based inhomogeneity corrections, TPS algorithms had not yet been fully tested in the small-field conditions characterizing dynamic MLC delivery, and the control systems and interfaces between the treatment planning systems and the treatment units were under constant revision [27,48,50]. The uncertainty in the tools available for early IMRT techniques motivated proponents to contend that it was vital to do full 3D dose assessment of the deliveries during a clinic's IMRT commissioning and, perhaps, even for every patient before treatment [46,51] (and encouraged the development of the IC3Ddose community). With time the treatment planning algorithms were better validated, the treatment unit control systems improved, and more efficient alternate patient specific QA was developed using detector arrays [52,53] or Electronic Portal Imaging Devices (EPIDs) [54–58] so that the requirement for regularly performed full 3D dosimetry diminished. Currently such patient specific delivery validation is considered part of a tiered RT QA environment often illustrated as a multi-layered pyramid (see figure 2) [59,60]. The various QA tests at each level are to ensure that all individual components required for the RT process work as intended individually and in their interactions. And external agency guidance documents, national technical control guidelines, and internal treatment and QM procedures have been developed to set requirements for appropriate unit

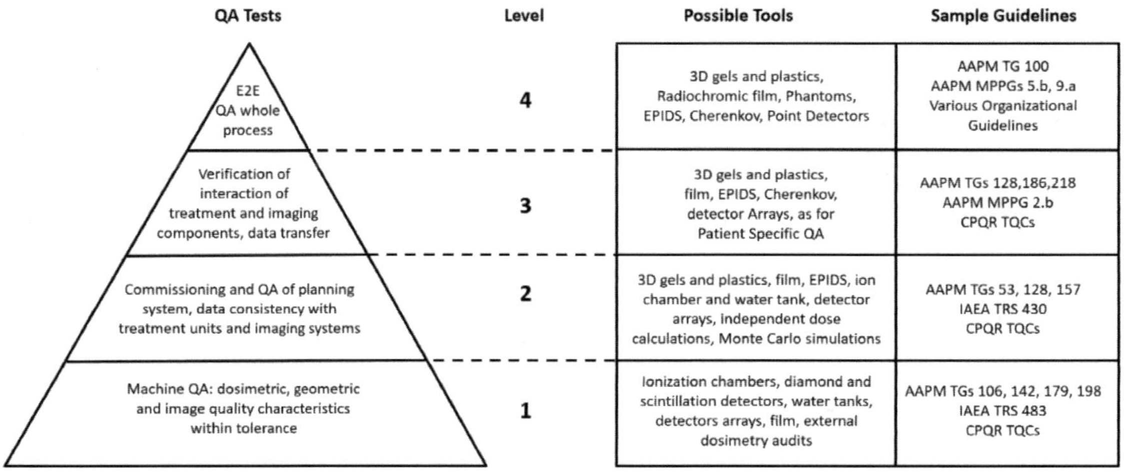

Figure 2. A representation of the quality assurance testing environment associated with radiation therapy. The left most pyramid gives a build-up of the QA program required for safe patient treatment. The tests progress from unit tests of individual components at level 1 to higher level tests that validate the operation as units work together with data exchange between those units, to eventual quality management of the whole process. The different levels build on the conceptual pyramid proposed by De Wagter [59]. The details of the tests at each level might be debated and one could move elements between the levels depending on definitions set (for example the Australian Clinical Dosimetry Service (ACDS) describes the hierarchy using 3 levels [60]); the figure is intended to illustrate that RT is a complicated process with many elements that need to be tested individually and also as they work together. End-to-end testing for some treatment schema acts at the fourth level (ACDS level 3). Some sample tools and reference guidelines proposed by various bodies in the radiation oncology and medical physics community are listed to illustrate the complexity of the QA testing needed to ensure quality radiation delivery [see text for references].

testing [61–68] of the equipment (level 1 testing in figure 2), the validation of the interaction of select adjacent stages [68–74] in the IGART process (level 2 and 3) and finally to guide the more comprehensive QM of the complete IGART process [17,32,33,37,38,75–77]. In this schema, the machine-specific QA unit testing is specified to be implemented at specific stages of the IGART process (illustrated in figure 1b) to ensure that the equipment functions as required within specific tolerances [17] (for example, validating the radiation output of the linear accelerator, the couch motion on treatment unit, the positioning of the EPID, etc.). The further level 2 and 3 type testing is stipulated to ensure the interaction of units, and data transfer between units, function as expected. A common clinical implementation at these levels is the patient-specific QA practiced in cancer centres to validate that the treatment prescribed for a particular patient can be delivered as planned on the treatment unit. Since the clinical scenarios specific to the individual patients are used, such delivery QA is a type of end-to-end testing (see below), albeit of only one select component of the IGART process (encompassing dose calculation in the treatment planning system, through determination of treatment unit's control parameters with transfer to the delivery unit, through to dose delivery). Level 2 and 3 testing is also directed at other components of IGART particularly addressing data transfer and interconnection of different systems in the IGART chain. The functioning together of the CT or MR imaging components with the treatment planning systems, of the planning systems and the radiation delivery units, and of the delivery units and on-board imaging modules add additional QA requirements with appropriate specifications to ensure that the systems function appropriately together and that their interaction does not add failure modes not apparent when the components are tested alone [17,27]. The guidelines at these levels may include additional unit tests with slightly different specifications and tighter tolerances to validate that adjacent components function well together [27]. The main tactics for QA directed to the highest level in the pyramid fall into two approaches: firstly, risk-based analyses of the radiotherapy processes within a radiation service's particular setting to identify efficient, effective, and safe treatment delivery; and secondly, end-to-end testing of the complete process. Risk based approaches to radiotherapy process QM have evolved in the last 15 years [78] reaching fruition in the report of the American Association of Medical Physicist's Task Group 100 (TG100) [76]. The TG100 report represents a departure from previous TG reports in that it does not provide guidelines for specific testing (with set tolerance levels and frequencies) but rather presents a framework for designing QM activities that work through the full RT planning and delivery process. The analysis uses engineering tools such as process mapping, failure modes and effects analysis (FMEA), and fault tree analysis (FTA) to set and prioritize a program's QM activities based on estimates of the probability of identified failures in the process with measures of the impact of these failures on clinical outcome [78]. Examples of specific RT process analysis in TG100 (restricted to IMRT) and reports of clinical applications of FMEA RT analysis all conclude that human factors have a considerable impact on the quality of the radiation delivery. The analysis of the complex interaction of equipment, RT processes and people on RT QM has led a number of guidelines to suggest that end-to-end testing should be more regularly practiced to achieve quality radiation delivery [17,77,79–81].

3. End-to-end testing

End-to-end testing (E2E) originated as a computer science methodology to assess if a software application, built up from linked distinct software routines, performed as required in the field: the application's performance would be tested from beginning to end under real-world working scenarios. In computing, E2E tests are devised to validate that the software routines forming the application work well together, that the application communicates and interacts with hardware, networks, databases, and other applications as required for proper functionality, and that the application delivers the output expected by users. E2E testing for radiation therapy has the same intent, with the testing serving to verify equipment and processes (for example as shown in figure 1) rather than software applications.

3.1 E2E testing in radiation therapy

In the RT context, E2E testing is to confirm that some planning/delivery/adaptation process works from start to finish correctly so that the desired radiation dose is delivered accurately to the intended spatial location in the irradiated volume [81]. As such, E2E testing operates at the higher levels of the testing pyramid from figure 2. Since the intent is to verify dose delivery, RT E2E testing is always performed using phantoms containing some dose measuring system(s). In effect, the testing uses the phantom as a patient surrogate and the application of the RT process being tested is evaluated by having staff perform the specific process protocol to this surrogate (see figure 3). The ability to verify the delivered dose provides a quantifiable metric for the evaluation. Thus, E2E testing enables a quantitative indication that i) the RT process is performing as required and that ii) the team applying the process is doing so correctly. This builds confidence in the procedure within the clinic.

3.2 Application of E2E testing in standard radiation therapy practice

To date E2E QA has been practiced intermittently in RT programs, most commonly in three settings: *Setting i)* during the creation of new radiotherapy devices and techniques in some development environment (in industry or a clinical research setting); *Setting ii)* when a clinic is implementing a new technique, especially if the technique is more advanced than the clinic's standard practice to date, or *Setting iii)* during external auditing under some multi-institutional program [81–90]. This latter application of E2E testing through some external audit is usually performed to help benchmark a clinic's execution of a specific IMRT or IGART delivery process against performance in the wider community or to credential a cancer centre's ability to perform a specific delivery protocol when applying to join a cooperative multi-centre clinical trial using that technique. A universal feature of the external E2E audits is that the service is designed to probe the particular cancer centre's ability to deliver the specific RT technique. For audits performed for clinical trial credentialling, the intent is to verify that the cancer centre can deliver the RT protocol required in some arm of a cooperative multi-centre clinical trial. This serves to improve the RT influenced data in the trial by confirming quality

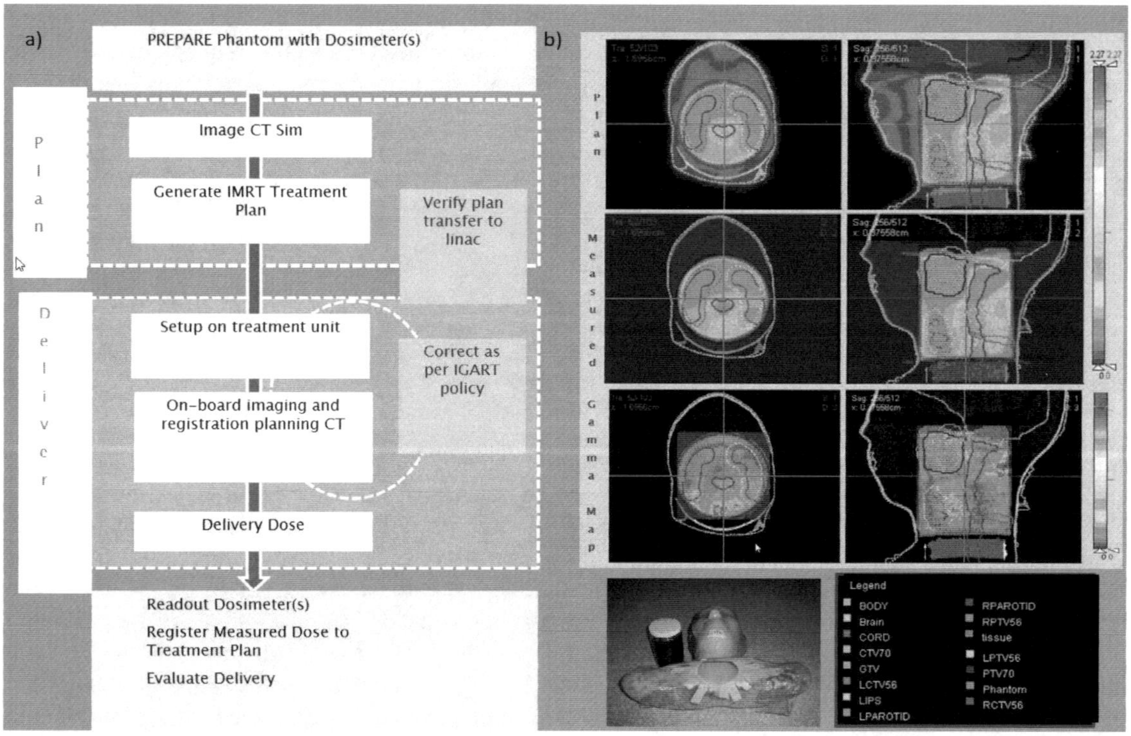

Figure 3. An example of end-to-end testing as applied to radiation therapy. a) The workflow for the E2E test that was run to validate the specific IGART process that was being implemented in Kingston during the transition from 3D conformal delivery to full IMRT treatment [19,21]. b) The E2E test used an in-house purpose-built head phantom containing a 3D xylenol-orange Fricke gel dosimeter [21] (bottom photograph). The phantom was irradiated using the workflow under the clinic's head-and-neck (H&N) IMRT treatment protocol. The resulting performance evaluation showing planned and measured dose distributions in two orthogonal views through the treatment volume. The top distributions are from the Varian-Eclipse generated treatment plan; the middle dose distributions are from the optical-CT measurements of the irradiated gel dosimeter. The lower images show the gamma evaluation maps (3%/3mm) indicating the spatial dose agreement between intent and delivery. Although this was an E2E test on a phantom, target and organ at risk contouring was performed by a

radiation oncologist (labels at bottom right) as per the H&N treatment protocol being implemented, and all dose distributions and the resulting gamma comparisons were mapped on the anatomical contours to flag if any delivery failures indicated by the gamma evaluations were clinically relevant (rather than just rely on gamma pass rates which can mask discrepancies [47,74,91–93]). Providing contours also enabled comparison of measured versus planned dose volume histograms for regions of interest in the irradiated volume [19,94–96]. The E2E validation during IMRT implementation before the first patient was treated provided considerable confidence to the implementation team that our IMRT protocol was within our program's capabilities.

radiation delivery from single centre(s) and better ensuring consistent RT delivery across the centres participating in the trial [81,84,97].

3.3 Lessons learned from past radiation therapy E2E testing

The objective of the E2E testing in the developmental Setting i) above is to validate that the technical components (equipment, hardware and software) required for some subset of the system under development are operating individually and together as intended [27] (thus, complementing level 1 and 2 testing). Such E2E testing for the novel treatment system being put together does not necessarily cover the full RT process, rather it concentrates on a select limited segment of the system to ensure that the segment is not perturbed by unforeseen complications introduced as it is assembled. The testing in this setting is most often performed by specialists, or by select development team members, rather than the individuals in the clinic who will eventually execute the process during treatment delivery. Such work is not widely reported in the open literature; it is mentioned in this review for completeness but is not considered further.

There have only been a few reports of RT E2E testing performed when novel treatment techniques are first introduced in the clinic (Setting ii above). A barrier for widespread reporting of such testing has been that such work is often interpreted as regular clinical practice and not innovative publishable research. The literature that is available usually presents developments of novel dosimeters or phantoms for particular E2E clinical applications or describes particular E2E protocols for specific RT techniques. Some of these Setting ii) applications have been published in past IC3Ddose conferences, often to illustrate the application and possible advantages for 3D dosimeters in clinical E2E QA (well-reviewed in ref. [98]). Some specific developments and applications for E2E QA to evaluate specific clinical treatment protocols are referenced in Table 1 below (which presents a list of vendors that supply various commercial tools and services to aid in-house E2E testing) and in the most recent IC3Ddose proceedings from the virtual meeting in 2021 [99–103]. These and other [21,53,96,104–112] clinical reports span a range of investigations into the feasibility of RT E2E testing, nonetheless they do make some common observations. The papers indicate that clinically relevant E2E QA can be performed with a wide range of phantom designs that may contain various point, 2D or 3D dosimetry systems (see discussion below). A common feature in this work is that the phantoms and dosimeters must be well characterized and tailored to the RT delivery technique being tested. Select studies acknowledge that there are extra demands in the data processing for E2E RT QA, starting with a requirement for tools that can manage the TPS planned dose distributions and the

Table 1. A list of commercial (and open source) service providers offering tools enabling in-house E2E QA including phantoms, 2D and 3D dosimeters, optical-computed tomography (opt-CT) scanners for volumetric dose readout, software tools for dose calibration and mapping (including modules for registration and comparison of measured and planned dose distribution data). The table includes the MDADL laboratory which provides a commercial IROC-H phantom service for select IMRT and dynamic RT technique E2E clinical testing outside of trial credentialling.

Provider	Vendor Website	Select Literature reporting use of vendor products for E2E QA
GeVero, Poland	www.polygevero.com/home	[95,114–118]

Heuris Pharma, USA	www.presage3d.com	[96,119,120]
MGS Research Inc., USA	www.mgsresearch.com	[121–124]
MD Anderson Dosimetry Laboratory	https:// IROC-Hhouston.mdanderson.org/ MDADL/home.htm	[125]
Modus QA, Canada	www.modusqa.com	[103,126–130]
RT Safe, Greece	www.rt-safe.com	[94,106,131–134]
Slicer-RT, Canada	www.slicerrt.org/	[113,135–137]

measured delivered dose data sets well and that can ensure that they are well registered before evaluating agreement [96,113]. Of relevance to this review, the literature investigating clinical E2E testing positively reports that establishing in-house E2E QA is feasible and that the tools (phantoms, dosimeters, analysis systems) for a cancer centre's E2E testing program can be prepared locally with reasonable effort or be purchased from existing service providers.

The most comprehensive literature describing the benefits of RT E2E testing comes from the experience from external audits (*Setting iii* above). Basic historical reviews [81,83,84] summarize E2E and dosimetry audits in general, giving an overview of E2E auditing by an external agency. Other reports describe the E2E activities of particular bodies such as the Australian Clinical Dosimetry Service (ACDS) [86], the Institute of Physics and Engineering in Medicine in the United Kingdom [90], the International Atomic Energy Agency (IAEA) [88,89,138], the Imaging and Radiation Oncology Core QA Center in Houston (IROC-H) in North America [83,84,139,140], or in a multicentre collaboration such as in the United Kingdom [130].

These reviews outline well how external audit RT E2E testing proceeds: the auditing body prepares a phantom containing dose measuring tools appropriate to the specific technique under audit. The objectives of the test (e.g., required given doses to the specific targets) are set and communicated to the centre performing the run, as are the criteria that determine if the observed E2E delivery passes or fails expectations. The phantom is delivered to the cancer centre and the RT clinic staff is directed to run the phantom through all steps and manipulations as would be used on a patient undergoing the RT process under investigation (for example with a workflow similar to that shown in figure 3). Once the radiation delivery has been completed, the phantom is returned to the auditing centre and the dosimetry systems are analysed to determine whether the delivery was as intended.

There are some common observations from the experience gained over the large number of external E2E audits cited above. While most of the centres coming into the external audits had well-established processes for the IMRT and IGART techniques being audited, and had well-formed in-house physics QA programs, a significant fraction of centres did not pass the initial audit runs. This result is reported consistently across all auditing bodies. One encouraging trend observed by IROC-H, which has the longest reported follow-up for audits on multiple phantom sets designed for different RT delivery tests (e.g., head and neck (H&N), lung, liver, etc.), is that the passing rate for E2E testing on the various phantoms has improved with time. For example, Ibbott [141] reported that ~30% of institutions had failed the 250 IROC-H H&N IMRT E2E audits performed before 2008, by 2013 only ~19% failed (over 1139 tests for ~760 institutions) [140], and by 2016 the failure rate dropped further to ~10% [142]. Unfortunately, the pass rates for external E2E audits have not yet approached 100%. Also, a not insignificant number of centres fail the E2R audit again on subsequent runs on the same RT protocol, an indication perhaps that the failure was not the result of random error but rather was related to some problem with the radiation delivery process in use by that centre [143].

An instructive literature has grown which reports on investigations of the patterns of failures observed over the IROC-H experience with external E2E audits [52,93,140,142–148]. Throughout the IROC-H audits, program failures have consistently been identified with issues in the physics domain, even daily output variations have contributed to unsuccessful E2E phantom irradiation [144]. About 1 in 5 failures in IROC-H H&N phantom tests could be attributed to TPS errors [149]. The errors included incorrect data entered into the TPS (e.g., output factors, percent depth doses, multi-leaf collimator (MLC) defining parameters, etc.) and inaccurate beam modelling within the TPS (which led to pencil beam algorithm models being disallowed in National Cancer Institute sponsored trials [144]). Some of the failures were well identified with known modelling issues at the time (MLC leaf gap modelling in a particular TPS [150]) and were corrected by the medical physics community; but similar problems then reappeared more recently resulting in unsuccessful E2E tests in centres using another TPS [145]. With lung and spine E2E audit phantom irradiations [146] the failures resulting from dose calculation errors were more evident with the highly modulated spine phantom irradiations. These results indicate clearly that E2E testing (in which the whole treatment chain is tested from planning to delivery) can reveal problems with basic physics modelling data and suboptimal TPS calculation algorithms that are missed during the unit specific evaluations in conventional physics QA. This observation has led to the recommendation that E2E testing be a part of TPS commissioning process [77,145,151].

One observation from the IROC-H H&N phantom audits [93] worth note is that audit failure does not correlate well with the results of the patient specific IMRT QA used by the institution performing the phantom irradiation. The discordance may be attributed in part to the fact that the patient specific IMRT QA test might use some surrogate measure of the delivery (with fluence measurement, or dose to a detector array positioned outside of the phantom compared against a TPS prediction). And the in-house IMRT QA might measure the delivery from a single gantry angle rather than going through the full gantry motion in the actual radiation delivery. Furthermore, the pass criteria in the patient specific QA might use too simple a metric (e.g., gamma pass rate rather than gamma maps or other comparator) to score success, and this may can mask failed delivery [47,74,91–93]. The poor correlation between patient specific and E2E QA validation of treatment delivery suggests that E2E testing should also be part of any commissioning of internal E2E patient specific QA techniques. This would be a type of Setting ii) E2E testing.

Some E2E audit failures have been attributed to incorrect positioning of the phantom during irradiation specifically because the RT clinic's delivery protocols, which specify how patients are to be set up, have their position verified, and corrected if needed, were not followed. In this case the E2E testing is catching noncompliance with the clinic's RT delivery protocol. As noted above, the auditing bodies instruct centres to have staff handle the phantom as if treating a patient. But often external audit testing is performed solely by medical physicists and may not involve the radiation therapists who treat the patient. This can lead to suboptimal performance during the phantom irradiation, especially when the test requires careful phantom set-up; E2E test failures have been attributed directly to the physicists not following set-up procedures as per centre policy [140,147]. Such errors have been detected because the IROC-H phantoms contain multiple dosimetry systems, in particular point detectors and film dosimeters for localization of the delivered dose distribution [152–154]. In many of the E2E runs the auditors have been able to use the point and film detectors to distinguish between systematic failures leading to the *given dose* (the dose value at points in the irradiated volume [155]) being out of the 7% acceptance criteria, and localization errors in which the correct given dose was delivered but to a shifted position perhaps because of mistaken phantom set-up or perhaps due to incorrect accommodation of motion (in IGART E2E testing of radiation delivery under motion). The advantages of dual detectors for the breakdown analysis of E2E test failures have been reported for head and neck [140,142], lung and spine [143] and liver phantom [147] audits.

The auditing E2E QA literature clearly establishes that E2E QA can identify deficiencies in the implementation of RT techniques that are not detected in the cancer clinic's conventional physics and radiation therapy QA nor apparent in the design of the protocols that set the delivery processes for that technique. In an IMRT uptake audit [88], the IAEA identified that clinics with more IMRT delivery experience did have higher pass rates in their audit. But no similar significant difference was observed in the pass rates for small and large cancer centres in IROC-H H&N IMRT audits [140]. Audit success has also been reported to be unaffected by the number of treatment linear accelerators (linacs) in a centre or the number of physicists per linac affected [140]. More broadly, the full IROC-H clinical trial credentialling experience shows that even well-established and experienced cancer centres fail phantom irradiation tests. It should be noted that the passing criteria for the IROC-H tests are generous (for a variety of historical and practical reasons, delivered given dose must be within 7% of plan and distance to agreement within 4 mm); if the criteria were tightened the failure rates would increase [142].

In summary, external audit experience shows that E2E testing adds value to a cancer centre's QM program. Deficiencies in treatment delivery can be identified which are otherwise missed in conventional clinic QA testing. Success in a particular RT technique's E2E test adds confidence to the centre that the radiation program can deliver the technique as intended. Running the test provides a learning experience for the team involved and can help fine-tune the policies and procedures defining the technique's delivery process. The results from the test can provide benchmark data for future reference useful to establish continued compliance in the delivery. These benefits to RT QM do not require that the E2E testing be moderated by an external auditing centre, they would apply equally to in-house E2E implementation.

4. Additional Observations on RT QM

While QM has been a constant concern in RT practice from its very beginning, QM programs have evolved as RT has become more complex, in part by adopting tools and approaches from engineering and industry [78]. Incident reporting, FMEA risk analysis, and human factors analysis have all been employed in the last years to advance safe clinical practice. The lessons learned from these developments have become as important a part of RT practice as are fundamental medical physics and dosimetry knowledge and practice. The idea that human factors (i.e., factors related to human behaviour, abilities, limitations, and relationships to the physical and organizational work environment) influence RT processes is now well established. And it is well understood that protocols and procedures for specific RT treatments must be well defined and laid out with appropriate training to minimize these human factors perturbing safe practice. The results of incident reporting programs have established that regular review of clinical performance with reports of these reviews to cancer centre staff can improve protocol compliance and reduce future incidents. FMEA can be used to scrutinize a program so that riskier steps in the RT process come under more scrutiny. Analysis may also enable QM programs to better allocate resources so that a program's efforts are devoted to the most beneficial QA by moving from tests that give little value to tests that better limit failure.

But the literature also indicates that these interventions and analyses cannot eliminate all policy non-compliance. For example, incident reporting [39], FMEA [78], and human factors analyses [18,24,42] accounts have indicated that the patient set-up immediately prior to RT dose delivery often remains subject to operating procedure non-compliance because of human behaviour, even after a process has had steps and workflow improved. The protocol deviations in patient setup have been explained as being the result of staff not being suitably trained or not fully understanding the situation they are encountering (because of limited information or communication, or because they are working under stress). And various corrective recommendations follow, for example suggesting the adoption of using checklists [156], or time-outs, etc., to provide more stable operating conditions. But it could be that the variation of performance is not only the result of poor training or judgement or imperfect process design, but rather is a feature of how individuals manage tasks and decisions in a process and how they make decisions.

For one thing individuals have variable risk tolerance. This is observed in our day-to day lives and regularly in the clinic, for example, during linac output calibration where one physicist might adjust an output that is within the QA tolerance while others will leave it be. This example suggests an illustration of the influence of variable risk tolerance on a QA test, see figure 4. The figure is a modification of a classic representation of how action and tolerance levels [17,157] are used to define the outcome of some QA test and how measurement uncertainty may influence the course of action. The modification is the addition of another row at the bottom to illustrate that decisions made during the QA are also influenced by the risk tolerance of the person evaluating the measurement. This added factor determining action introduces variability to the QA outcome;

different risk tolerance would similarly influence the completion of any RT processes since the processes are comprised of multiple decision points that define subsequent action. Investigation of the influence of risk tolerance on decision making in medicine is in early stages [158–161], and multiple factors such as training, past clinical experience, personality traits, and tolerance for uncertainty, have been identified as affecting an individual's decision making. Some models for decision making [161] propose that a person's reasoning is not exclusively based on analysis of quantifiable facts and measures; rather their information processing also involves simplified categorization of situations into more qualitative and intuitive framework (termed "gist") influenced

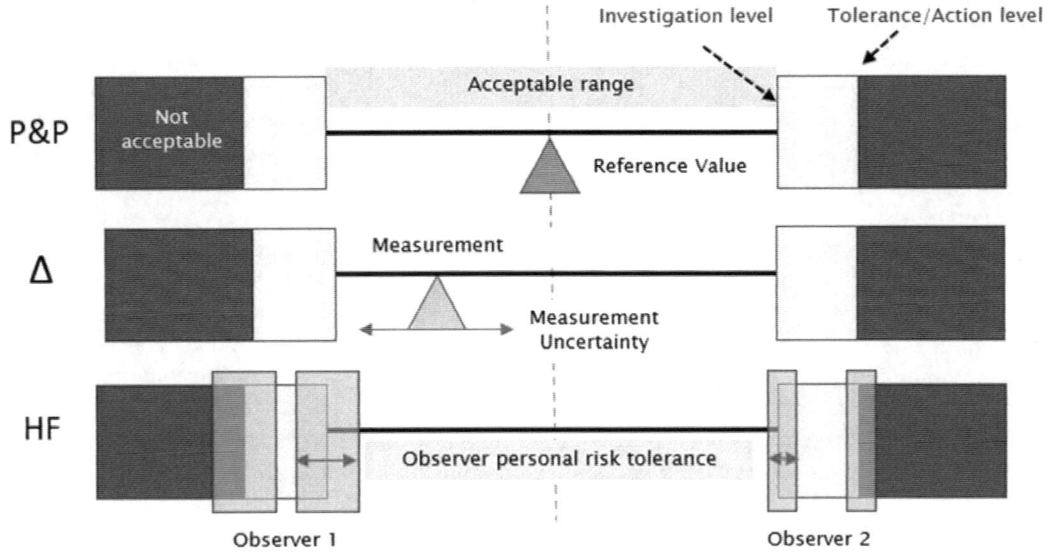

Figure 4. A common illustration (rows P&P and Δ) of a QA test in which a measurement result must be within an acceptable range defined by some tolerance specified by investigation and action levels used to trigger a specific action after testing [17,157]. A QA policy and procedure will use the investigation and action levels to mark when the system might only require examination after the test (for example, a repeat measurement later in the day in daily linac output measurement) or perhaps generate a more critical action (taking treatment unit out of service until recalibrated). QA protocols accommodate measurement uncertainty (Δ) by setting these levels appropriately [17,157]. Row HF is intended to show that human factors such as individual observer risk tolerance may blur the decision to undertake a specific action after some observation. This may lead to variation of practice even when a process protocol has been well defined and laid out. For example, not all staff might reposition a patient whose set-up images do not align with the planning image within limits of the treatment protocol even though they are working to the same treatment policy.

by past experience and past encounters of similar situations. And as individuals become more experienced in their roles, their decision making becomes more gist weighted [161]. This is all to say that variations of performance in an RT process cannot be completely controlled by external factors since they may be basic consequences of human behaviour.

In a seminal paper on risk management [162] Rasmussen suggests that top-down process design using traditional approaches may not work in a dynamic environment with changing technology and a high degree of integration and coupling of systems (as encountered in RT programs). Traditional design might attempt to account for the whole process being more than the sum of its uncoupled parts by adding task instructions (i.e., protocols or performance criteria) to better ensure the whole process proceeds as intended. But such an approach misses that the tasks may not just proceed in sequence but may be active simultaneously. And the individuals doing the task (the actors) may start to not follow the task instructions to the letter but rather adopt practical modifications to better achieve the outcome intended. The paper points out that this is the reason "working to rule" is an effective labour action since work often slows down when actors do things exactly as written in policy. Therefore, modeling of human behavior in terms of a stream of acts with task instructions to control processes may not be a reliable standard for judging actual behavior in work.

A recent analysis of the variability in practice in medical physics RT treatment plan review (i.e., plan or chart checking) across 15 cancer centres in Ontario, Canada, provides some interesting observations [163] on these points above. The investigators observed that standardized checklist use within a centre did not automatically reduce checking variability. They note individual physicists may not have been following checklists to the letter and suggest this indicates the importance of checklist development guidelines [156] to avoid ineffective checklists. The paper also highlights a regular aspect of planning checking, when to initiate a replan. It notes that there was consensus in all centres that plans that deviated from the centre protocols (e.g., did not meet planning dose-volume histogram (DVH) objectives) would be sent for replanning. However, sub-optimal plans (which met the program's DVH criteria but could still be improved) generated variable action, not only because of workload or a desire to get the patient on treatment as soon as possible, but also because individual medical physicists tolerated differently the level of sub-optimality and the need for replanning. The authors comment that variation in practice is not inherently a bad thing but that in principle uniformity of practices should be desired and that identifying factors that cause variation in practice can lead to improved processes and ultimately higher quality RT.

The nature of RT delivery in a process with human decision making throughout, means that there will be variation of delivery within a centre and between institutions. External audit E2E QA has been established to evaluate and control the variation between institutions. It seems advisable for cancer centre RT programs to use similar E2E methodologies to monitor RT delivery in-house.

5. In-house E2E Testing in the Radiotherapy Clinic

The discussion above provides the foundation for the main argument of this review: that more regular in-house E2E QA is doable and provides an important component for every comprehensive cancer treatment facility's QM program. But adopting E2E testing into an RT QM program is challenging, especially in the *Setting iii)* type testing, a setting which the author advocates needs wider application in local clinics. E2E testing is resource (material, and human) intensive and must be implemented prudently so as not to interfere with a clinic's main purpose, patient treatment. Therefore, establishing E2E testing requires forward planning to allocate the time and resources that best advance the clinic's QM needs. Some personal insights on how one might best fit E2E into a clinical QM program are presented in the next sections.

5.1 Implementing in-house E2E testing into a clinical QM program

It might be practical, in the spirit of TG100 [76], to initiate an E2E program by first applying a test to a simple commissioning or delivery protocol, and to then expanding the program more widely through the institution's QM. One entry point for clinics, especially those new to E2E testing, is suggested by a universal observation from the external audits in the IROC-H, the IAEA, the ADLS and UK experience: many phantom irradiation tests fail because of issues at the physics level of the IGART process (issues with incorrect beam modelling, MLC parameter entry, TPS algorithms, etc.). This has motivated calls for E2E testing to complete TPS commissioning [77,145,151]. The E2E test would involve generating a planned delivery, for example, for some more complex dynamic IMRT technique that challenges the algorithms and radiation delivery and irradiating the phantom under that plan. This is a form the *Setting i)* E2E QA described earlier. Such testing might be limited to the cancer centre's physics team, as most of the set-up tasks would be simple, although guidance from the radiation therapists might be needed at some stage if the IMRT application involved shifts of the phantom or motion. The E2E testing could be performed on simple phantom sets that are already in the centre's inventory, although one requirement for such an E2E test might be that the phantom contain multiple dosimetry systems that could measure both the delivered given dose and the spatial dose distribution (depending on the IMRT technique). A similar *Setting i)* E2E testing scenario might occur with the commissioning of some established RT treatment on a treatment unit new to the clinic or on which that technique has not yet been used. Implementing such *Setting i)* E2E QA would add an important component to the cancer centre's QM program since it would close the path for full physics commissioning of the involved hardware, software, or equipment. And if the centre is new to E2E methodologies it would enable staff to bring on the skill sets required to implement E2E method through a well-controlled scenario.

An alternate initial E2E test, or a next step in further implementation of an E2E program, would be in a *Setting ii)* application, for example, during the roll-out of a new RT technique within the centre. In this case the team involved with the testing would have to expand beyond just physicists (adding radiation therapists and radiation oncologists) to ensure that design, implementation, and evaluation of the test is appropriate to the RT technique. This would involve everyone on the test team understanding the policies and procedures required to deliver the technique so that the E2E testing is running exactly as the technique's implementation team has laid out. The people performing the test would likely still be a subset of the full radiation clinic staff, and the E2E test might not be as broad as in a full *Setting iii)* test (perhaps leaving out the planning stages by using a standard special test plan to a standard CT dataset for the phantom). Adopting *Setting ii)* type tests in the clinic would have the benefit of ensuring that a new technique has been well commissioned before irradiating the first patient, and it would help the staff performing the tests to gain a full understanding of the treatment process with all its steps and interactions. It would also expand the clinic staff involved with the E2E testing program beyond the medical physics department.

The approach to adopting E2E testing for the clinic in the two settings above is guided in a large part by a specific development in the clinic (a new TPS, a different treatment unit, or novel IGART technique). Adding type *Setting iii)* E2E testing is more complicated since in any given centre E2E testing could likely be applied to multiple RT deliveries. One could consider piloting *Setting iii)* E2E by adopting the testing for a simple IMRT process (starting simple, as TG-100 [76] suggests). On the other hand, one might direct the initial E2E adoption by using information from the centre's incident reporting program or using risk-based FMEA/FTA methods to determine which of the IGART techniques practiced in the clinic would benefit most from the increased scrutiny the E2E methodology provides. Such analysis will be cancer centre specific and there are no reports in the literature the author is aware of that provide guidance (although the TG-100 practical guides appendices provide a possible starting point [76]). Some guiding principles from computing science would recommend selecting RT deliveries whose failure would cause most harm and to avoid exception testing (testing of seldom encountered scenarios) by focussing on deliveries regularly used in the clinic.

Initial implementation of *Setting iii)* E2E QA may also benefit from an FMEA to assess specific technical requirements for the E2E test on the specific RT delivery to be useful. Such an analysis might show that a certain E2E test is not yet feasible in some clinic given the tools and resources available in the clinic at that time, while another test could be more readily started. This can be illustrated using two examples: i) E2E testing for treatment of multiple metastatic brain lesions in a Fractionated Stereotactic Radiation Therapy (FSRT) program, and ii) E2E testing of an adaptive RT treatment protocol involving target motion guidance or the correction for tissue deformation over a multi-fraction course of treatment.

The FMEA for the FSRT treatment for multiple brain metastases might establish that the main motivation for the E2E testing is to validate the spatial integrity of the delivery (confirming that the high dose is being delivered only to the multiple small targets that need treatment) [78]. Analysis may determine that there is only a small risk that incorrect doses would be given to a target since there was rigorous and complete commissioning of the TPS calculation for the small radiation fields that will be used and because the brain is a homogeneous medium well handled by the TPS. The risk analysis may also determine that the regular physics and radiation therapist QA performed on a particular treatment unit, along with patient specific portal imaging fluence based verification prior to treatment, are sufficient to ensure that the radiation delivery will be as predicted by the TPS. Therefore, the main risk established from the FMEA might be a targeting miss due to errors in image registration and set-up correction prior to irradiation, a step that involves human intervention. Such analysis might conclude that an effective FSRT E2E QA could have relaxed requirements for dose determination but increased requirements for high resolution spatial localization of the delivered dose. These objectives would release some constraints on the tools required for FSRT E2E testing, particularly relaxing calibration concerns in determining the delivered given dose when using film or 3D gel or radiochromic dosimeters for the spatial assessment. Efficient E2E testing could be performed using a phantom containing radiochromic film with transmission scanner readout, or a 3D NIPAM polymer gel dosimeter with CBCT readout [112,164,165], soon after the test irradiation and before all radiation-induced reactions have finished; or with polymer gels read out in an optical CT system where scatter perturbation might affect given dose measurement but not localization [126,166,167]; or using MR measurements based in relaxation time weighted imaging sequences [168–170] rather than more lengthy quantitative relaxation rate determining sequences [171]. Alternatively, the QA could incorporate a vendor supplied 3D gel dosimeter which is prepared remotely and sent to the clinic through regular shipping and delivery (that is, in an uncontrolled environment) with in-house or remote readout. The FMEA would also inform the design of the phantom required for the E2E QA. Since the risk mitigation motivating the FSRT E2E QA is associated with geographic misalignment of the patient during treatment it might be beneficial to use an opaque anthropomorphic head phantom that can be accommodated in the patient immobilization system. The phantom would have to appear patient-like on CT imaging to mimic the registration task during pre-irradiation setup to fully align the E2E test with the FSRT patient treatment. In the end, the FMEA analysis of the requirements for valid FSRT E2E QA may present options that make implementing the QA more tenable and hence a good starting point for a *Setting iii)* application.

FMEA of radiation deliveries under motion guidance, or of future adaptive processes with multi-fraction deliveries with dose correction schemas to correct for target deformation, would likely flag additional requirements for the E2E testing since the validation would require improved dosimetry with more exact calibration to verify that the delivered given dose is measured accurately. This is in part because these specific adaptive applications incorporate complex IGART processes with segments of the process that may still be under some development. For example, a particular test may be to validate adaptations incorporating on-line dose recalculation using weaker characterised on-line CBCT image sets than simulation CT scans [172], or using dose calculations with novel deformation algorithms that have yet to be fully characterized against measurement. The FMEA for IGART techniques involving motion management or tumour tracking adaptation will likely lead to recommendations that the E2E testing be designed specifically to detect any failure of the motion management strategies [173]. In all these cases, the dosimeters in the test phantoms will need to be able to measure the delivered given dose and the spatial dose distribution in the irradiated volume precisely and accurately which may require multiple detectors (as used in many of the external audit E2E QA phantoms).

Risk-based analysis could also be used to determine the frequency for any *Setting iii)* E2E testing within the cancer centre's QM program. The test frequency would depend on the technical aspects of the RT delivery, on the complexity of the interchanges within the delivery process, and perhaps on staff dependent human factors; and it would not necessarily be the same for all RT deliveries in the E2E program. It may be that a particular well-established IMRT treatment need only be tested annually to provide a regular assessment of performance. The E2E testing for the FSTR delivery (described above) may be scheduled as a regular quarterly test on each FSRT treatment unit because it is more easily performed, it confirms that dose delivery is maintained for a technique demanding accurate targeting, and it maintains a regular application of E2E methodology in the clinic

QM culture. In another possible situation, analysis of incident reports for a RT technique involving complicated set-up verification with patient shifts during the treatment may show that incidents reports are submitted regularly after new staff rotate onto the treatment unit delivering that technique. The RT program may decide that it would benefit by performing more regular E2E QA for this technique a week or so after staff rotation. Such decisions can be formed by the risk-based approaches now being recommended for QM design. With time E2E QA frequency might also be determined through observed pass/failure data with added scrutiny given to the RT techniques showing higher failure rates. The scheduling will need to balance the benefit of testing with the demands on clinic resources but as a clinic establishes its E2E QA program the testing should become more efficient, and more testing could become manageable.

An E2E program will require additional software and equipment infrastructure to proceed effectively. As indicated by experience, many different phantom configurations can be used for effective E2E testing [153,154,174]. Regular shaped physics phantoms [88,108–110,130,175–177] with well-defined geometry and appropriate locations for dosimeters are sufficient for many of the *Setting i)* and *Setting ii)* testing required to validate, for example, that the TPS has been commissioned correctly and that the TPS dose calculation algorithms model well the conditions of the RT delivery (e.g., accounting for highly modulated deliveries under dynamic MLC and gantry motion). Similar simple phantoms would work well also for some *Setting ii)* E2E testing of new delivery techniques being introduced into the clinic. There may be additional requirements to commission some treatments in a *Setting ii)* application or as a clinic introduces *Setting iii)* testing. If the delivery under evaluation involves motion management, then the phantom will have to be able to be integrated into some commercial or in-house built motion testing system [129,130,178,179]. For some RT deliveries the E2E QA phantoms may need to be more anthropomorphic in external appearance (e.g., if the delivery requires phantom set-up in some patient immobilization system with set-up validation and repositioning) or to simulate internal structure heterogeneity with organ mimicking structures (e.g., if the E2E testing is being applied to RT delivery to specific anatomy and the assessment includes target conformity) [89,102,104,106,111,180,181]. The material for the phantom need not be exactly tissue or water equivalent although it should be similar to any organs of interest and should be well characterised by CT, and perhaps MR imaging, before use [174]. The phantom must be able to accommodate the required dosimetry systems (3D dosimeters, radiochromic film, detector arrays and point detectors) to enable measurement of delivered doses. Experience with E2E testing in Kingston and published reports from external audit agencies indicate that the use of both point and 2D or 3D detectors improves the dose measurements required to assess delivery [102,127]. The ability to establish delivered given dose with well positioned point detectors can relax the dosimetry requirements on the spatial detectors (2D radiochromic film or gel or plastic 3D dosimeters) which typically provide accurate relative dose data, but which may experience preparation history dependent perturbation of dose sensitivity that limit precision when measuring given dose. Finally, the E2E testing system needs specialized software that can import planning image sets, the TPS predicted dose distributions, and measured spatial dose distributions for the analysis [113,118,137,182]. For some analysis it can be helpful for the E2E software to be able to import the set-up imaging data from the treatment unit. The software must be able to register these image data with dose data sets easily; this can be facilitated by incorporating fiducial markers into the phantom during preparation [96,102,113]. If the delivered dose distribution measurements are being determined in-house, the software must be able to convert data from the dosimeter readout system (e.g., optical CT, X-ray CBCT or MR data) into dose data via some calibration data. And the software must be able to calculate, and display, planned and delivered dose distribution comparison metrics such as gamma maps and conformity indices. These requirements have been reported in multiple IC3Ddose proceedings in this journal over the years. Such E2E capable analysis software is available commercially, including from most vendors listed in table 1.

5.2 In-house implementation of Setting iii) E2E testing

Since *Setting iii)* E2E QA requires that a phantom be processed through all steps and manipulations that would be used on a patient undergoing the RT treatment, the testing should not be limited solely to medical physicists, but rather it should involve radiation therapists and radiation oncologists as they would be during patient treatment (see figure 5) [21,183]. The various staff would contribute to the running of the E2E test as it proceeds through the delivery's stages:

i) The phantom would be prepared by a member of the medical physics department. This would involve selecting the appropriate phantom from the clinic's inventory and inserting the appropriate dose measurement systems (either prepared in-house or purchased from a vendor) into the phantom. At the end of this stage the phantom should be able to act as a patient to undergo the next stages of the test.

ii) The phantom would be delivered to the planning team and imaged according to departmental treatment simulation policy, and the images would then be contoured by the planning CT/MR therapists and, ideally, a radiation oncologist. The contoured images would then be sent for treatment planning.

iii) Treatment planning would be completed by dosimetrists/planning therapists using the planning protocol for the RT technique under test. Once approved, the treatment plan and phantom images would be sent to the treatment unit as in clinical practice. The TPS dose distributions would also be sent to the physics team so that the planned dose distribution data could be prepared in the E2E software system for subsequent comparison with measured dose data at the end of the irradiation.

iv) The RT delivery would be executed as closely to treatment protocol as possible at the treatment unit by the treating radiation therapists. (For an E2E test of FSRT delivery the phantom set-up would include CBCT imaging at the unit with registration, comparison with planning images, and shifts initiated as required by the RT techniques positioning policy as shown in figure 5). Once the phantom is deemed to be in the correct position the phantom would be irradiated.

v) The phantom would be returned to the medical physics team who would remove the test dosimeters and read out the dose data (although the centre may have established E2E testing protocols to image dose directly on the treatment unit using on-board CBCT or MR imaging, in this case the dose images would need to be transferred to the E2E data analysis software). Alternatively, if using commercial services, the irradiated dosimeters might be shipped back to the vendor for measurement with the dose data subsequently sent back to the centre for analysis. The physics team would compare the measured dose data to the planned dose delivery and evaluate if the E2E test has been passed or not.

vi) Once the E2E evaluation of the delivery is complete, the results would be reviewed, initially by the medical physicists to confirm the integrity of the analysis, and then by all members of the E2E test team. Regular test review sessions would be a requisite feature of a well-established E2E program. And the E2E program would send reports summarizing test results periodically to the cancer centre's QM program for its review.

Adopting a framework for *Setting iii)* E2E testing along the lines outlined above would ensure that the full radiation program staff would be involved in the testing and be aware of the test outcomes. It would lead to confidence within the RT program that it provides quality radiation delivery, thereby benefiting the QM culture within the cancer centre.

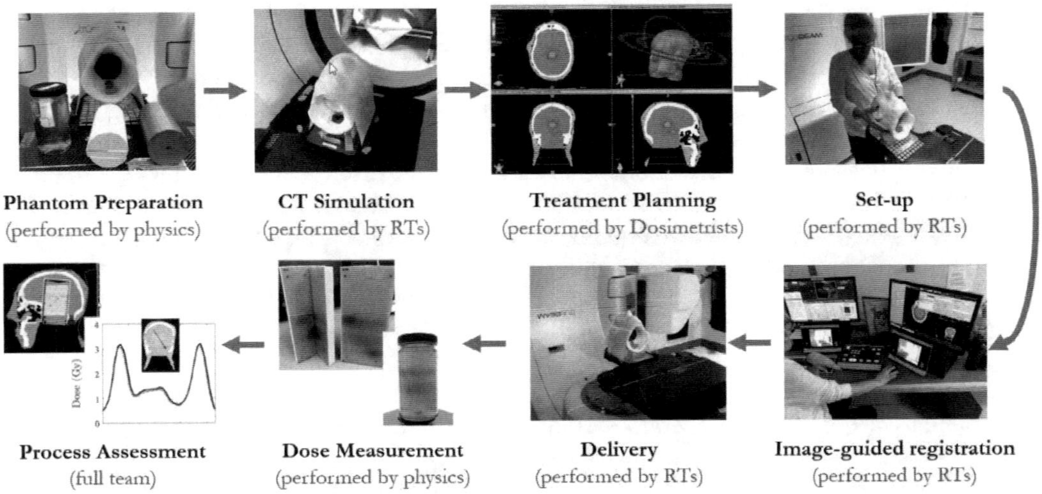

Figure 5. Clinical implementation of *Setting iii)* E2E QA validation of an FSRT treatment protocol [102]. The E2E test is performed by the whole treatment team (medical physicists, radiation therapists and radiation oncologists) with staff responsible for specific steps in the patient treatment process performing the same tasks during the E2E validation. The QA is designed to replicate the treatment protocol as faithfully as possible given the tools available (e.g., in this figure the phantom cannot replicate patient motion after set-up so that variable cannot be tested). The final assessment should reviewed by the whole team so that process validation is comprehensive, provides opportunity for process review, and offers learning opportunities to staff.

5.3 Benefits and Caveats

The benefits of end-to-end QA have been described throughout this review and they need not all be repeated here. One point that can be reiterated is that performing an E2E QA phantom irradiation does require that all the personnel involved in the test review and become familiar with delivery protocol and have the necessary understanding to complete the process as intended. Therefore, E2E QA not only tests the technical components of the RT delivery, but it also provides an assessment of the staff preparation. And the E2E testing can help identify the segments of the process that need greater vigilance during performance [143]. It has been suggested that such identification of failure points can help inform risk-based FTA/FMEA review [130].

There are limitations to E2E methodology that should be flagged as implementation is being brought into the local clinic. E2E QA cannot and should not replace all the conventional QA undertaken in the RT program. When an E2E test does fail, the cause is usually not immediately obvious. Troubleshooting and tracking where the failure originated requires that investigators understand how individual segments are functioning, and this is informed in part from the unit testing QA history. Furthermore, adding a new level of resource intensive testing to the RT QM program may be unsustainable. Therefore, cancer centres will have to also undertake FMEA review of their QA programs. There has been debate for some time [23,184,185] that not all conventional QA performed by medical physicists adds value to the centre's QM program. This does need attention, and some of the procedures long held sacrosanct may need to be reassessed (for example, spending weekends on annual water tank measurements of multiple linacs that have shown insignificant beam data variation over the years) to enable adoption of QA that better influences quality control.

Establishing an in-house E2E testing program with wider involvement of clinic staff in the testing will require a change in clinic QM culture. There may be initial challenges to get individuals on board who have had limited experience performing QA. Therefore, there will have to be a considerable effort on education and change management. Ensuring that the outcomes of testing are regularly reviewed in an open QM forum will help this transition.

There is a caveat associated with moving E2E QA from external audits by an independent body into a local setting. One of the main advantages of external audits is that the testing performance and analysis is directed by a single body that provides a homogeneous testing framework which has been a major benefit of external audit QA to the radiation therapy community. Moving E2E QA in-house removes this benefit. Therefore, such a move will have to be supported by the various bodies in the United Kingdom, at the IAEA, at IROC-H and the ACDS, and within the various professional bodies that help guide practice management to standardize how in-house E2E QA should proceed. Technical quality control documents and medical physics practice guidelines will need to be developed to standardize approaches.

5.4 Future directions

The observations from the RT literature and recommendations from medical physics practice guidelines motivate increased E2E testing adoption in RT clinics in-house. This indicates that the community should continue to develop improved tools for E2E testing. There is room to improve dose measurement, for example by reducing the variation of dosimeter response due to post-preparation history, or by developing novel point dose dosimeters that do not perturb measurement when inserted into 3D gel dosimeters. And there are still opportunities to advance phantom systems for the RT environment. For example, developing a fleet of phantoms able to accommodate interchangeable dosimeter inserts between phantoms would improve phantom assembly and disassembly and ease of use. Other work could be devoted to developing inserts to provide the image contrast to enable structure contouring during planning and to enable testing of on-board imaging on the treatment unit. Research to better integrate TPS planning systems and E2E analysis software might enable more efficient planned versus delivered dose comparisons and improve the E2E software environment. This is work well suited to the IC3Ddose community. And it will require consultation and collaboration with commercial vendors to develop phantoms and dosimeter systems that are more exchangeable and transferable between suppliers and within clinics.

The literature reviewed in this article also suggests that there are two additional applications for E2E testing methodologies in the clinic. As has been noted, the success of a phantom irradiation test during an E2E audit has not always been well predicted by the conventional QA or patient specific IMRT QA performed in-house. Perhaps E2E testing should become an integral part of the commissioning of internal QA techniques, in particular patient specific IMRT validation that uses surrogate tests (such as fluence measurements on detector arrays) to validate dose delivery. This would be a type of *Setting ii)* E2E testing. The second application would be in a role using full *Setting iii)* approaches. Experience has shown that E2E testing requires the individuals performing the test to fully understand all the components and steps that define the given RT delivery. And performing the E2E QA on a phantom involves going through all the steps of the process that would be used on a patient; in effect it serves as a simulation run of the patient delivery. E2E methodologies therefore provide a learning tool to the clinic staff. It may be beneficial to a cancer centre's QM program to consider how E2E methodologies could be used as simulation training in an educational setting. Simulation training has been shown to add benefit in RT because it provides hands-on application of learned concepts in real clinical scenarios involving team members from different professions working together [186,187]. Care would have to be given to ensure testing and education objectives are clearly defined and kept separate. But as E2E testing becomes more tenable, an expanded role for staff education seems well indicated.

6. Conclusion

The quality management considered appropriate for good radiation therapy delivery has changed considerably over the last three decades as radiotherapy processes have become more complex. Guidelines for quality assurance testing have moved beyond protocols for unit testing of equipment to recommendations for more comprehensive testing of interacting links in each clinic's delivery system, with risk-based analysis of the RT environment to establish points in the process that may require particular attention for testing or improvement. Incident reporting and human factor studies have shown that health care personnel can influence the quality of RT delivery, and protocol review and revision strategies to make processes more robust and less perturbed by persons have become commonplace. But such strategies cannot eliminate all the variables that might cause an RT protocol to deviate from plan. Therefore, testing strategies that evaluate the complete process are required. End-to-end testing provides an appropriate methodology to achieve this goal. E2E testing has been shown to detect deficiencies in the delivery process not detected by the conventional QA performed by medical physicists and radiation therapists in the clinic, thereby improving delivery accuracy. And E2E QA can give the RT team confidence that the entire RT treatment delivery process being evaluated can deliver the intended dose to the intended location. If there are faults detected during testing on phantoms, these can be corrected and re-verified before treating patients. And as the E2E testing is performed it provides opportunities for learning by increasing team members' awareness of the details and requirements at various stages of the protocols they are using. The tools enabling in-house E2E testing have become available in the last years. More regular E2E testing would provide an important tool for any cancer clinic's RT QM so that it can confidently provide its patients quality radiation delivery.

7. Acknowledgments

The ideas presented in this address have been formed through research funded through Canadian CIHR project MOP-115101, CHRP projects CIHR CPG 151964 and NSERC CHRP 508528-17, and through years of interaction with many colleagues in the IC3Ddose community. I would like to thank those who provided me with examples of IGART protocols from their clinic for my presentation on this topic and colleagues from IROC-H Houston, especially Andrea Molineu, who shared ideas as I was forming my thoughts for this review especially in the discussions in sections 3 and 5.

8. References

A version of this reference list with titles is available from the author on request.

[1] Trump JG *et al* 1961 *Radiol.* **76** 275
[2] Jaffray D A *et al* 2002 *Int. J. Radiat. Oncol. Biol. Phys.* **53** 1337-49
[3] Raaymakers BW *et al* 2017 *Phys. Med. Biol.* **62** L41-50
[4] Ibbott GS 2018 *Ann. ICRP* **47** 160-76
[5] de Crevoisier R *et al* 2005 *Int. J. Radiat. Oncol. Biol. Phys.* **62** 965-73
[6] van Herk M 2007 *Semin. Radiat. Oncol.* **17** 258-67
[7] Bujold A *et al* 2012 *Semin. Radiat. Oncol.* **22** 50-61
[8] McPartlin AJ *et al* 2016 *Radiat. Oncol.* **119** 371-80
[9] Dewan A *et al* 2016 *Asian Pacific Journal of Cancer Prevention* **17** 985-92
[10] Sonke JJ *et al* 2019 *Semin. Radiat. Oncol.* **29** 245-57
[11] Schaly B *et al* 2021 *J. Appl. Clin. Med. Phys.* **22** 168-74
[12] Paganelli C *et al* 2018 *Phys. Med. Biol.* **63** 22TR03
[13] Kontaxis C *et al* 2015 *Phys. Med. Biol.* **60** 7485-97
[14] Mackie TR *et al* 2003 *Int. J. Radiat. Oncol. Biol. Phys.* **56** 89-105
[15] Green OL *et al* 2019 *Semin. Radiat. Oncol.* **29** 219-27
[16] Bertholet J *et al* 2020 *Radiat. Oncol.* **153** 88-96
[17] Kron T *et al* 2022 *J. Med. Imaging Radiat. Oncol.* **66** 279-90
[18] Kalapurakal JA *et al* 2013 *Int. J. Radiat. Oncol. Biol. Phys.* **86** 241-8
[19] Schreiner LJ 2009 *J. Phys.: Conf. Ser.* **164** 012001

[20] Schreiner LJ and Olding T 2009 ed D W Roger and J E Cygler *AAPM Medical Physics Monograph: Clinical Dosimetry Measurements in Radiotherapy* vol 34, (Madison, WI: Medical Physics Publishing) pp 979–1026
[21] Schreiner LJ *et al* 2011 *Proc. Internat. Symp. Standards, Applications and Quality Assurance in Medical Radiation Dosimetry (IDOS) Book of Extended Synopses Vol 2* (Vienna, Austria: IAEA) pp 197–205
[22] Knöös T 2015 *J. Phys.: Conf. Ser.* **573** 12001
[23] Amols HI *et al* 2011 *Med. Phys.* **38** 5835-7
[24] Chan AJ *et al* 2012 *Radiat. Oncol.* **102** 142-7
[25] Marks LB *et al* 2007 *Int. J. Radiat. Oncol. Biol. Phys.* **69** 1579-86
[26] Ganesh T 2014 *J. Med. Phys.* **39** 203-5
[27] Williamson JF *et al* 2008 *Int. J. Radiat. Oncol. Biol. Phys.* **71** S2-12
[28] Thwaites D *et al* 1995 *Radiat. Oncol.* **35** 61-73
[29] Palta JR *et al* 2008 *Int. J. Radiat. Oncol. Biol. Phys.* **71** S13-7
[30] Leer JWH *et al* 1995 *Radiat. Oncol.* **35** 75-81
[31] Kehoe T and Rugg LJ 1999 *Radiat. Oncol.* **51** 281-90
[32] Canadian Partnership for Quality Radiotherapy 2019 (Available from: https://www.cpqr.ca/wp-content/uploads/2020/03/QRT2019-12-04.pdf)
[33] IAEA 2019 *Introduction of Image Guided Radiotherapy into Clinical Practice* vol 16 (Vienna, Austria: IAEA)
[34] IAEA 2008 *Setting Up a Radiotherapy Programme: Clinical, Medical Physics, Radiation Protection and Safety Aspects* (Vienna, Austria: IAEA)
[35] The Royal College of Radiologists 2008 *Towards Safer Radiotherapy.* (London)
[36] Kron T *et al* 2015 *Clin. Oncol.* **27** 325-9
[37] Yamada Y *et al* 2012 *Brachytherapy* **11** 20-32
[38] Strnad V *et al* 2018 *Radiat. Oncol.* **128** 411-20
[39] Clark BG *et al* 2010 *Radiat. Oncol.* **95** 344-9
[40] Pawlicki T *et al* 2017 *Clin. Oncol.* **29** 562-7
[41] Arnold A *et al* 2022 *J. Med. Imaging Radiat. Oncol.* **66** 291-8
[42] Chan AJ *et al* 2010 *Radiat. Oncol.* **97** 596-600
[43] Schreiner LJ 1999 ed. LJ Schreiner *Proc.1st Internat. Workshop Radiation Therapy Gel Dosimetry* (AB: Canadian Organization of Medical Physicists) pp 1-10
[44] Huq MS *et al* 2008 *Int. J. Radiat. Oncol. Biol. Phys.* **71** S170-3
[45] Schreiner LJ 2011 *J. Med. Phys.* **36** 189-91
[46] Thwaites D 2013 *J. Phys.: Conf. Ser.* **444** 12006
[47] Feygelman V and Nelms BE 2013 *J. Phys.: Conf. Ser.* **444** 12012
[48] Low DA 2015 *J. Phys.: Conf. Ser.* **573** 012009
[49] Low DA 2019 *J. Phys.: Conf. Ser.* **1305** 012001
[50] Ezzell GA *et al* 2003 *Med. Phys.* **30** 2089-115
[51] S.G. Scheib E and Lazaridis SG 2001 ed C Baldock *Proc.2nd Internat. Workshop Radiation Therapy Gel Dosimetry* 130–2
[52] McKenzie EM *et al* 2014 *J. Appl. Clin. Med. Phys.* **15** 241-51
[53] Lim SB *et al* 2022 *J. Appl. Clin. Med. Phys* **23** e13747
[54] Bojechko C and Ford E C 2015 *Med. Phys.* **42** 6912-8
[55] Mijnheer B *et al* 2013 3D *J. Phys.: Conf. Ser.* **444** 012011
[56] Spreeuw H *et al* 2016 *Med. Phys.* **43** 3969-74
[57] Mijnheer B *et al* 2018 *Physica Medica* **54** 137-45
[58] Kutuzov I *et al* 2021 *J. Appl. Clin. Med. Phys.* **22** 110-20
[59] de Wagter C 2004 *J. Phys.: Conf. Ser.* **3** 4-8
[60] Williams I *et al* 2012 *Australas. Phys. Eng. Sci. Med.* **35** 407-11
[61] Das IJ *et al* 2008 *Med. Phys.* **35** 4186-215

[62] Klein EE *et al* 2009 *Med. Phys.* **36** 4197-212
[63] Kirkby C *et al* 2018 *J. Appl. Clin. Med. Phys.* **19** 22-8
[64] Lagmago Kamta G 2018 *J. Appl. Clin. Med. Phys.* **19** 18-25
[65] Vandervoort E *et al* 2018 *J. Appl. Clin. Med. Phys.* **19** 29-34
[66] IAEA 2017 *Dosimetry of Small Static Fields Used in External Beam Radiotherapy, Technical Reports Series No. 483* (Vienna, Austria: IAEA)
[67] Hanley J *et al* 2021 *Med. Phys.* **48** e830-e885
[68] Bissonnette J-P *et al* 2012 *Med. Phys.* **39** 1946-63
[69] IAEA 2004 *Commissioning and Quality Assurance of Computerized Planning Systems for Radiation Treatment of Cancer, Technical Reports Series No. 430* (Vienna, Austria: IAEA)
[70] Beaulieu L *et al* 2012 *Med. Phys.* **39** 6208-36
[71] Villarreal-Barajas J E 2018 *J. Appl. Clin. Med. Phys.* **19** 35-8
[72] Pfeiffer D *et al* 2008 *Med. Phys.* **35** 5471-89
[73] McCullough SP *et al* 2021 *J. Appl. Clin. Med. Phys.* **22** 73-81
[74] Miften M *et al* 2018 *Med. Phys.* **45** e53-83
[75] IAEA 2008 *Transition from 2-D Radiotherapy to 3-D Conformal and Intensity Modulated Radiotherapy, IAEA-TECDOC-1588* (Vienna, Austria: IAEA)
[76] Huq MS *et al* 2016 *Med. Phys.* **43** 4209-62
[77] Geurts MW *et al* 2022 *J. Appl. Clin. Med. Phys.* **23** 14-34
[78] Rath F 2008 *Int. J. Radiat. Oncol. Biol. Phys.* **71** S187-90
[79] American Society for Radiation Oncology 2019 *Safety is no accident: A framework for quality Radiation Oncology and* Care (available from https://www.astro.org/ASTRO/media/ASTRO/Patient%20Care%20and%20Research/PDFs/Safety_is_No_Accident.pdf)
[80] Halvorsen PH *et al* 2017 *J. Appl. Clin. Med. Phys.* **18** 10-21
[81] Followill DS *et al* 2017 ed. BJ Mijnheer in *Clinical 3D dosimetry in modern radiation therapy* (CRC Press, Boca Raton, Fl) Chapter 9
[82] Gillis S *et al* 2005 *Radiother. Oncol.* **76** 340-53
[83] Ibbott GS *et al* 2006 *Technol. Cancer Res. Trea.t* **5** 481-7
[84] Ibbott GS and Thwaites D I 2015 *J. Phys:. Conf. Ser.* **573** 012002
[85] Palmer AL *et al* 2017 *Radiother. Oncol* **125** 453-8
[86] Lehmann J *et al* 2018 *Phys. Imaging Radiat. Oncol.* **6** 5-11
[87] Commissie N and Stralingsdosimetrie V 2018 Report 28 of the Netherlands Commission on Radiation Dosimetry (available from https://radiationdosimetry.org/?worker=add_footer&text=This+NCS+report+has+been+&file=files/documents/0000107/305-ncsreport28imrtvmataudit.pdf)
[88] Wesolowska P *et al* 2019 *Acta Oncol* **58** 1731–9
[89] Kazantsev P *et al* 2020 *Acta Oncol* **59** 141–8
[90] Clark CH *et al* 2015 *Brit. J. Radiol.* **88** 20150251
[91] Schreiner LJ *et al* 2013 *J. Phys.: Conf. Ser.* **444** 012016
[92] Nelms BE *et al* 2013 *Med. Phys.* **40** 1-15
[93] Kry SF *et al* 2014 *Int. J. Radiat. Oncol. Biol. Phys.* **90** 1195-201
[94] Nierer L *et al* 2022 *Z Med. Phys.* **32** 312-25
[95] Kozicki M *et al* 2020 *Physica Medica* **69** 134-46
[96] Wang YF *et al* 2021 *J. Appl. Clin. Med. Phys.* **22** 24-35
[97] Weber DC *et al* 2011 *Radiat. Oncol.* **100** 150-6
[98] Schreiner LJ 2017 *J. Phys.: Conf. Ser.* **847** 012001
[99] Kairn T *et al* 2022 *J. Phys.: Conf. Ser.* **2167** 012012
[100] Venning A *et al* 2022 *J. Phys.: Conf. Ser.* **2167** 012016
[101] Elter A *et al* 2022 RSC: *J. Phys.: Conf. Ser.* **2167** 012020
[102] Alexander K *et al* 2022 *J. Phys.: Conf. Ser.* **2167** 012022

[103] Colnot J *et al* 2022 *J. Phys.: Conf. Ser.* **2167** 012017
[104] Zakjevskii VV *et al* 2016 *J. Appl. Clin. Med. Phys.* **17** 497-510
[105] Elter A *et al* 2019 *Phys. Med. Biol.* **64** 225003
[106] Choi Y *et al* 2021 *Progress in Medical Physics* **32** 99-106
[107] Matrosic CK *et al* 2021 *Physica Medica* **84** 159-67
[108] McKenna JT 2021 *Med. Phys.* **48** 105-3
[109] Poder J *et al* 2018 *J. Appl. Clin. Med. Phys.* **19** 99-108
[110] Grams MP and de los Santos LEF 2018 *Physica Medica* **50** 59-65
[111] Dimitriadis A *et al* 2017 *Br J Radiol* **90** 20170053
[112] Adamson J *et al* 2019 *Int. J. Radiat. Oncol. Biol. Phys.* **103** 1271-9
[113] Alexander KM *et al* 2018 *BioMed. Phys. Eng Express* **4** 045041
[114] Kozicki M *et al* 2009 *J. Phys.: Conf. Ser.* **164** 012062
[115] Kozicki M *et al* 2014 *Phys. Med. Biol.* **59** 4111-36
[116] Kozicki M and Maras P 2015 *J. Phys.: Conf. Ser.* **573** 12070
[117] Maras P *et al* 2021 *Radiat. Phys. Chem.* **189** 109730
[118] Kozicki M *et al* 2015 *Introduction to polyGeVero® software for 3D radiotherapy dosimetry* (available from https://www.researchgate.net/publication/283711431_INTRODUCTION_to_polyGeVeroR_software_for_3D_radiotherapy_dosimetry)
[119] Guo P *et al* 2006 *J. Phys.: Conf. Ser.* **56** 191-4
[120] Sakhalkar HS *et al* 2009 *J. Phys.: Conf. Ser.* **164** 012058
[121] Avery S *et al* 2015 *J. Phys.: Conf. Ser.* **573** 012020
[122] Wuu C *et al* 2004 *J. Phys.: Conf. Ser.* **3** 297-300
[123] Scheib SG *et al* 2004 *J. Phys.: Conf. Ser.* **3** 228-31
[124] al Kafi MA *et al* 2021 *Radiat. Meas.* **143** 106568
[125] Davidson SE *et al* 2008 *Med. Phys.* **35** 5434-9
[126] Olding T and Schreiner LJ 2011 *Phys. Med. Biol.* **56** 1259-79
[127] Olding T *et al* 2013 *J. Phys.: Conf. Ser.* **444** 012052
[128] Olding T *et al* 2015 *J. Phys.: Conf. Ser.* **573** 12019
[129] Olding T *et al* 2013 *J. Phys.: Conf. Ser.* **444** 012073
[130] Palmer AL *et al* 2017 *Radioth. Oncol.* **125** 453-8
[131] Ziro ne L *et al* 2022 *Gels* **8** 481
[132] Han EY *et al* 2020 *J. Appl. Clin. Med. Phys.* **21** 278-85
[133] Han EY *et al* 2022 *J. Appl. Clin. Med. Phys.* **23** e13518
[134] Pappas E *et al* 2019 *Phys. Med. Biol.* **64** 225009
[135] Alexander KM *et al* 2015 *J. Phys.: Conf. Ser.* **573** 012042
[136] Alexander KM *et al* 2017 *J. Phys.: Conf. Ser.* **847** 012061
[137] Alexander KM *et al* 2015 *IFMBE Proceedings* vol 51 (Springer Int. Publ.) pp 521–4
[138] Gershkevitsh E *et al* 2014 *Acta Oncol.* **53** 628-36
[139] Ibbott GS 2010 *J. Phys.: Conf. Ser.* **250** 012001
[140] Molineu A *et al* 2013 *Med. Phys.* **40** 1-8
[141] Ibbott G S *et al* 2008 *Int. J. Radiat. Oncol. Biol. Phys.* **71** 71-5
[142] Carson ME *et al* 2016 *Med. Phys.* **43** 6491-6
[143] Edward SS *et al* 2020 *Pract. Radiat. Oncol.* **10** 372-81
[144] Kry SF *et al* 2013 *Int. J. Radiat. Oncol. Biol. Phys.* **85** e95-100
[145] Koger B *et al* 2020 *J. Appl. Clin. Med. Phys.* **21** 82-8
[146] Edward SS *et al* 2020 *Med. Phys.* **47** 4502-8
[147] Taylor PA *et al* 2021 F *Pract. Radiat. Oncol.* **11** e322-8
[148] Hughes J *et al* 2021 *Med. Phys.* **48** 6184-97
[149] Kerns JR *et al* 2017 *Int. J. Radiat. Oncol. Biol. Phys.* **98** 1197-203
[150] Cadman P *et al* 2002 *Phys. Med. Biol.* **47** 3001

[151] IAEA 2008 *Commissioning of Radiotherapy Treatment Planning Systems: Testing for Typical External Beam Treatment Techniques* (Vienna, Austria: IAEA)
[152] Molineu A *et al* 2005 *Int. J. Radiat. Oncol. Biol. Phys.* **63** 577-83
[153] Followill DS *et al* 2007 *Med. Phys.* **34** 2070-6
[154] Followill DS 2014 eds. DeWerd L A and Kissick M The Phantoms for Medical Physics Medical Physics pp 39-51
[155] Podgorsak EB 2005 *Radiation Oncology Physics: A Handbook for Teachers and Students* (Vienna, Austria: IAEA)
[156] Fong de los Santos E *et al* 2015 *J. Appl. Clin. Med. Phys.* **16** 37-59
[157] Malkoske KE *et al* 2018 *J. Appl. Clin. Med. Phys.* **19** 44-7
[158] Dinhofer DS and Agarwal S 2021 *EBEM* **12** 113-25
[159] Probst MA *et al* 2014 *Am. J. Emerg. Med.* **32** 645-50
[160] Saeed NA *et al* 2022 *Adv. Radiat. Oncol.* **7** 1-7
[161] Reyna VF and Lloyd FJ 2006 *J. Exp. Psychol. Appl.* **12** 179-95
[162] Rasmussen J 1997 *Saf. Sci.* **27** 183-213
[163] Salomons G *et al* 2021 *Physica Medica* **89** 293-302
[164] Pant K *et al* 2020 *Med. Phys.* **47** 927-36
[165] Pant K *et al* 2022 *J. Phys.: Conf. Ser.* **2167** 012007
[166] Olding T *et al* 2009 *J. Phys.: Conf. Ser.* **164** 012031
[167] Olding T *et al* 2010 *Phys. Med. Biol.* **55** 2819-40
[168] Schreiner LJ *et al* 1994 *Magn. Reson. Imag.* **12** 901-7
[169] Chu WC *et al* 1998 *Med. Phys.* **25** 2326-32
[170] De Deene Y *et al* 2020 *Phys. Med. Biol.* **65** 225031
[171] De Deene Y 2020 ed. De Deene Y NMR and MRI of Gels (Royal Society of Chemistry, London UK) pp 275–357
[172] Bae Y-K *et al* 2020 *Sci Rep* **10** 473
[173] Keall PJ *et al* 2021 *Med. Phys.* **48** e44-64
[174] Tino R *et al* 2020 *Internat. J. Extreme Manufacturing* **2** 012003
[175] Stark LS *et al* 2020 *Phys Imaging Radiat Oncol* **16** 109-12
[176] Iijima K *et al* 2021 *Med. Phys.* **48** 7541-51
[177] McCulloch J *et al* 2020 *J. Appl. Clin. Med. Phys.* **21** 107-15
[178] Lempart M *et al* 2017 *Australas. Phys. Eng. Sci. Med.* **40** 717-27
[179] Liu X *et al* 2021 *Med. Phys.* **48** 5479-89
[180] Olding T *et al* 2010 *J. Phys.: Conf. Ser.* **250** 012003
[181] Marot M *et al* 2021 *Med. Phys.* **48** 5501-10
[182] Alexander KM *et al* 2015 *J. Phys.: Conf. Ser.* **573** 012042
[183] Lehmann J *et al* 2012 *Spectrum* **19** 14-7
[184] Smith J C *et al* 2011 *Med. Phys.* **38** 553-5
[185] Zygmanski P and Halvorsen P 2021 *Med. Phys.* **48** 4715-8
[186] Giuliani M *et al* 2014 *Radiat. Oncol.* **9** 1-7
[187] Winter JD *et al* 2020 *Int. J. Radiat. Oncol. Biol. Phys.* **107** 844-9

Journal of Physics: Conference Series **2630** (2023) 012008 doi:10.1088/1742-6596/2630/1/012008

Evolution of 3D X-Ray Dose Computation Algorithms

J. Battista[1], J. Chen[1], S. Sawchuk[1] and G. Hajdok[1,2]
[1]Department of Oncology and Department of Medical Biophysics, Western University, London, Ontario, Canada
[2]Department of Medical Physics, Walker Family Cancer Centre, St. Catharines, Ontario, Canada

E-mail: George.Hajdok@niagarahealth.on.ca

Abstract. Radiation treatment planning of individual cancer patients relies on the accurate computation of dose distributions in irradiated tissue. Inaccurate dose maps have the potential to mislead clinical decision-making and compromise the balance between effective tumour control and side effects in surrounding normal tissue. In the context of this conference, 3D dosimetry is important for the experimental validation of computed dose distributions. Dose computation methods for external beams of high energy x rays have evolved over the past decade with computer simulation models more closely aligned with the fundamental physics of x-ray scattering and absorption in heterogeneous tissue. In this article, we first present a historical review from a Canadian perspective, followed by an introductory intuitive description of contemporary algorithms used in clinical treatment planning: (1) Convolution-superposition algorithm fundamentally based on the Green's function method; (2) Stochastic Monte Carlo simulation of x-ray interactions in tissue, and (3) Deterministic numerical solution of a system of Boltzmann transport equations. In principle, all these methods solve the same problem of predicting x-ray scattering and absorption in heterogeneous tissue. However, the mathematical tools differ in their approach and approximations to achieve sufficient speed for routine clinical application. In the conclusion of this article, the evolution of 3D x-ray dose computation is summarized, in terms of accuracy and computational speed.

1. Historical Perspective

Dose computations used for treatment planning have advanced with several "giant leaps" forward: (a) splitting of radiation dose into primary and scatter components (1960s); (b) incorporation of *in vivo* tissue densitometry by x-ray computed tomography (CT) (1970s); and (c) exponential gains in cost-effective computing power. The evolution of algorithms occurred within a backdrop of concurrent technological advances in medical physics, as illustrated in Figure 1. Early computer methods (1960s) focused on automating the treatment planning process through rapid recall of stored measured beam data, replacing the manual overlay of transparent isodose curve charts and dose calculations performed with a hand-held electronic calculator.

The decoupling of primary and secondary components of dose enabled better prediction over a wider range of beam configurations, beyond those that were measured. Semi-empirical models of the 1970s considered irregularly-shaped fields including a basic form of beam intensity-modulation [1]. These fields were produced with shielding blocks and tissue compensating filters, before the commercial advent of multi-leaf collimators (MLCs). Dose distributions were first computed on the assumption of a patient composed entirely of water-like tissue, and were then adjusted with secondary "inhomogeneity correction factors" for tissue density variations [2-5]. The tissue space was segmented by internal contours of regions with non-aqueous density, typically done in a single central slice of the patient (i.e. 2D planning). In the 1970s, the introduction of x-ray computed tomography (CT) and patient-specific tissue densitometry, caused a sudden and major paradigm shift from 2D vector-based to 3D pixel-based dose computations [6-8]. Historically, it is interesting that a precursor pixel-based method [8] using differential Scatter-Air Ratios (SARs) [9] accounted for the influence of an isolated voxel on the scattered radiation field. Dr. John R. ("Jack") Cunningham was a key figure in such developments and their implementation on computer workstations (e.g. Theraplan, Oncentra); we dedicate this review article to his memory [10]. Later experiments measured the influence of an air-filled toroid on central axis dose points in a water phantom [11, 12]. Two counteracting radiation effects were observed: (a) reduced scattering from the air void *per se*, and (b) enhanced transmission through the toroid, causing greater intensity of scatter produced in underlying voxels. This observation accentuated the need to develop dose computation algorithms with full consideration of scattering effects in 3D [6, 13-15].

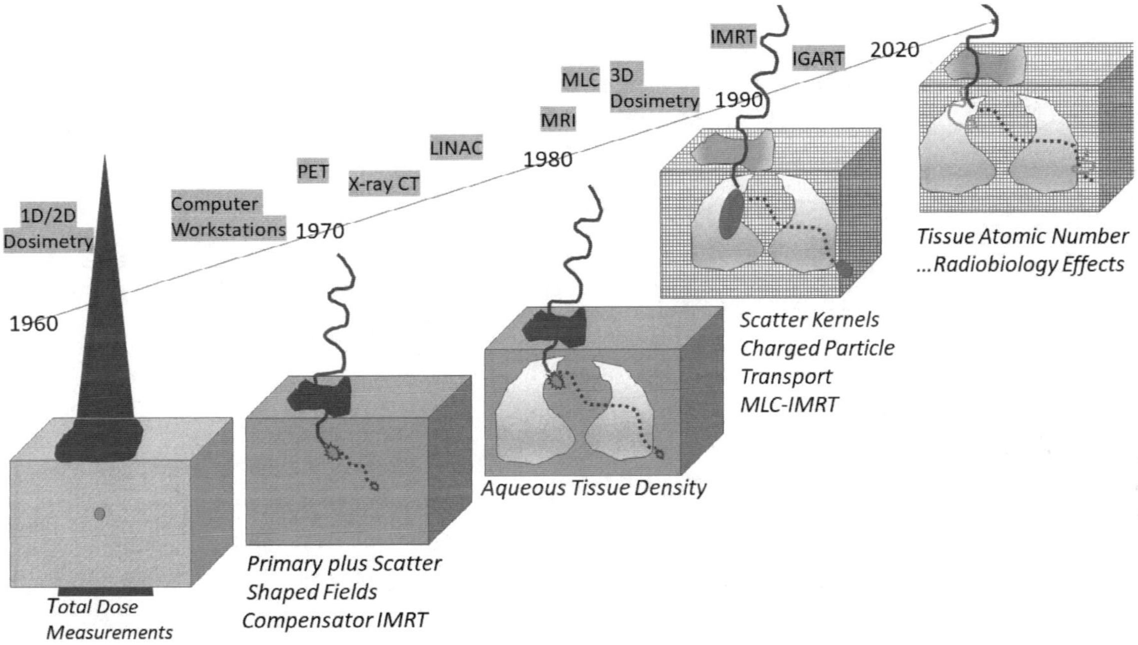

Figure 1. Chronology of major advances in medical physics and their influence on dose computation.

During the era of Cobalt-60 machines, energy transferred from x rays to secondary knock-on electrons was generally assumed to be deposited *on-the-spot,* or equivalently, charged particle equilibrium (CPE) was assumed to be guaranteed at all dose voxel locations. However, this assumption weakened for higher energy x-ray beams delivered by linear accelerators that soon replaced Cobalt units (1970s). The energetic x rays liberate electrons with longer ranges in porous tissue such as lung. Localized dose perturbations had also been observed experimentally in the buildup region of depth-dose curves, beam penumbrae, and near tissue interfaces with metallic implants. It became clear that *electron migration away from primary photon interaction* sites had to be considered in the next generation of multi-particle algorithms [16]. The convolution method, introduced in the mid-1980s [17] was a major driving force in reawakening interest in photon-electron transport; it sparked the rapid deployment of a new generation of 3D computational methods [15, 18-20].

This article focuses on algorithms (Types B and C, or 2nd and 3rd generation) that are used for contemporary clinical treatment planning in radiation oncology (Figure 2) [3, 21]. These techniques

Journal of Physics: Conference Series **2630** (2023) 012008 doi:10.1088/1742-6596/2630/1/012008

solve a common problem - modeling the propagation of radiation through a heterogeneous tissue space. However, each technique approaches the situation from a different perspective, and applies unique mathematical tools and approximations during practical implementation: (a) Convolution-superposition algorithms are based on Green's impulse-response concept; (b) Monte Carlo techniques simulate a large number of *individual* particle trajectories and scores energy deposition in voxels in real space; and (c) Deterministic techniques solve a system of coupled differential (Boltzmann transport) equations that constrain the flow and energy of particle cohorts across a multidimensional *phase space.* The goal of this article is to introduce the reader to these different methods in an intuitive manner. Mathematical details are found elsewhere [22-27].

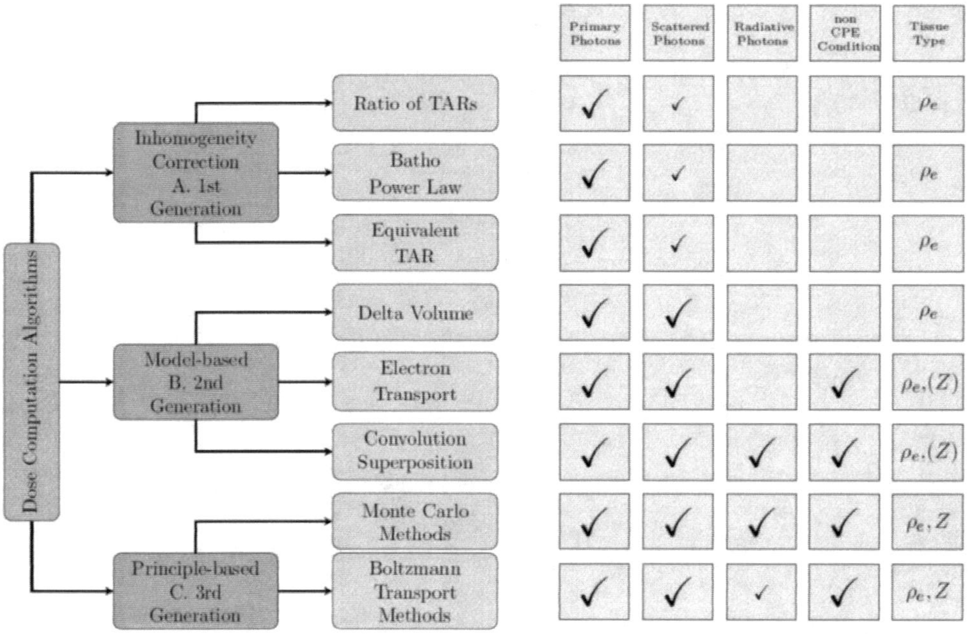

Figure 2. Categories of algorithms and corresponding generations. *CPE* denotes charged particle equilibrium. ρ_e denotes electron density (electrons per cm^3). *Z* denotes consideration of atomic number effects. *(Z)* denotes partial consideration.

2. Convolution & Superposition Methods

Convolution-superposition algorithms offer a good compromise between accurate dose modelling and efficient computation. This algorithm has become the most popular method in commercial treatment planning systems. However, implementations vary considerably from 2D (i.e., pencil beams) to consideration of the 3D shower of scattered radiation launched by primary photons. The concept is easy to comprehend intuitively; each photon interaction site becomes an epicentre (i.e., kernel) of nanoscale "explosions" that produce a debris field of scattered energy. The mathematical foundation of convolution is the Green's function method that has a rich history, well before applications to medical physics. The impulse-response approach uses a data base of energy deposition *kernels* that describe the debris field. Kernels are pre-computed for a homogeneous water absorber exposed to various incident energies using a one-time Monte Carlo simulation [28]. X-ray interactions are artificially forced to occur at an isolated point in a large homogeneous water absorber and the emerging particles create the scattered dose deposition pattern. In this approach, attenuation of primary radiation is cleanly decoupled from the spreading of energy by secondary particles. For an external beam of x rays, each kernel is weighted by the local energy released by primary photons at a grid of launch sites. The magnitude of each energy burst is quantified by a special quantity called TERMA (total energy released per unit mass) [29]. The *attenuated* TERMA is computed by raytracing through voxels lying along primary rays, from the radiation source to each interaction point in the patient space. Radiological pathlengths across heterogeneous tissue are determined from megavoltage linear attenuation coefficients at the treatment beam energy, inferred from CT image data usually obtained with kilovoltage beams. The kernels are intensity-modulated by the attenuated TERMA within the irradiated space and each kernel disperses the released energy to surrounding voxels. The sum of all energy deposited to each dose voxel forms the total dose distribution. When the in-water kernel is applied "as is" throughout an irradiated water volume, the process is a *convolution with a spatially-invariant kernel,* as illustrated in Figure 3. This process resembles the method used for computing a brachytherapy dose distribution from a discrete array of radioactive sources placed within homogeneous tissue. The dose pattern around each source is the kernel.

For heterogeneous tissue exposed to a more realistic poly-energetic divergent x-ray beam, the situation becomes more complex because the kernels must be modified at each launch point. Each kernel must be rotated slightly in order to align with divergent primary beam rays, and consideration must also be given for the local energy spectrum and tissue type, plus the external border of the patient [15, 30]. The spectral changes are due to x-ray "beam hardening", but *tissue inhomogeneity poses the biggest challenge.* The calculation involves raytracing throughout tissue along pairs of interaction-deposition voxels within the kernel domain. Variations in electron density (electrons per cm^3) alter the mean free path and range of Compton-scattered photons and Compton electrons, respectively.

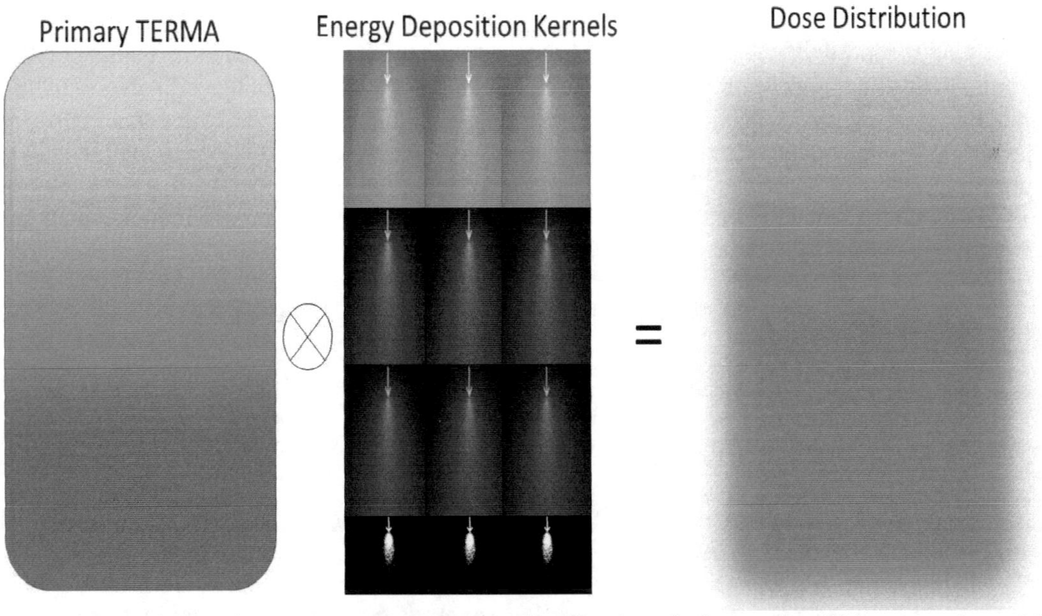

Figure 3. Convolving the attenuated TERMA distribution (left panel) with energy deposition kernels (middle panel) produces the blurred dose distribution (right panel).

Dose spread kernels are available for a large all-water medium, categorized by each type of energy-spreading mechanism (i.e. primary electrons, first-scattered photons, multiply-scattered photons, radiative photons) [28]. The kernels for heterogeneous tissue, however, can only be approximated, assuming water-like atomic composition. For primary Compton electrons and first-scattered photon kernels, the water-equivalent pathlength along fan lines from the kernel origin to each dose destination point is used to look up in-water kernel values. The kernels effectively become distorted by a radial expansion or contraction along these fan lines. This approximation assumes that primary electrons and first-scattered photons travel paraxially along these direct flight paths. Fortunately, the majority of dose is deposited by these two dominant kernels. For other kernels, the diversity of pathways requires density averaging over a wider volume [17][30]. When *spatially-variant kernels* become necessary, the convolution integral transforms into a *superposition integral*.

We now introduce the mathematical formalism for 3D convolution-superposition. Consider energy launch sites at source positions ($\vec{s}$). The vector $\vec{r}$ specifies the location of a dose-receiving point in 3D. The fraction of energy that is liberated and subsequently deposited is defined for each type of energy-spreading process (i). The kernels are normalized as fractions of the total energy released (i.e., TER = TERMA × mass) at location $\vec{s}$, then deposited in a surrounding voxel at location $\vec{r}$, per unit volume of the receiving voxel. Primary kernels are shaped by short-range secondary electrons (primarily Compton electrons for MV x rays), with a high concentration of distributed energy near the kernel origin. Other kernels reflect Compton-scattered or radiative photons that travel far from the interaction site with a much longer mean free path. The partial dose contribution is governed by the differential energy spectrum of the TERMA (T_E), the kernel $k_i(E)$ for i-type of interactions and incident photon energy, E:

$$\boldsymbol{D_i^{con}(\vec{r}) = [1/\rho(\vec{r})] \int_E \iiint_V T_E(\vec{s})\, \rho(\vec{s})\, k_i(E, \vec{r} - \vec{s})\, \mathrm{d}^3 s\, \mathrm{d}E} \tag{1}$$

The passage of energy between a pair of interaction-deposition voxels in an all-water medium is wholly defined by the vector line that joins these voxels directly, (i.e., $\vec{r} - \vec{s}$). The leading density term in equation (1) yields the dose absorbed in tissue of the receiving voxel, assuming water-like atomic composition.

As mentioned previously, realistic conditions of beam divergence, poly-energetic x-ray beams, tissue inhomogeneity, and finite borders of the patient negate the use of the idealistic convolution integral (Equation 1). This integral *must* then be replaced by the more general *superposition* integral. The kernel values for a pair of interaction-deposition voxels considers radiological pathlengths along ray lines as previously described. The short-hand notation $(\vec{\boldsymbol{r}}; \vec{\boldsymbol{s}})$ is intended to draw attention to the scattering *entourage* around $\vec{\boldsymbol{r}}$ within the scope of the kernel placed with its origin at $\vec{\boldsymbol{s}}$. The following superposition integral replaces the convolution integral Eq. (1):

$$\boldsymbol{D_i^{sup}(\vec{r}) = [1/\rho(\vec{r})] \int_E \iiint_V T_E(\vec{s})\, \rho(\vec{s})\, k_i(E, \vec{r}; \vec{s})\, \mathrm{d}^3 s\, \mathrm{d}E} \tag{2}$$

The total dose distribution is the sum of energy deposited by all energy-spreading mechanisms (i):

$$\boldsymbol{D(\vec{r}) = \textstyle\sum_i D_i^{con/\,sup}(\vec{r})} \tag{3}$$

Various strategies have been developed to alleviate the computational burden of superposition algorithms because they involve a large set of ray tracings for the primary TERMA calculation and for kernel reshaping in heterogeneous tissue. Pratx and Xing [31] described the use of graphical processor units (i.e., GPUs) in medical physics. GPU-aided superposition can be accomplished in 3D with impressive speed gains of approximately 50-fold for the TERMA calculation step and 940-fold for the -dose calculation step [32]; 3D dose distributions are computed within seconds! Efficient software has also been developed based on the Fast Fourier transform (FFT). Unfortunately, the spatially-*variable* kernel limits the practical application of this otherwise attractive strategy [33][34][35]. Reducing the dimension of scatter integration from 3D to 2D is another obvious accelerant, but this approach compromises flexibility and accuracy of such algorithms (e.g., pencil beam, SVD - singular value decomposition, AAA – Anisotropic Analytical Algorithm). This strategy can be applied during dose optimization cycles that necessitate high speed, but their inaccuracies could lead to IMRT fields with suboptimal optimization.

The most practical compromise between dose accuracy and speed of computation has been achieved with the Collapsed Cone Convolution (CCC) approximation [36][37]. The technique is, in fact, a superposition rather than convolution model. The energy released at a photon interaction point is propagated into a set of small cones and projected onto the cone axes (akin to closure of an umbrella). This approximates the lateral transport of scattered energy, but greatly reduces the computational burden by limiting the ray tracing burden and capitalizing on an efficient recursive expression that was developed to be applied along the cone axes. The misplacement of lateral energy causes a moderate loss in dose accuracy for most clinical situations. The CCC algorithm has become the most prevalent algorithm on commercial workstations with products such as Pinnacle (Philips), Tomotherapy (Accuray), RayStation (RaySearch Laboratories), Monaco (Elekta CMS), Oncentra (Nucletron), and Mobius3D (Mobius Medical Systems).

There are inherent assumptions and limitations of the superposition methods. The electron density used in calculating water-equivalent pathlengths for the primary TERMA calculation and kernel modifications assumes 100% Compton events and "soft" electron collisions in *all* tissues at the treatment beam energy; atomic number effects are not normally considered. If there are obvious regions of compact bone or foreign prosthetic materials observed in the CT image data, more accurate *ad-hoc* attenuation coefficients can be "looked up" for the TERMA calculation. However, the approximate distortion of the kernels, using only electron density data will produce less accurate dose at interfaces of tissues with different atomic composition. The following methods have the potential to account for atomic number effects more accurately.

3. Monte Carlo Simulation – Stochastic Model

The Monte Carlo method is the random sampling method named after the famous gambling casino in Monaco. The technique has been applied to a wide range of fields including mathematics, economics, simulation of games and lotteries, and radiation sciences [38]. It has become an invaluable and flexible tool for uncovering weak (and often tacit) assumptions of more approximate dose computation models. In medical physics, the Monte Carlo simulation method has often been regarded as the de facto *gold standard* technique for decades, with a strong record of corroborating experimental validation. The execution of this method can be viewed as a virtual radiation experiment evolving in slow motion. Furthermore, the experiment and internal physics can be manipulated to isolate key physical effects most important to accurate dosimetry. The method has also been used to design experimental apparatus or to clarify the interpretation of signals generated in radiation dosimeters [39]. The method has yielded essential data on radiation emerging from linear accelerators [40] and energy deposition kernels [28] shown in Figure 4. As a point of historical interest, these kernels were originally computed in 1987, running 1 million histories per kernel for a total of 0.5 years of computational time on a cluster of VAX-11 computers (Digital Equipment Corporation) across participating institutions in Canada. Today, the entire database can be regenerated in less than 1 hour using a personal computer!

The general applicability of stochastic solutions fundamentally stems from the ability to evaluate the integral of any function, including those that may be difficult to express analytically. The integral can be evaluated by counting the fraction of random points that land under the surface of a multi- dimensional integrand. The disadvantage of the Monte Carlo method is the computational burden of tracking millions or billions of individual particle "histories" to reach a required precision of a few percent in high dose regions. The reduction in statistical uncertainty of dose values is notoriously sluggish because of an *inverse square root* relationship with the number of events contributing dose to scored voxels. The reader is reminded that precision does not guarantee accuracy. The accuracy of a Monte Carlo run is fundamentally dependent on the quality of the underlying radiation physics infrastructure: *you cannot fool the physics!*

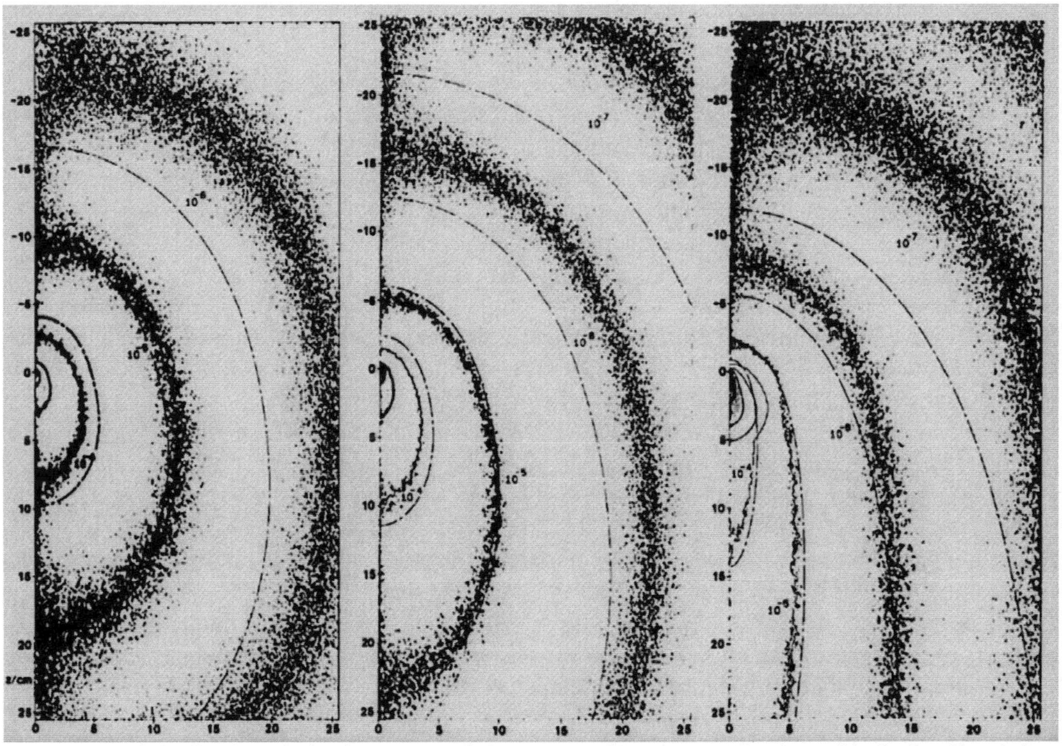

Figure 4. Total (points) and singly scattered Compton photon contributions (dash-dot lines) kernels for monoenergetic incident photon energies of 0.4, 1.25, and 10 MeV, from left to right. The kernel units are fractions of energy released at the origin (TER) subsequently deposited in neighbouring voxels per unit volume. Reproduced with permission [29].

An advantage of the Monte Carlo concept is that it is also easy to understand intuitively. The processing of one photon at a time, and the single-event structure facilitates graphical visualization [41] and debugging of software. The disadvantage is lengthy computation times with a clear trade-off between dose noise level and spatial resolution (i.e., dose voxel size). In addition, wise choices of run-time parameters (e.g., energy or geometric cutoffs) and applying bias to promote more relevant radiation events can shorten simulation times; artificial bias can be corrected *a posteriori*. Efforts are ongoing to reduce the computation time using variance reduction strategies, highly specialized codes for treatment planning and faster processors (i.e., CPUs or GPUs). The Monte Carlo method is very well suited to parallel processing because batches of particle histories or types of interactions can be distributed to sub-processors. Monte Carlo computer codes that are used clinically are summarized in Table 1.

The programs are classified according to their practical application: treatment planning or quality assurance of simpler algorithms applied to unusual clinical cases [42]. The clinical programs are usually descendants of parent codes used previously in nuclear physics, such as DOSXYZnrc coupled to EGS developed at Stanford University and enhanced and repackaged for medical physics applications as EGSnrc by the Canadian National Research Council (NRC).

Journal of Physics: Conference Series **2630** (2023) 012008 doi:10.1088/1742-6596/2630/1/012008

Table 1. Monte Carlo codes used clinically [42].

System	Application		Monte Carlo Software			
	Treatment Planning	QA*	LINAC Head	Patient Dose	Self-Contained*	Software Platform
CARMEN	✓		BEAMnrc	DOSXYZnrc		MATLAB®
CERR		✓		VMC++/DPM		MATLAB®
Corvus	✓		PSF	PEREGRINE	C	
Eclipse	✓		PSF	MMC	C	
eIMRT		✓	BEAMnrc	DOSXYZnrc	✓	
iPLAN RT	✓		Model	XVMC	C	
IsoGray	✓		PENELOPE	PENFAST	✓	
MCDE	✓		BEAMnrc	DOSXYZnrc		GRATIS
MCDOSE	✓		Model	EGS4		FOCUS
MCV		✓	EGS4	EGS4		Pinnacle
MCVS		✓	BEAMnrc	DOSXYZnrc		CERR
MMCTP		✓	BEAMnrc	EGSnrc		
Monaco	✓		Model	XVMC	C	
MSKCC		✓	Model	EGS4	✓	
Oncentra	✓		Model	VMC++		
Pinnacle	✓			DPM	✓	
PlanUNC		✓	EGSnrc	EGSnrc/MCNP		
PRIMO		✓	PENELOPE	PENELOPE	✓	
RTGrid	✓		BEAMnrc	EGSnrc		Globus
SMCP	✓		Model	EGSnrc/VMC++		Eclipse
VIMC		✓	BEAMnrc	DOSXYZnrc/VMC++		Eclipse
XiO	✓		Model	XVMC	C	

* QA = quality assurance; C = commercial product

4. Solving Boltzmann Transport Equations – Deterministically

The deterministic approach has been used to model many natural phenomena including electrodynamics, fluid dynamics, and radiation transport in outer space. In modelling dose distributions within the patient, significant progress has recently been made in developing efficient "solvers" of a large system of differential (Boltzmann transport) equations [23][26][43][44]. Boltzmann and Monte Carlo techniques can generally use similar input databases of atomic cross sections that prescribe the possible interactions of multi-particle systems. Dose distribution results calculated by either method are in excellent agreement, provided that input physics and patient-related data, and run-time parameters are well matched. This is confirmed by several studies [45][46][47][48][49], including tests performed for difficult clinical situations such as pelvic regions with hip prostheses [50] and small lung lesions treated by stereotactic radiosurgery [51][52][53]. Boltzmann dose distributions are based on expectation values of dose without statistical noise associated with Monte Carlo results. The efficiency of deterministic problem solving has been compared with stochastic simulation. The relative speed is highly dependent upon the complexity of the problem (i.e. the "curse of dimensionality") [38]. With fast execution on a commercial software platform and evolving familiarity with the method in the medical physics community, clinical adoption of Boltzmann techniques is expected to rise in the coming decade.

The core accounting principle underpinning the Boltzmann equations is conservation of mass-energy. The collection of radiation particles is treated as a *fluid*. The radiation field is described by vector quantities that specify particle fluence intensity *and direction*. When using deterministic methods, cohorts of particles are tracked in a virtual phase space. The re-distribution of particles within this hyperspace is governed by atomic cross sections and kinetics of interactions constrained by mass-energy conservation. Ultimately, the solution to the Boltzmann equations yields fluences of charged particles that are directly responsible for depositing energy in voxels *via* their stopping power.

4.1 Introduction to Phase Space

A significant shift in mentality is required to understand hyper-dimensional *phase space* [54][55][56]. Three coordinates specify voxel locations in 3D space and two spherical coordinates (i.e., polar and azimuthal angles) specify particle direction. Adding the particle's instantaneous energy, the phase space becomes six-dimensional for each particle type (i.e., photon or electron). Before we develop a mathematical formulation, we define important quantities used in radiation transport theory (**Table 2**) [26][57]. Quantities on the left specify the density and flow rates for particles travelling with a specific direction ($\widehat{\Omega}$) and energy (E) at a fixed 3D location ($\vec{r}$) sampled at a moment in time (t). The upper-case quantities in the middle column are the results of integrating left-side quantities over all particle directions and energies. Quantities on the right side are further integrated over radiation exposure time. They then become scalar in nature, more often seen in familiar radiation dosimetry.

4.2 Counting Particles in Phase Space

The book-keeping principle embodied in the Boltzmann equation is that mass-energy must be conserved as particles migrate across phase space "bins". Changes in bin occupancy (*n*) are cross-checked from two perspectives: (1) net flow of particles through dose voxel borders, and (2) particle interaction events that can occur within the voxel. The equations impose the equality of results obtained by both viewpoints, accounting for internal energy losses, and removal, creation, or mass-energy transformations of particle types (e.g., pair production). The particle occupancy in phase space bins evolves as particles enter, start/stop in, transform, or exit a phase space voxel. For example, particle fluences are reduced in a photon phase space bin by a photelectric absorption. Scattering may yield the same type of emerging particle (e.g., Compton-scattered photons) with a switch to bins tagged by lower energy and new directional indices. In addition, interactions can spawn new particle types (e.g., photons releasing electrons). Bin counts are then transferred across two particle phase spaces that must be tracked for the case of megavoltage photon-electron showers. For example, *bremsstrahlung* due to charged particles may spawn a fresh contribution to the photon phase space bins. Particles that pass straight through a voxel without interaction retain their incoming energy and orientation, and simply have an updated emerging 3D location. The Boltzmann transport equations enforce a strict *auditing and cross-checking process* as particle counts are transferred across phase space bins and domains.

In Figure 5, a simple numerical example introduces the reader to the accounting principle in a photon phase space. Imagine a voxel at 3D location ($\vec{r}$) exposed to a bi-energetic bi-directional fluence of photons over a fixed time. Each arrow in Figure 5 represents the flow of 10 photons over the irradiation period. The two starting populations (in blue and green) initially occupy different bins in phase space due to their inbound orientation and energy difference. Sixty (60) of the incoming photons have initial conditions $(\vec{r}, \widehat{\Omega}_1, E_1)$ – coordinates of the phase space bin of interest in this calculation. Another 20 photons have a higher inbound energy ($E_2 > E_1$) and different orientation ($\widehat{\Omega}_2$). On passing through the voxel, 20 photons with incoming energy E_1 undergo total photoelectric absorption and another 10 photons are 'out-scattered' from their original phase space bin with a degraded energy $E_3 < E_2$ and re-orientation to $\widehat{\Omega}_3$. In this context, *out-scattering* refers to a photon count being transferred out of the phase space bin of interest $(\widehat{\Omega}_1, E_1)$. The remaining 30 photons with incoming energy E_1 are transmitted through the voxel without interacting in the voxel. Twenty (20) photons with greater starting energy E_2 are *scattered into* the phase space bin of interest. This obvious coincidence was forced to illustrate the basic concept of accounting for 'in-scattering'. The book-keeping is shown in the inset of Figure 5, with a loss of 30 counts due to full absorption (-20) combined with out-scattering (-10), while the gain from in-scattering is 20. The net change in the count at phase space bin $(\widehat{\Omega}_1, E_1)$ is therefore -10. This change in occupancy is alternatively predictable from the photon *divergence* which is the difference between incoming and outgoing particles; 60 photons entered the voxel with attributes $(\widehat{\Omega}_1, E_1)$ but only 50 exited *with the same attributes*, yielding a net loss (-10) in bin $(\widehat{\Omega}_1, E_1)$.

Table 2. Radiometric Quantities used in transport theory. Particles with direction $(\hat{\Omega})$ and energy (E) are at location $(\vec{r})$ in 3D space for a moment in time (t) [26].

number density distribution (scalar) $n(\vec{r}, \hat{\Omega}, E, t)$ $[\mathrm{cm^{-3}\,sr^{-1}\,MeV^{-1}}]$	$\int dE \int d\Omega$ $\longrightarrow$	**number** density (scalar) $N(\vec{r}, t)$ $[\mathrm{cm^{-3}}]$		
$v\hat{\Omega}$ $\downarrow$				
current density distribution (vector) $\vec{j}(\vec{r}, \hat{\Omega}, E, t)$ $[\mathrm{cm^{-2}\,sr^{-1}\,MeV^{-1}\,s^{-1}}]$	$\int dE \int d\Omega$ $\longrightarrow$	**current** density (vector) $\vec{J}(\vec{r}, t)$ $[\mathrm{cm^{-2}\,s^{-1}}]$		
$\frac{1}{\hat{\Omega}}$ $\downarrow$				
flux density distribution (scalar) $\dot{\varphi}(\vec{r}, \hat{\Omega}, E, t)$ $[\mathrm{cm^{-2}\,sr^{-1}\,MeV^{-1}\,s^{-1}}]$	$\int dE \int d\Omega$ $\longrightarrow$	**fluence** rate (scalar) $\dot{\Phi}(\vec{r}, t)$ $[\mathrm{cm^{-2}\,s^{-1}}]$	$\int dt$ $\longrightarrow$	**fluence** (scalar) $\Phi(\vec{r})$ $[\mathrm{cm^{-2}}]$
E $\downarrow$				
energy flux density distribution (scalar) $\dot{\psi}(\vec{r}, \hat{\Omega}, E, t)$ $[\mathrm{cm^{-2}\,sr^{-1}\,s^{-1}}]$	$\int dE \int d\Omega$ $\longrightarrow$	**energy fluence** rate (scalar) $\dot{\Psi}(\vec{r}, t)$ $[\mathrm{MeV\,cm^{-2}\,s^{-1}}]$	$\int dt$ $\longrightarrow$	**energy fluence** (scalar) $\Psi(\vec{r})$ $[\mathrm{MeV\,cm^{-2}}]$

In Figure 5, a simple numerical example introduces the reader to the accounting principle in a photon phase space. Imagine a voxel at 3D location $(\vec{r})$ exposed to a bi-energetic bi-directional fluence of photons over a fixed time. Each arrow in Figure 5 represents the flow of 10 photons over the irradiation period. The two starting populations (in blue and green) initially occupy different bins in phase space due to their inbound orientation and energy difference. Sixty (60) of the incoming photons have initial conditions $(\vec{r}, \widehat{\Omega}_1, E_1)$ – coordinates of the phase space bin of interest in this calculation. Another 20 photons have a higher inbound energy ($E_2 > E_1$) and different orientation ($\widehat{\Omega}_2$). On passing through the voxel, 20 photons with incoming energy E_1 undergo total photoelectric absorption and another 10 photons are 'out-scattered' from their original phase space bin with a degraded energy $E_3 < E_2$ and re-orientation to $\widehat{\Omega}_3$. In this context, *out-scattering* refers to a photon count being transferred out of the phase space bin of interest $(\widehat{\Omega}_1, E_1)$. The remaining 30 photons with incoming energy E_1 are transmitted through the voxel without interacting in the voxel. Twenty (20) photons with greater starting energy E_2 are *scattered into* the phase space bin of interest. This obvious coincidence was forced to illustrate the basic concept of accounting for 'in-scattering'. The book-keeping is shown in the inset of Figure 5, with a loss of 30 counts due to full absorption (-20) combined with out-scattering (-10), while the gain from in-scattering is 20. The net change in the count at phase space bin $(\widehat{\Omega}_1, E_1)$

is therefore -10. This change in occupancy is alternatively predictable from the photon *divergence* which is the difference between incoming and outgoing particles; 60 photons entered the voxel with attributes $(\widehat{\Omega}_1, E_1)$ but only 50 exited *with the same attributes*, yielding a net loss (-10) in bin $(\widehat{\Omega}_1, E_1)$.

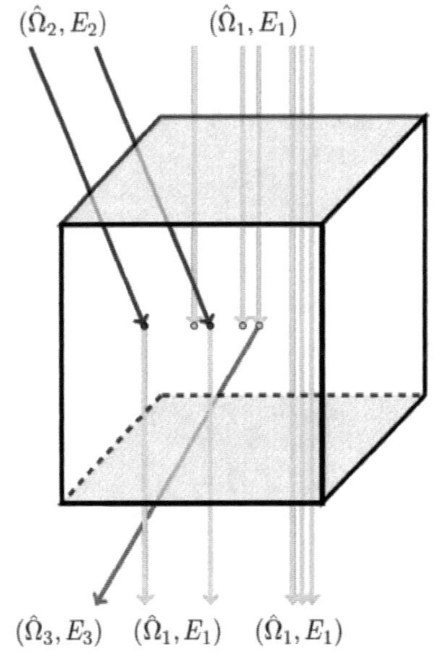

Phase Space Bin	Incoming	Outgoing	Bin Change $(\hat{\Omega}_1, E_1)$
$(\hat{\Omega}_2, E_2)$	20	0	+20 In Scatter
$(\hat{\Omega}_3, E_3)$	0	10	−10 Out Scatter
Absorbed	-	-	−20
$(\hat{\Omega}_1, E_1)$	60	50	**-10**

Figure 5. Voxel exposed to a fluence of bi-energetic bi-directional photons. Book-keeping of the phase space bin counts is shown in the inset on the right side.

The equality between divergence and net losses in a voxel forms the basis of the steady-state Boltzmann equation. On the left hand side of equation (4), the leading divergence term is notated as a dot product. This term tracks the *net* passage of particle fluences through all voxel boundaries. All the gains and losses of particles *within* the voxel are listed on the right hand side of the equation. For a given type of particle (photon or electron), the following equality must hold true, ignoring internal radioactive sources:

$$\widehat{\Omega} \cdot \vec{\nabla}\varphi(\vec{r}, \widehat{\Omega}, E) = \int \mathrm{d}E' \int \mathrm{d}\widehat{\Omega}' \; \Sigma_s(\vec{r}, \widehat{\Omega}' \rightarrow \widehat{\Omega}, E' \rightarrow E)\; \varphi(\vec{r}, \widehat{\Omega}, E') \; - \; \Sigma_t(\vec{r}, E)\; \varphi(\vec{r}, \widehat{\Omega}, E) \qquad (4)$$

Rearranging of terms yields the more common version of this equation [26]:

$$\left[\widehat{\Omega} \cdot \vec{\nabla} + \Sigma_t(\vec{r}, E)\right]\varphi(\vec{r}, \widehat{\Omega}, E) = \int \mathrm{d}E' \int \mathrm{d}\widehat{\Omega}' \; \Sigma_s(\vec{r}, \widehat{\Omega}' \rightarrow \widehat{\Omega}, E' \rightarrow E) \;\; \varphi(\vec{r}, \widehat{\Omega}, E') \qquad (5)$$

where $\varphi(\vec{r}, \widehat{\Omega}, E)$ is the flux density distribution integrated over the exposure time. The Σ_t term refers to the total macroscopic cross section for photon interactions that remove photon counts from a phase space bin through total absorption or out-scattering or re-orientation. The Σ_s term refers to the scattering cross section that can convert an incoming photon's direction $\widehat{\Omega}' \rightarrow \widehat{\Omega}$ *and* energy $E' \rightarrow E$, causing an entrance move to the phase space bin of interest at $(\vec{r}, \widehat{\Omega}, E)$.

In practice, a large set of coupled Boltzmann transport equations must be solved to predict the flow of photons and charged particles across a 3D matrix of dose voxels. The ultimate solution is the charged particle fluence, $\Phi^{(cp)}$, that reaches geometric voxels in real patient space coordinates. Once this fluence is determined, the dose can be calculated using the local mass collisional stopping power,

$(S/\rho)_c$, specific to the voxel's atomic composition. The following equation is simplified for monoenergetic charged particles and assumes delta-ray equilibrium (δCPE) within a voxel:

$$D(\vec{r}) \overset{\delta\text{CPE}}{=} \Phi^{(cp)}\,(S/\rho)_c(\vec{r}) \tag{6}$$

In practice, a polyenergetic variant of this equation is actually used. Note that absorbed dose is the concentration of energy deposited *in a specific absorber*. The flippant use of the word "dose" without specifying the absorbing medium has led to ambiguity in dose results reported by different algorithms [58-60].

As an example of a nuclear physics code that was adopted for medical physics applications, the Atilla code (http://silverfirsoftware.com/attilads.html) from the Los Alamos National Laboratory has been recast as the Acuros® XB algorithm on Varian's Eclipse treatment planning system [61]. Simplifications and assumptions are required to achieve practical turnaround times: (a) the normal Cartesian voxel space of the patient anatomy is restructured into a lattice of tetrahedrally-shaped finite elements; (b) scattering cross sections are approximated by a finite series of Legendre polynomials; (c) energy levels are quantized *via* the multi-grouping method, and (d) particle directions are quantized *via* discrete ordinates. The accuracy of dose results remains competitive with that of Monte Carlo simulations, and the process can be completed in practical times for routine radiotherapy planning.

To recapitulate, solving the linear Botzmann equations with boundary conditions of clinical beams of x rays incident on a finite heterogeneous patient space has now become practical. Advances in numerical methods and computer hardware have produced practical solutions for clinical applications of ever increasing complexity. The biggest disadvantage at present is unfamiliarity with commissioning a new computational technique for clinical application. A primer on Boltzmann concepts and equations is available for readers facing the challenging learning curve [23].

5. Conclusion: State of the Art

The progression of algorithms has been driven by an interplay across the fields of nuclear physics, numerical computation, 3D medical imaging, and experimental dosimetry – the theme of this conference [62]. The accuracy of Monte Carlo or Boltzmann results is becoming difficult to distinguish because results often lay within the range of experimental uncertainties. Table 3 summarizes the performance of various contemporary algorithms for typical clinical situations. Direct head-to-head comparison across algorithms is complicated by variations in computer hardware, run-time parameters, and debated tolerance metrics for judging acceptable clinical accuracy. Remarkable progress has been achieved in the development of stochastic and deterministic algorithms. Both methods have recently "caught up" with convolution-superposition algorithms in terms of speed of execution, and potentially offer better accuracy over a wider scope of clinical conditions. The next evolutionary step for these algorithms will likely be application to rapid dose optimization cycles and adpative radiotherapy at the treatment machine console while the patient remains in treatment position.

Progress in computer technology will continue to have an impact not only on treatment planning but also on multi-modality imaging, automated dose optimization, and on-line dose reconstruction for adaptive radiotherapy. The GPUMCD code was implemented on a NVIDIA GTX480 processor to achieve a speed gain on the order of 1000 over the general-purpose EGSnrc/DOSXYZ code running on a CPU (Intel i7 860) [63], reducing execution times to a matter of seconds. As another clinical example, a multi-core CPU-based system (PhiMC code, Xeon V3) produces clinical IMRT treatment plans, as shown in Figure 6, in about 10 seconds [64]. Cloud-based shared computing resources are also under investigation [65]. Dose accuracy will continue to evolve, with opportunities for interactive treatment re-optimization at the treatment console for dose-adaptive 4D radiotherapy [66-69]. With assessment of acute and long-term clinical outcomes obtained under such tightly-controlled dosimetry, radiobiological models of human tumour and normal tissue response will take a quantum leap forward [70].

Journal of Physics: Conference Series **2630** (2023) 012008 doi:10.1088/1742-6596/2630/1/012008

Table 3. Contemporary dose computation algorithms [26].

Target Site	Algorithm	System	Technique	Dose Matrix (voxels)	Voxel (mm)	Time	Remarks
Head and Neck	SVD	Pinnacle	2 Arcs; 182 Beams	100 × 100 × 100	2.5	30 s	8-core Intel 2.9 GHz Xeon
	CCC	Pinnacle	2 Arcs; 182 Beams	100 × 100 × 100	2.5	4.5 m	8-core Intel 2.9 GHz Xeon
	CCC	RayStation	2 Arcs; 182 Beams	100 × 100 × 100	2.5	18.2 s	1 server with a GPU
	AAA	Eclipse	2 Arcs; 182 Beams	100 × 100 × 100	2.5	35 s	13 frame agent server, 30 cores
	Acuros	Eclipse	2 Arcs; 182 Beams	100 × 100 × 100	2.5	1.9 m	1 frame agent, 2X 8-core, 2.9 GHz
	EGSnrc	Stand-Alone	2 Full Arcs	240 × 240 × 159	2.5	39 m	16 x 8-core Intel Xeon 2.0 GHz (2 threads/core)
	PhiMC-DPM	Royal Marsden	IMRT; 9 Beams	256 × 256 × 116	1.6	8.2 - 18.8 s	0.9-0.6% precision; Xeon V3
Lung	SVD	Pinnacle	2 Arcs; 108 Beams	100 × 100 × 100	2.5	21 s	8-core Intel 2.9 GHz Xeon
	CCC	Pinnacle	2 Arcs; 108 Beams	100 × 100 × 100	2.5	2.6 m	8-core Intel 2.9 GHz Xeon
	CCC	RayStation	2 Arcs; 108 Beams	100 × 100 × 100	2.5	13 s	1 server with GPU
	AAA	Eclipse	2 Arcs; 108 Beams	100 × 100 × 100	2.5	20 s	13 frame agents, 30 cores were used
	Acuros	Eclipse	2 Arcs; 108 Beams	100 × 100 × 100	2.5	47 s	1 frame agent, 2X 8-core, 2.9 GHz
	EGSnrc	Stand-Alone	2 Partial Arcs	259 × 259 × 180	2.5	60 m	16 x 8-core Intel Xeon 2.0 GHz (2 threads/core)
Prostate	SVD	Pinnacle	2 Arcs; 182 Beams	147 × 118 × 86	3	35 s	8-core Intel 2.9 GHz Xeon
	CCC	Pinnacle	2 Arcs; 182 Beams	147 × 118 × 86	3	5.1 m 3 s	8-core Intel 2.9 GHz Xeon
	CCC	RayStation	2 Arcs; 182 Beams	147 × 118 × 86	3	19 s	1 server with GPU
	AAA	Eclipse	2 Arcs; 182 Beams	147 × 118 × 86	3	26 s	13 frame agents, 30 cores were used
	Acuros	Eclipse	2 Arcs; 182 Beams	147 × 118 × 86	3	1.33 m	1 frame agent, 2X 8-core, 2.9 GHz
	GPUMCD	Monaco	IMRT; 7Beams	344 × 286 × 172	2	2.7 m	2% precision, B-Field ON
	GPUMCD	Monaco	IMRT; 7Beams	324 × 265 × 152	2	1 m	2% precision, B-Field OFF
	EGSnrc	Stand-Alone	Full Arc	240 × 240 × 175	2.5	60 m	16 x 8-core Intel Xeon 2.0 GHz (2 threads/core)
	PhiMC-DPM	Royal Marsden	IMRT; 9 Beams	256 × 256 × 234	2	10.8-26.7 s	1.1-0.7%; Xeon V3

* Data courtesy of Odette Cancer Centre (GPUMCD), London Regional Cancer Program (SVD, CCC, AAA, Acuros), Vancouver Cancer Centre (EGSnrc) (Popescu et al. 2015), and published worl (PhiMC-DPM) (Ziegenhein et al. 2015). Algorithms are used for treatment planning, except for EGSnrc used for quality assurance. Typical computation times.

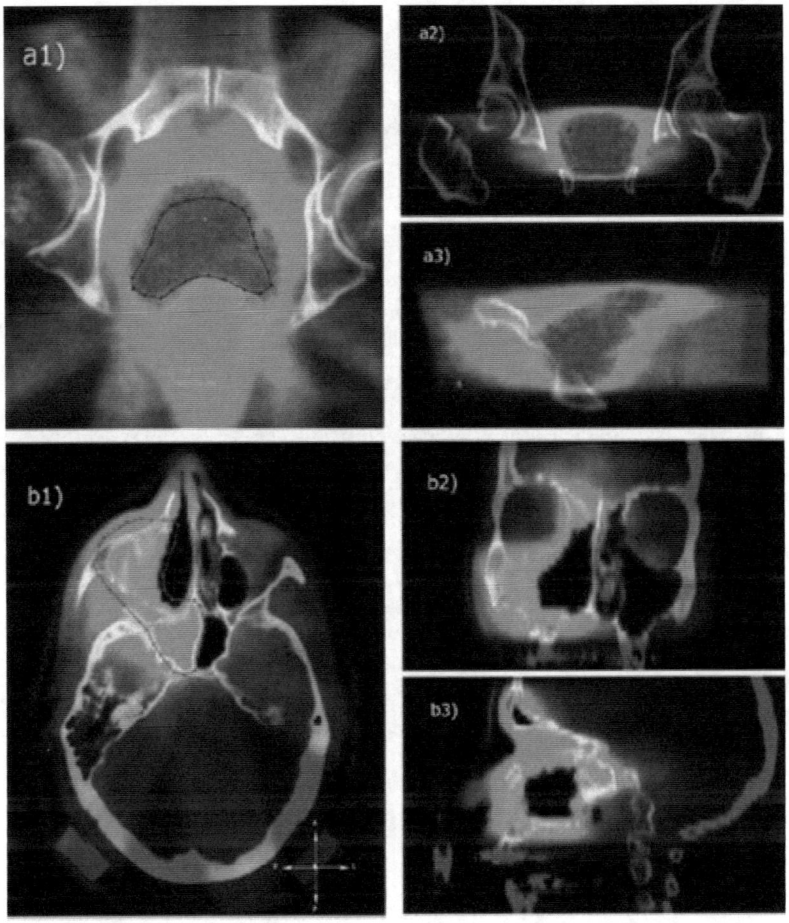

Figure 6. High-resolution dose distributions with 1% precision for prostate (top panels) and head & neck (bottom panels) IMRT cases produced with the Phi Monte Carlo code in 10 seconds [64]. Reproduced with permission of the Creative Commons license.

6. References

[1] Johns H E *et al* 1983 *The Physics of Radiology (Fourth Edition)*. (Charles C. Thomas, Springfield Illinois) 336-381
[2] Purdy J A *et al* 1987 *Front. Radiat. Ther. Oncol.* **21** 4-24
[3] Papanikolaou N *et al* 2004 Report of the AAPM Radiation Therapy Committee Task Group 65. 1-142
[4] Rinks M 2012 *J. Med. Radiat. Sci.* **59** 1-4
[5] Wong J W *et al* 1990 *Med. Phys.* **17** 807-14
[6] Sontag M R *et al* 1978 *Radiology* **129** 787-4
[7] Wong J W *et al* 1984 In *Proceedings of the 8th International Conference on the Use of Computers in Radiation Therapy*. (Computer Society Press, IEEE, Silver Spring, Maryland) 26-30
[8] Cunningham J R *et al* 1974 In *International Congress of the XIIIth International Congress of Radiology*. Madrid, Spain. (Excerpta Medica, Amsterdam) **2** 653-7
[9] Cunningham J R 1972 *Phys. Med. Biol.* **17** 42
[10] Van Dyk J *et al* 2020 *Med. Phys.* **47** 1417
[11] Yu C X *et al* 1987 *Med. Phys.* **14** 78-83

[12] Wong J W *et al* 1981 *Med. Phys.* **8** 783-91
[13] Wong J W *et al* 1981 *Med. Phys.* **8** 775
[14] Wong J W *et al* 1983 *Med. Phys.* **10** 199-208
[15] Battista J J *et al* 1992 *Australas. Phys. Eng. Sci. Med.* **15** 159-178
[16] Yu C X *et al* 1995 *Med. Phys.* **22** 1157- 65
[17] Mackie T R *et al* 1984 *Med. Phys.* **12** 188-96
[18] Knöös T 2017 *J. Phys.: Conf. Series* **847** 1-10
[19] Lu L 2013 *Int. J. Cancer. Ther. Oncol.* **1** 01025-8
[20] Ahnesjö A *et al* 1999 *Phys. Med. Biol.* **44** R99-R155
[21] Battista J J *et al* 1997 In *Proceedings of the Xiith International Conference on the Use of Computers in Radiation Therapy*. Salt Lake City 39-42
[22] MacDermott P N 2016 In *Tutorials in Radiotherapy Physics*. CRC Press, Taylor & Francis Group 120-69
[23] MacDermott PN. Deterministic Radiation Transport: a Rival to Monte Carlo Methods. In: *Tutorials in Radiotherapy Physics*. CRC Press, Taylor & Francis Group, 2016, pp. 170–216.
[24] Andreo P 2017 In *Fundamentals of Ionizing Radiation Dosimetry*. Weinheim, Germany: Wiley-VCH 349-96
[25] Seco J *et al* 2016 *Monte Carlo techniques in radiation therapy* (CRC press, Taylor & Francis Group)
[26] Battista J *et al* 2019 *Introduction to megavoltage x-ray dose computation algorithms* (CRC press, Taylor & Francis Group)
[27] Vassiliev ON 2016 *Monte Carlo Methods for Radiation Transport*. (Springer, New York)
[28] Mackie T R *et al* 1998 *Phys. Med. Biol.* **33** 1-20
[29] Ahnesjö A *et al* 1987 *Acta. Oncol.* **26** 49-56
[30] Sharpe M B 1997 *A unified method of calculating the dose rate and dose distribution for therapeutic x-ray beams (Doctoral Thesis)*. University of Western Ontario, London, Canada.
[31] Pratx G *et al* 2011 *Med. Phys.* **38** 2685-97
[32] Hissoiny S *et al* 2010 *Med. Phys.* **37** 1029-37
[33] Boyer A L 1984 *Med. Phys.* **11** 552-4
[34] Wong E *et al* 1996 *Med. Phys.* **23** 1511-21
[35] Field G C *et al* 1987 In *Proceedings of the IXth International Conference on the Use of Computers in Radiation Therapy*. ICCR 103
[36] Ahnesjö A 1989 *Med. Phys.* **16** 577-92
[37] Ahnesjö A *et al* 2017 *Comput. Methods Programs Biomed.* **139** 17-29
[38] Bielajew A 2016 In *Monte Carlo Techniques in Radiation Therapy*. CRC Press, Taylor & Francis Group
[39] Rogers D W O 2006 *Phys. Med. Biol.* **51** R287-R301
[40] Rogers D W O *et al* 2021 NRCC Report PIRS-0509 (A) Rev L, Ottawa, Canada: National Research Council of Canada (NRCC)
[41] Verhaegen F *et al* 2006 *Phys. Med. Biol.* **51** N157-N161
[42] Brualla L *et al* 2017 *Strahlentherapie und Onkol* **193** 243-59
[43] Kan M W K 2013 *BioMed Res. Int.* 1-10
[44] Seuntjens J 2017 In *Fundamentals of Ionizing Radiation Dosimetry*. Wiley-VCH, Weinheim, Germany 279-313
[45] Ojala J 2014 *Int. J. Cancer Ther. Oncol.* **2** 1-12
[46] Bush K *et al* 2011 *Med. Phys.* **38** 2208-21
[47] Hoffmann L *et al* 2018 *Med. Phys.* **45** 3909-15
[48] Hoffmann L *et al* 2012 *Acta Oncol.* **51** 376-85
[49] Vassiliev O N *et al* 2010 *Phys. Med. Biol.* **55** 581-98
[50] Ojala J J *et al* 2014 *J. Appl. Clin. Med. Phys.* **15** 162-76
[51] Fogliata A *et al* 2017 *Physica Medica* **44** 157-62

Journal of Physics: Conference Series **2630** (2023) 012008 doi:10.1088/1742-6596/2630/1/012008

[52] Ojala J J *et al* 2014 *J. Appl. Clin. Med. Phys.* **15** 4-18
[53] Zhou C *et al* 2017 *Radiat. Oncol.* **12** 1-11
[54] Wareing T A *et al* 2001 *Nucl. Sci. Eng.* **138** 256-68
[55] Kan M W K *et al* 2013 *Biomed. Res. Int.* **692874** 1-10
[56] Vassiliev O N *et al* 2008 *Int. J. Radiat. Oncol. Biol. Phys.* **72** 220-7
[57] Andreo P *et al* 2017 *Fundamentals of Ionizing Radiation Dosimetry*. Wiley-VCH, Weinheim, Germany
[58] Andreo P 2015 *Phys. Med. Biol.* **60** 309-37
[59] Smilowitz J B *et al* 2015 *J. Appl. Clin. Med. Phys.* **16** 14-34
[60] Kry S *et al* 2020 *Med. Phys.* **47** e52-64
[61] Failla G A *et al Varian White Pap RAD 10156A*
[62] Mijnheer B 2018 *Clinical 3D Dosimetry in Modern Radiation Therapy*. CRC Press, Taylor & Francis Group
[63] Hissoiny S *et al* 2011 *Med. Phys.* **38** 754-64
[64] Ziegenhein P *et al Phys. Med. Biol.* **60** 6097-111
[65] Ziegenhein P *et al* 2017 *Phys. Med. Biol.* **62** 437589
[66] Popescu I A *et al* 2015 *J. Phys.: Conf. Series* **573** 1-11
[67] Spreeuw H *et al* 2016 *Med. Phys.* **43** 3969-74
[68] Colvill E et al 2018 In Mijnheer B (ed) *Clinical 3D Dosimetry in Modern Radiation Therapy*. Boca Raton, Florida: CRC Press, Taylor & Francis Group 281-299
[69] Hill B *et al* 2002 *Phys. Med. Biol.* **47** 4233-246
[70] El Naqa I 2018 A Guide to Outcome Modeling in Radiotherapy and Oncology: Listening to the Data. CRC Press, Taylor & Francis Group

EPID-based *in vivo* dosimetry – new developments and applications

B M C McCurdy[1,2,3,4]
[1]Medical Physics Department, CancerCare Manitoba, Winnipeg, MB, Canada
[2]Department of Physics and Astronomy, University of Manitoba, Winnipeg, MB, Canada
[3]Department of Radiology, University of Manitoba, Winnipeg, MB, Canada
[4]CancerCare Manitoba Research Institute, Winnipeg, MB, Canada

E-mail: bmccurdy2@cancercare.mb.ca

Abstract. *In vivo* dosimetry has been shown to be a powerful quality assurance method in modern radiation therapy. The most common tool used for *in vivo* dosimetry is the electronic portal imaging device (EPID) which can quantitatively image the therapeutic beam fluence exiting the patient during treatment delivery. Since the last major literature review on this topic was published five years ago, the radiation oncology community has shown continued strong interest in this subject. Commercial options have become more widely available, with a related increase in validation efforts and sensitivity testing, while new applications continue to be explored. Work has been done to understand and increase the accuracy of the EPID for dosimetric applications, as well as continued efforts to provide practical, quantitative experiences from clinical implementation of *in vivo* dosimetry systems. This review examines the published literature related to *in vivo* EPID dosimetry from January 2017 to February 2022. The literature is classified into three main topical areas: (1) new or improved algorithmic developments including validation work, (2) applications of the *in vivo* EPID dosimetry method, and (3) error identification and error sensitivity analyses.

1. Introduction

Modern radiation therapy continues to witness a rapid pace of technological development which allows advances in clinical procedures delivering higher dose in fewer fractions and to increasing levels of accuracy and precision. However, differences between the intended delivery and the actual delivery can still occur. *In vivo* dosimetry (IVD) is a valuable tool that allows a measurement verification of the actual dose delivered to the patient. A recent review paper by an ESTRO task group of experts provided a standard definition of *in vivo* dosimetry: "IVD is a radiation measurement that is acquired while the patient is being treated containing information related to the absorbed dose in the patient." [1]. Note that this definition excludes related and potentially useful but often misperceived techniques such as monitoring the constancy of transmission images as well as log-file based analysis. The *in vivo* dosimetry approach is able to detect errors in machine/hardware delivery, errors in dose calculation, errors in patient setup, and changes in patient anatomy between planning and delivery. It has been demonstrated to be one of the most important quality control checks in radiation therapy [2], and is recommended by several organizations to increase patient safety (e.g. WHO [3], ESTRO [1], and AAPM [4]).

In the 1980's and 1990's, there was interest in using electronic portal imaging devices (EPIDs) for *in vivo* dosimetry measurements. When the current generation amorphous-silicon style EPIDs became widely available in the 2000's, interest grew dramatically due to their excellent dosimetric properties [5, 6] which were significantly improved over previous EPID technologies (both camera-based and liquid ionization chamber designs). The EPID's broad availability on modern linear accelerators, convenience of use, large 2D detector format, and digital image nature ensure that EPID-based *in vivo* dosimetry is significantly more attractive compared to alternative, manually placed, single point detector methods.

Several excellent reviews of a variety of aspects of EPID based *in vivo* dosimetry are available in the literature. Van Elmpt *et al* provided a comprehensive literature review in 2008 [7], which was updated to late 2016 [8]. In 2020, an ESTRO task group on *in vivo* dosimetry provided a broad assessment of the current state of *in vivo* dosimetry clinical implementation, as well as describing areas requiring more development, and suggested approaches to improve clinical acceptance [1]. There have also been recent reviews of EPID-based *in vivo* dosimetry focusing on SBRT delivery techniques including for lung treatment, where available techniques and clinical dosimetry results were summarized in 2017 [9], and for all SBRT treatment sites, providing an overview of accuracy and sensitivity of published results in 2020 [10]. Other related reviews include a report on clinical usage of *in vivo* dosimetry methods in the UK [11] as well as *in vivo* dosimetry methods in brachytherapy [12].

The purpose of the present review of EPID based *in vivo* dosimetry is to update the literature reviews of references [1] and [8] to the period of January 2017 to February 2022, focusing on *in vivo* dosimetry. Three main categories will be examined: (1) new or improved algorithmic developments including validation work, (2) new applications of the IVD method, and (3) error identification and error sensitivity analyses. It is recognized that some papers qualify in multiple categories, in which case a decision was made as to the best single category placement.

2. New or improved predictive algorithms for *in vivo* dosimetry

2.1. Predictive methods or algorithms

Several groups have contributed additional empirical methods aimed at estimating dose at a 2D plane within the patient, generally located at mid-plane. An existing empirical method to estimate 2D midplane dose in the patient was improved through use of a more accurate off-axis correction method, incorporating an asymmetric 2D Gaussian kernel to account for beam penumbra effects [13], with the MatLab code (MathWorks, Natick, CT) made available to researchers. Hashemi *et al* [14] developed an empirical method for homogeneous phantoms accounting for beam hardening, patient scatter, and off-axis EPID response, and verified it using a 6MV, 8x8 cm^2 field on a 20 cm thick water equivalent slab comparing reconstruction to a treatment planning system (TPS) with a gamma pass rate (3%, 3 mm DTA) of 94.9% for points at mid-plane. This method was then extended with a water-equivalent pathlength concept to account for patient CT data [15] and tested on a pelvis phantom with a conformal field and a 4-segment step-and-shoot IMRT field, with gamma results (3%, 3 mm DTA) within 97.0% and 93.5% respectively. Anvari *et al* [16] proposed an empirical method with attenuation correction and off-axis correction terms to estimate 2D mid-plane dose and also EPID plane dose and tested it with a homogeneous phantom using a fluoroscopic style EPID as well as a planar dose array, observing a gamma pass rate (3%, 3 mm DTA) of 96% for fields from pelvic plans.

A 3D backprojection method originally developed by Berry *et al* [17] was improved by the addition of a correction factor to remove dependence on source-imager-distance, and by incorporating a model of the treatment couch [18]. The method was tested with six IMRT plans and seven VMAT plans on a homogenous slab phantom, RT01 pelvis phantom, and anthropomorphic pelvis phantom, with gamma pass rates (3%, 3 mm DTA) of >95% for all the IMRT fields and >95% for 23 of 24 VMAT arcs.

Results for a 2D forward projection approach using a Monte Carlo method have been recently published by Feng *et al* [19] using a fast simulation method based on Jia *et al* [20], with average gamma pass rates (3%, 3 mm DTA) for 268 fractions over 20 IMRT patients of 87.8% (±14.0%) using the planning CT data set and 93.6% (±8.2%) using the daily CBCT data set.

There is continued interest in utilizing commercial treatment planning systems (TPS) to calculate dose to the EPID plane. Deshpande *et al* [21] used Pinnacle (v9.1) and introduced a water-equivalent virtual EPID into the TPS and compared predictions to measurements using a commercial EPID that they physically converted to a water-equivalent EPID by replacing the metal/phosphor buildup layers with solid water. Gamma pass rates (3%, 3 mm DTA) of >94% were achieved for IMRT and VMAT fields applied to inhomogeneous and anthropomorphic phantoms. Martinez-Ortega *et al* [22] used Pinnacle (v8.0m) and introduced a water-equivalent virtual EPID model to the TPS but also implemented a photon spectral correction based on radiological thickness of the patient along individual raylines. They compared predictions to a-Si EPID measurements that were converted to be water-equivalent, and achieved gamma pass rates (5%, 3 mm DTA) of >80.3% for three square fields on an anthropomorphic phantom (cranial, thorax, and pelvis sites) and >94.6% for five IMRT fields on the pelvis site. Antoniou and Penfold developed a script to add a virtual EPID model consisting of a 5cm thick water slab backed by a 4 cm thick lead slab, into the patient calculation volume using Raystation (v8B) [23]. Mean differences with a measured integrated EPID image for a VMAT delivery were within 9.0%, with the mean difference across the entire image of 1.5%, although the investigation was more focused on establishing correlations between transit fluence changes and DVH metrics.

Guo *et al* developed a new hybrid method for estimating the patient-generated photon scatter fluence entering the EPID [24] combining three different algorithms suited for different photon scatter types: analytical for singly scattered photons, hybrid Monte Carlo for multiply scattered photons, and pencil beam for bremsstrahlung and positron annihilation photons. The authors also optimized it's sampling for speed and accuracy, achieving the accuracy of full Monte Carlo simulation at a fraction of the time [25]. That algorithm can be used as part of the CancerCare Manitoba (CCMB) inhouse 3D dose reconstruction technique to improve overall accuracy. The Netherlands Cancer Institute (NKI) group developed a more accurate version of their inhouse algorithm specific to small field dosimetry (ie. stereotactic body radiotherapy or SBRT) by removing the influence of the ion chamber volume effect from the EPID scatter kernel, extending output factors to smaller fields, and updating scatter-to-primary ratios using smaller incident fields [26], changes which reduced dose discrepancies in tested brain VMAT fields from 6.8% to 0.8% at isocentre. They also extended their *in vivo* EPID dosimetry technique to an MR-linac (Elekta Unity, Elekta AB, Sweden), documented in several publications as follows. In 2017, Torres-Xirau *et al* demonstrated the feasibility of modifying the EPID dose prediction method to account for attenuation and scattering of the cryostat in an MR-linac, simulated with a conventional linac [27]. Once the MR-linac was available for testing, the authors provided a more detailed update of the NKI *in vivo* EPID model [28], specifically for EPID plane dose prediction. Twenty-five IMRT fields (two prostate plans, one rectum plan) were tested with gamma test results (3%, 3 mm DTA) all >95%. This was eventually extended to 3D patient dose reconstruction [29] with plans from five patients being tested comparing EPID reconstructed dose with in-phantom measured dose (Octavius, PTW, Frieburg, Germany), achieving a gamma pass rate (3%, 2 mm DTA) average of 91.9%. However, to improve agreement further especially for larger fields whose outer regions are more significantly affected by attenuation and scatter from gradient coils, Olaciregui-Ruiz *et al* [30] used a deep learning approach to correct *in vivo* reconstructed dose in these areas, demonstrating the average gamma pass rate (3%, 2 mm DTA) outside of the central region improved from 44.5% to 90.2%, using 45 large field rectum plans.

2.2. Validation of algorithms

The last five years has seen an increase in the interest in and use of commercial options for *in vivo* dosimetry, as demonstrated by numerous validation studies available in the literature during this time. Validation studies typically focus on examining *in vivo* dosimetry results measured under carefully controlled situations (i.e. phantom irradiation) and comparing those to patient/phantom dosimetry calculated in a treatment planning system.

Stevens *et al* [31] validated Dosimetry Check (Lifeline Software, Tyler, TX), a 3D patient dose reconstruction software, using PAGAT gel dosimetry in two phantoms using a single VMAT plan, and obtained gamma pass rates (3%, 3 mm DTA) of >92% and >94%. Esposito *et al* [32] validated Dosimetry Check with Octavius phantom measurements for 20 clinical VMAT plans, demonstrating a mean gamma pass rate (3%, 3 mm DTA) of 94.2% (±3.4%). They also evaluated error sensitivity to various setup errors and anatomical changes. Held *et al* [33] validated EpiGray (DOSIsoft, Paris, France) using solid water phantoms of three thicknesses and a variety of field shapes including IMRT. Dose reconstructed at a single point (typically isocentre) was found to be within 3-5% of the TPS, while off-isocentre points typically showed poorer agreement. The PerFraction software (Sun Nuclear, Melbourne, FL) was validated by Aziz-Sait *et al* [34] using an ion chamber and a 2D ion chamber array (MapCheck, Sun Nuclear) to validate single arc VMAT plans in several phantoms (homogeneous and inhomogeneous) and found 97.8% of 270 compared points were within 3% of expected. Olaciregui-Ruiz *et al* [35] validated the iViewDose software (Elekta AB, Stockholm, Sweden) with Octavius phantom measurements using 10 IMRT plans and 34 VMAT plans (2 arcs each), finding an average gamma pass rate (3%, 2 mm DTA) of 92.2% (±5.2%).

The CCMB inhouse method was validated for *in vivo* dose reconstruction on 4D CT data sets using a dynamic thorax phantom and lung tumour insert [36] showing a gamma pass rate (3%, 3 mm DTA) within the PTV of >99.3%. The method was also validated for non-coplanar beam configurations such as those used in stereotactic radiosurgery (SRS) brain treatments [37], using an IMRT QA phantom with irradiations delivered via XML scripting in research mode on a TrueBeam linac (Varian Medical Systems, Palo Alto, CA), achieving gamma pass rates (2%, 2 mm DTA) >97%.

2.3. *Infrastructure work*

This section describes development work that is not itself a type of *in vivo* dose calculation, but rather relates to required methodological infrastructure or foundational understanding which impacts the accuracy of an implemented *in vivo* dosimetry method.

For example, the pixel sensitivity matrix (PSM) is a technique used to correct for the variation in gain of individual pixels throughout the EPID panel, while allowing the image to still retain beam profile dosimetric information. Accounting for this effect improves the accuracy of EPID *in vivo* dosimetry applications by up to ~5%. Cai *et al* proposed an approach acquiring five, large-field overlapping irradiations where the EPID is shifted by a few mm between irradiations [38]. Beam on images were alternately acquired with dark field images, which were later subtracted in order to eliminate ghosting effects. Another proposed approach avoided shifting the EPID and used a regression model to differentiate the shape of beam fluence (i.e. low frequency signal) from the pixel-to-pixel variation (i.e. high frequency signal) [39], thus deriving the PSM in a single EPID image acquisition.

Abbasian *et al* [40] showed that one could measure the timing and amplitude of linac dose pulses and use this to modify a 2D EPID dose image prediction method to account for the spatial-temporal readout nature of the Varian EPID. Improved 2D image prediction for individual EPID frames was demonstrated, of interest for real-time image comparison applications.

Renaud *et al* [41] performed a systematic evaluation of the dosimetric uncertainty for a unique film-in-EPID dose calibration method for EPIDs (Elekta model examined).

The same authors also evaluated long term stability of the Elekta EPID panel and showed it to be dependent on supply voltage, internal operating temperature, and accumulated absorbed dose in the EPID [42]. Barnes *et al* [43] demonstrated the long-term drift of sensitivity using a Varian EPID and recommended regular (i.e. monthly) cross calibration with ion chamber measurement.

3. Applications of *in vivo* dosimetry

3.1. 2D forward dose projection in the EPID

Fuangrod *et al* [44] applied statistical process control methods to the comparison of measured and predicted transmission EPID images for IMRT and VMAT treatments to flag treatment discrepancies which required investigation. Control limits were generated using initial data from 90 IMRT and 10 VMAT patient deliveries. Fourteen patients were selected as case studies, and five patients had treatments flagged when verification exceeded the control limits. Investigation of the flagged treatments revealed that most errors were due to anatomical uncertainty.

3.2. Dose reconstruction in the patient for external beam treatments

There has been significant effort in publishing clinical experiences with *in vivo* EPID dosimetry programs over the literature review period. Note that many of these are using commercial products while only a few use inhouse developed techniques, a change from earlier literature reviews, which had very few results from commercial products.

The Softdiso system (Best Nomos, Pittsburgh, PA, USA) has been applied to a group of 15 lung SBRT patients (80 fxs, 2 arcs/fx) [45], and observed an out-of-tolerance rate of 5%. The system has also been used on a mix of 3DCRT and VMAT deliveries across breast, thorax, abdomen, pelvis, H&N, and brain (823 patients, 11357 fxs) [46], finding 21% of 3DCRT fractions and 6% of VMAT fractions out-of-tolerance, with the higher value for 3DCRT attributed mainly to less consistent patient setup and also attenuation through uncontrolled positioning of couch supports. The Softdiso system was also applied to a mix of 3DCRT (breast, lung, pelvis, abdomen), IMRT (pelvis, H&N), and VMAT deliveries (prostate, H&N) on a total of 386 pts, 8474 fxs, and observed 9%, 7%, and 9% out-of-tolerance rates respectively [47], with the results of initial fractions used to identify and correct errors in subsequent fractions to bring them within tolerance.

Nailon *et al* [48] applied Dosimetry Check to verify *in vivo* dose at the reference point for 2943 3DCRT and 842 VMAT patients and found an out-of-tolerance rate of 4%. Esposito *et al* [49] evaluated Dosimetry Check on abdominal and prostate SBRT patients (80 pts, 152 fxs), observing an out-of-tolerance rate of 7%. A pre-cursor to the iViewDose software, modified to work on the Unity MR-linac system (Elekta), was applied to 127 IMRT treatments (74 prostate, 19 rectum, 19 liver, and 15 lymph node oligomets) during 1207 adapted treatment fractions, with a 9% out-of-tolerance rate observed [50]. Bossuyut *et al* [51] used the PerFraction software for *in vivo* dosimetry (and pre-treatment QA) and reported *in vivo* results from 24011 fractions across 3671 patients (used on first three days then weekly thereafter) over a wide range of anatomical treatment sites using integrated EPID transmission images, as well as discussing error types and the development of clinical tolerances. For their *in vivo* dosimetry results they found a 16% out-of-tolerance rate, of which 6% were false positives and 10% were related to patient setup, with newer linacs associated with about half the false positive rate of older linacs.

Clinical results from non-commercial algorithms include McCowan *et al* [52] using the CCMB inhouse method on 602 fractions delivered to 117 lung, spine, and liver SBRT patients, and demonstrating an out-of-tolerance rate of 10% after EPID frame averaging was optimized. Meanwhile Peca *et al* [53] used their 2D midplane algorithm on 10 rectal patients treated with 4-field box plans and observed out-of-tolerance differences in seven patients due to issues with positioning of the belly-board, patient positioning on the belly-board, and bowel gas. Sterkx *et al* [54] used the 'in aqua' *in vivo* EPID dosimetry method [55], originally developed for lung patients, on prostate patients who were treated with a rectal balloon device to circumvent the negative impact of the presence of air or low density anatomical volumes to accurate patient dose reconstruction. Anvari *et al* [56] used their inhouse method, an empirical 2D mid-plane dose reconstruction [16], to verify small animal (ie. rat) irradiations showing agreement of TPS predicted and EPID measured central axis exit dose rates of 3.1% (on average).

The literature review also showed that many researchers are using available *in vivo* EPID dosimetry systems as a tool to investigate other aspects of radiation therapy, beyond the evaluation or quantification of the *in vivo* dosimetry system itself. For example, the Softdiso system was used to demonstrate and quantify improvements in lung SBRT treatment reproducibility in 10 patients when using the ABC breathing system [57], as well as to assess the treatment quality in 48 breast patients and correlate this to an improved patient setup procedure [58]. The EpiGray software was used to help quantitatively evaluate the dosimetric impact of high-density bone cement [59] during radiation therapy, of interest for post-surgical spine SBRT patients. The PerFraction software was applied to quantitatively evaluate the effectiveness of new patient preparation strategies for the bladder and rectum in prostate SBRT patients, and while was shown to detect changes in bladder and rectal filling, could not confirm the hypothesis [60]. Tan *et al* [61] used the non-commercial *in vivo* EPID dosimetry approach of Najem *et al* [18] to help develop a new quality assurance methodology to determine the target localization accuracy of phase gated treatments.

3.3. Dose reconstruction in the patient for brachytherapy treatments

Smith *et al* [62] verified dwell times and source positions for HDR treatments of two prostate patients using an external imaging panel (aS500, Varian) embedded in the treatment couchtop using the HDR source as an x-ray imaging source. They used a pre-treatment localization planar x-ray image to establish prostate position (using fiducials) and catheter positions (using x-ray dwell position markers). In a proof-of-concept study, Fonseca *et al* [63] also used an external EPID and combined the HDR source tracking with anatomical gamma ray imaging by the source itself, to remove the need for pre-treatment planar x-ray imaging, to determine source dwell times and positions for HDR treatments in a pelvic phantom.

4. Error identification and error sensitivity analyses

Some error sensitivity analyses are included in the initial validation work, but many publications solely focus on evaluating errors and developing a clinically relevant understanding of the types and magnitudes of errors detectable by *in vivo* EPID dosimetry.

Bedford *et al* [64] used the iViewDose software and introduced errors into thirteen clinical prostate VMAT plans and delivered these to a phantom in order to compare the results of back-projection *in vivo* dosimetry with forward projection *in vivo* dosimetry, demonstrating that while the two methods have their own error sensitivities, they are generally similar. That research group also investigated the use of a running sum of real-time EPID transmission images for comparison to predicted EPID transmission images for corresponding segments of a VMAT arc [65], in a forward projection framework. They introduced errors in MU's, field size, field position, and patient thickness for 13 clinical prostate plans and delivered to a phantom to determine error sensitivity. More recently they developed a composite difference metric to identify forward projection errors more quickly, combining central axis signal, mean image signal, and two different image difference metrics [66].

A pre-cursor to the iViewDose software was used by Van Der Bijl [67] to compare DVH based tolerances to multi-parametric gamma-based tolerances for pelvic VMAT patients (47 rectum, 37 prostate, 44 bladder pts; 387 fx total) and establish meaningful correlations with the goal of establishing more clinically intuitive tolerances. Also using a pre-cursor to the iViewDose software, Mijnheer *et al* [68] tested sensitivity to a variety of errors (ie. delivery, thickness, setup) physically introduced to treatments on an anthropomorphic phantom using four VMAT plans (prostate, lung, two head and neck). In addition to establishing more clinically relevant tolerance critieria, they also observed that the presence of patient setup errors reduced sensitivity to other types of errors. Again employing similar IVD software, Olaciregui-Ruiz *et al* [69] mapped the sensitivity for a variety of patient related errors using a virtual phantom approach, where errors are introduced into the CT data set or the treatment plan, and the dosimetry recalculated and compared to the non-errored dosimetry. They developed alert criteria specific to four anatomical treatment sites (using 104 VMAT plans from brain, rectum, head and neck, and lung) and the magnitude of errors they wanted to detect clinically.

The virtual phantom approach was also used by that research group to compare error detectability for *in vivo* dosimetry estimated on the patient's planning CT versus the patient's daily CBCT [70], demonstrating that the use of the daily CBCT was more sensitive in detecting patient positioning errors, which is logical considering it is a more accurate representation of the dose actually delivered to the patient. Yedekci *et al* [71] evaluated the sensitivity of the iViewDose software on a variety of errors (dose calibration, setup, collimator, MLC, and patient anatomy) introduced to ten prostate SBRT patient plans delivered to a pelvis phantom and showed that some positional errors could escape detection by *in vivo* dosimetry in that treatment site. Recently Mans *et al* [72] presented a procedure to perform long-term trend analysis of multi-year *in vivo* EPID results for prostate cancer patients to identify gradual changes over time, which led to the identification and reduction of underlying systematic dose uncertainties through implementation of new beam fits in the treatment planning system.

Esposito *et al* [73] performed an eight-centre study comparing clinical results (i.e. out-of-tolerance rates, errors identified) of three different commercial *in vivo* dosimetry systems (Softdiso, PerFraction, and Dosimetry Check), involving 32,276 fx and 2002 patients. They showed that the proportions of treatments flagged as out-of-tolerance, and the identified types of errors, were generally consistent between the three software systems.

Wolfs *et al* [74] used an inhouse, forward projecting EPID algorithm to generate integrated VMAT delivery images with 47 lung cancer patient data sets (59 tx plans and 175 beams). Anatomical, positional, and mechanical errors were introduced virtually via modifications of the patient CT data set and also the treatment plan. A convolutional neural network was trained to identify these errors and showed promising results (96%, 87%, and 73% classification accuracy for increasingly complex errors). While their inhouse *in vivo* EPID dosimetry method was not described in full detail, Li *et al* [75] performed a sensitivity assessment on it by introducing several errors (machine output, PTV deformation, MLC, and patient setup) to treatment delivery of one IMRT and one VMAT plan and evaluated on two simple geometric phantoms and one thoracic phantom.

5. Summary
A literature review focusing on recent EPID-based *in vivo* dosimetry is presented. 'Recent' is defined here as the approximate five-year period from January 2017 to February 2022, considering the last major comprehensive EPID dosimetry review [8] included publications to the end of 2016. The published literature was categorized into three main areas including: (1) new or improved algorithmic developments including validation work, (2) new applications of the IVD method, and (3) error identification and error sensitivity analyses. The number of publications annually on this topic indicates it maintains a strong interest from both the research and clinical communities in radiation oncology.

6. References
[1] Olaciregui-Ruiz Í *et al* 2020 *Phys. Imaging Radiat. Oncol.* **15** 8-16
[2] Bojechko C *et al* 2015 *Med. Phys.* **42** 5363-9
[3] World Health Organization 2008 *Radiotherapy Risk Profile Technical Manual* (Geneva: World Health Organization) p. 42.
[4] American Association of Physicists in Medicine (AAPM) Task Group No. 307 - The use of EPIDs for patient-specific IMRT and VMAT QA *in preparation*
[5] Greer PB *et al* 2003 *Med. Phys.* **30** 1618-27
[6] Winkler P *et al* 2005 *Med. Phys.* **32** 3095-105
[7] van Elmpt W *et al* 2008 *Radiother. Oncol.* **88** 289-309
[8] McCurdy BMC 2018 In *Modern Radiation Therapy* ed B Mijnheer (Boca Raton, FL: CRC Press, Taylor and Francis Group LLC)
[9] McCurdy BMC *et al* 2017 *Phys. Med.* **44** 123-130
[10] Esposito M *et al* 2020 *Radiother. Oncol.* **149** 158-167

[11] MacDougall ND *et al* 2017 *Br. J. Radiol.* **90** 20160915
[12] Fonseca GP *et al* 2020 *Phys. Imaging Radiat. Oncol.* **16** 1-11
[13] Peca S *et al* 2017 *Imaging Technol. Cancer Res. Treat.* **16** 944-955
[14] Hashemi SM *et al* 2019 *J. Biomed. Phys. Eng.* **9** 551-8
[15] Nasseri S *et al* 2022 *Appl. Radiat. Isot.* **182** 110116
[16] Anvari A *et al* 2019 *J. Cancer Res. Ther.* **15** 1011-7
[17] Berry SL *et al* 2012 *Med. Phys.* **39** 87-98
[18] Najem MA *et al* 2018 *Phys. Med.* **52** 143-53
[19] Feng B *et al* 2021 *Front. Oncol.* **11** 782263
[20] Jia X *et al* 2011 *Phys. Med. Biol.* **56** 7017-31
[21] Deshpande S *et al* 2018 *Med. Phys.* **45** 1266-75
[22] Martínez Ortega J *et al* 2018 *Med. Phys.* **44** 333-41
[23] Antoniou G *et al* 2021 *Phys. Eng. Sci. Med.* **44** 1121-30
[24] Guo K *et al* BMC 2020 *Phys. Med. Biol.* **65** 185008
[25] Guo K *et al* BMC 2021 *J. Appl. Clin. Med. Phys.* **22** 99-114
[26] Torres-Valderrama A *et al* 2021 *Phys. Med. Biol.* **66** 11NT01
[27] Torres-Xirau I *et al* 2017 *Phys. Med Biol.* **62** 6322-40
[28] Torres-Xirau I *et al* 2019 *Med. Phys.* **46** 4193-203
[29] Torres-Xirau I *et al* 2020 *Radiother. Oncol.* **146** 161-6
[30] Olaciregui-Ruiza I *et al* 2020 *Phys. Med.* **71** 124-31
[31] Stevens S *et al* 2018 *Phys. Med.* **45** 25-34
[32] Esposito M *et al* 2018 *Australas. Phys. Eng. Sci. Med.* **41** 1021-7
[33] Held M *et al* 2018 *Cureus* **10** e2139
[34] Aziz Sait A *et al* 2019 *J. Med. Phys.* **44** 16-20
[35] Olaciregui-Ruiz Í *et al* 2019 *J. Appl. Clin. Med. Phys.* **20** 79-90
[36] Aftabi S *et al* 2021 *Med. Dosim.* **46** 29-38
[37] Kutuzov I *et al* 2021 *J. Appl. Clin. Med. Phys.* **22** 110-120
[38] Cai B *et al* 2018 *J. Appl. Clin. Med. Phys.* **19** 73-85
[39] Ahmad M *et al* 2021 *Med. Phys.* **48** 4598-609
[40] Abbasian P *et al* 2020 *Med. Phys.* **47** 5301-11
[41] Renaud J *et al* 2022 *Med. Phys.* **49** 1238-47
[42] Renaud J *et al* 2022 *Med. Phys.* **49** 1231-7
[43] Barnes MP *et al* 2017 *J. Appl. Clin. Med. Phys.* **18** 139-50
[44] Fuangrod T *et al* 2017 *Int. J. Health Care Qual. Assur.* **30** 90-102
[45] Consorti R *et al* 2017 *Phys. Med.* **42** 157-61
[46] Piermattei A *et al* 2018 *Med. Biol. Eng. Comput.* **56** 1939-47
[47] Falco MD 2018 *Technol. Cancer Res. Treat.* **17** 1533033818779201
[48] Nailon WH *et al* 2019 *J. Appl. Clin. Med. Phys.* **20** 6-16
[49] Esposito M *et al* 2021 *Radiother. Oncol.* **154** 14-20
[50] Olaciregui-Ruiz Í *et al* 2021 *Radiother. Oncol.* **157** 241-6
[51] Bossuyt E *et al* 2020 *Phys. Imaging Radiat. Oncol.* **16** 113-29
[52] McCowan PM *et al* 2017 *Int. J. Radiat. Oncol. Biol. Phys.* **97** 1077-84
[53] Peca S *et al* 2017 *Technol. Cancer Res. Treat.* **16** 956-63
[54] Sterckx B *et al* 2019 *Phys. Imaging Radiat. Oncol.* **12** 7-9
[55] Wendling M *et al* 2012 *Med. Phys.* **39** 367-77
[56] Anvari A *et al* 2020 *Med. Phys.* **47** 1871-9
[57] Cilla S *et al* 2019 *J. Appl. Clin. Med. Phys.* **20** 37-44
[58] Kang S *et al* 2019 *J. Appl. Clin. Med. Phys.* **20** 43-52
[59] Dumas J-L 2021 *Phys. Med.* **85** 42-9
[60] Fiagan YAC *et al* 2020 *Tech. Innov. Patient Support Radiat. Oncol.* **16** 65-9
[61] Tan HQ *et al* 2022 *J. Appl. Clin. Med. Phys.* e13560

[62] Smith RL *et al* 2018 *Brachytherapy* **17** 111e121
[63] Fonseca GP *et al* 2021 *Phys. Med. Biol.* **66** 104001
[64] Bedford JL *et al* 2018 *Phys. Med. Biol.* **63** 025008
[65] Bedford JL *et al* 2019 *Phys. Med. Biol.* **64** 205009
[66] Bedford JL *et al* 2021 *Br. J. Radiol.* 94 20201014
[67] van der Bijl E 2017 *Radiother. Oncol.* **125** 405-10
[68] Mijnheer B *et al* 2018 *Phys. Med.* **54** 137-45
[69] Olaciregui-Ruiz Í *et al* 2019 *Med. Phys.* **46** 45-55
[70] Olaciregui-Ruiz Í *et al* 2020 *Med. Phys.* **47** 171-180
[71] Yedekci Y *et al* 2019 *Phys. Med.* 2019 **64** 204-09
[72] Mans A *et al* 2022 *Phys. Imaging Radiat. Oncol.* **21** 6-10
[73] Esposito M *et al* 2021 *Strahlenther Onkol.* **197** 633-43
[74] Wolfs CJA *et al* 2020 *Radiother. Oncol.* **153** 243-9
[75] Li Y *et al* 2019 *PLoS One* **14** e0218803

Journal of Physics: Conference Series **2630** (2023) 012010 doi:10.1088/1742-6596/2630/1/012010

FLASH radiotherapy and the associated dosimetric challenges

S Ceberg[1], A Mannerberg[1], E Konradsson[1], M Blomstedt[1], M Kügele[2], M Kadhim[1,2], A Edvardsson[2], S Å J Bäck[1,2], K Petersson[2,3], C Jamtheim Gustafsson[2] and C Ceberg[1]

[1]Department of Medical Radiation Physics, Lund University, Lund, Sweden

[2]Radiation Physics, Department of Hematology, Oncology and Radiation Physics, Skåne University Hospital, Lund, Sweden

[3]Oxford Institute for Radiation Oncology, Department of Oncology, University of Oxford, Oxford, UK

E-mail: sofie.ceberg@med.lu.se

Abstract. At Lund University and Skåne University Hospital in Lund, Sweden, we have, as the first clinic, modified a clinical Elekta Precise linear accelerator for convertible delivery of ultra-high dose rate (FLASH) irradiation. Whereas recently published reviews highlighted the need for standardised protocols for ultra-high dose rate beam dosimetry to be able to determine the true potential of FLASH irradiation, several dosimetry studies as well as *in-vitro* and *in-vivo* experiments have been carried out at our unit. Dosimetric procedures for verification of accurate dose delivery of FLASH irradiation to cell cultures, zebrafish embryos and small animals have been established using radiochromic films and thermo-luminescent dosimeters. Also, recently the first experience of electron FLASH radiotherapy (FLASH-RT) in canine patients in our clinical setting was published. Our research facilities also include a laboratory for 3D polymer gel manufacturing. Recently, we started investigating the feasibility of a NIPAM polymer gel dosimeter for ultra-high dose rate dosimetry. Furthermore, in the bunker of the modified Elekta linear accelerator, a Surface Guided Radiotherapy (SGRT) system is accessible. The Catalyst^{TM} system (C-Rad Positioning, Uppsala, Sweden) provides optical surface imaging for patient setup, real-time motion monitoring and breathing adapted treatment. Aiming at treating patients using ultra-high dose rates, a real-time validation of the alignment between the beam and the target is crucial as the dose is delivered in a fraction of a second. Our research group has during the last decade investigated and developed SGRT workflows which improved patient setup and breathing adapted treatment for several cancer patient groups. Recently, we also started investigating the feasibility of a real-time motion monitoring system for surface guided FLASH-RT. Both FLASH related studies; 3D polymer gel dosimetry and surface guided FLASH-RT are to our knowledge the first of their kind. Following an introduction to the field of FLASH and the associated dosimetric challenges, we here aim to present the two ongoing studies including some preliminary results.

1. Introduction

The paradigm of radiotherapy is to irradiate the tumour or treatment volume with high absorbed dose and simultaneously avoid healthy surrounding tissues as much as possible. Due to the last decades of technical development within the radiotherapy workflow, we can aim at the tumour more accurately and thus decrease the absorbed dose to the surrounding healthy tissues. Examples of these technical developments are multi-diagnostic imaging for tumour and organs at risk definition, intensity modulated beams and arcs, real-time image guidance to continually follow the target during beam-on, and deformable motion models that enable daily adaptive treatment. These technical advances have not only decreased the acute and long-term radiation-induced side effects but also allowed treatments that otherwise would result in intolerable normal tissue complications for the patients. State-of-the-art treatments such as gated stereotactic body radiation therapy (SBRT) liver treatment in an MR-linac workflow [1] and ventricular tachycardia radiotherapy, delivering single 25 Gy fraction during cardiobreathing motion [2], are notable results from the latest generation of technical and methodical developments. However, even with a perfect treatment technique using the current clinical accelerators, the tumour will still be surrounded by healthy tissue. To further improve radiotherapy, we need to increase the therapeutic window by considering other treatment concepts, i.e., to decrease the probability of normal tissue complication and increase/maintain the probability of tumour control.

The novel ultra-high dose rate FLASH irradiation is such a suggestion. The term FLASH was put on the map by Favaudon et al. 2014, presenting experiments carried out using a high-dose-per-pulse 4.5 MeV electron beam with an average dose rate larger than 40 Gy/s, which is at least hundred times higher than conventionally used in radiotherapy [3]. Pre-clinical FLASH studies have shown promising results, indicating a possibility to increase the therapeutic window [3]. The evidence for this so-called FLASH effect has started to accumulate and according to recent reviews, over 25 studies showed data on less normal tissue toxicity than conventional radiotherapy, of which over 10 studies also investigated and showed data on maintained similar tumour effect [4-6]. The in-vivo studies involved irradiation of zebrafishes, small animals, and mini pigs, but also veterinary canine and feline patients. The overall result showed that short beam-on time (<200 ms), and large single doses (≥ 10 Gy) can protect normal tissue by 20-40 % compared to conventional radiotherapy [3,7-15], without compromising the intended treatment effect [3,9,11,13]. The results showed less DNA damage and cell death in the normal tissue as well as less chronic inflammation and vascular damages compared to conventional RT. Furthermore, the studies show preserved function in the skin but also in cognitive functions. And in the meantime, according to another recent published review, a maintained similar tumour control (tumour growth) was demonstrated [16].

These results might enable possibilities for treatments of patients with more radioresistant tumours that are not cured today, or not even considered for radiotherapy, due to the risk of intolerable normal tissue complications.

2. Challenges in ultra-high dose rate beam dosimetry

Today the fundamental underlying physiochemical and radiobiological mechanisms for the FLASH effect are not yet fully understood and are currently under investigation [17]. Several laboratories have established the biological response and showed robustness of their experiments, however differences in the magnitude of the FLASH effect between laboratories have been reported [17]. In the two recently published special issue papers; Schüler et al. 2022 and Romano et al. 2022, the authors proclaim that we need to establish standardised protocols for ultra-high dose rate (UHDR) beam dosimetry to be able to determine the true potential of FLASH irradiation and thus finally move towards clinical translation [17,18]. The suggested required parameters are, among others, beam energy and time structure, total dose, mean dose rate, intra-pulse dose rate, pulse repetition frequency, dose-per-pulse, and pulse width [17]. Also, established methods for absorbed dose measurement and real-time output monitoring of UHDR beams need to be properly addressed [18]. An initiative to develop and improve dosimetry standards for FLASH radiotherapy was recently carried out by the European Joint Research Project UHDpulse [19]. However, currently, no recommendation for UHDR dosimetry is available, as TRS 398 and TG51 are for dosimetry of conventional radiotherapy.

Conventionally used measurement methods and detector systems have shown to be heavily affected by the dose rate and dose-per-pulse characteristic for UHDR irradiation [18]. For instance, ionization chambers, considered the gold standard detector for conventional radiotherapy dosimetry, suffer from significant ion recombination effects at FLASH irradiation dose rates, resulting in a dose rate dependence [20-22]. The Si-based semi-conductor detectors already suffer from dose rate dependence using conventional dose rates, and also beam angular dependence, long term irradiation damage and non-tissue equivalence [20]. Radiochromic film has on the other hand tissue equivalent dosimetric properties and high spatial resolution, however, the detector shows an LET dependence and cannot provide real-time measurements [18]. Nevertheless, these conventionally used detector systems available in a radiotherapy clinic have been shown to also be feasible for UHDR dosimetry, however, with careful consideration regarding corrections and calibrations. There are however numerous promising novel studies presenting other detector candidates for dosimetry of UHDR beams [20].

2.1 Charge-based dosimeters
The function of most charge-based detectors is based on the principle that the irradiation creates ion pairs or charges, which are collected and then correlated to absorbed dose. Ionisation chamber, Faraday Cup, and the solid-state detectors such as diamonds and diodes are examples on charge-based detectors. The advantages, having UHDR beam dosimetry in mind, are the detectors' high time resolution, in the range of ns-ms, which makes them suitable for real-time dosimetry, however, they are limited to point-dose measurements. Typical unfavourable characteristics for charge-based detectors are their dose rate- and energy dependence and non-tissue equivalence. Several promising modifications to meet the demands of the UHDR beam dosimetry have however recently been investigated for ion chambers and diamond detectors [23-26]. For instance, ion recombination correction model has been suggested by Cavallone et al. 2022 for online dosimetry using a Nano Chamber, and Gómez et al. 2022 demonstrated the ability to extend the dose rate operating range of an ultra-thin parallel plate ionization chamber by reducing the spacing between electrodes [23,24]. Furthermore, novel approaches for the feasibility of a diamond detector for UHDR beam dosimetry have also been carried out recently. Kranzer et al. 2022 showed that the linear range can be extended towards a higher dose-per-pulse range by reducing the active volume and thus the sensitivity [25]. Marinelli et al. 2022 showed the same but by fine-tuning other physical properties of the diamond detector such as the resistance [26].

2.2 Luminescent dosimeters
Any system which utilises optical photon generation upon irradiation and where the measured light correlates to the absorbed dose, belongs to the group of luminescent dosimeters. Examples of luminescent dosimeters are thermo-luminescent detectors (TLD), optically stimulated luminescent detectors (OSLD), and the Cherenkov detector [18,20]. Some crystals absorb and trap energy due to added impurities in the crystal lattice when exposed to irradiation. When heated (TLD) or optically stimulated (OSLD), the crystal releases the trapped energy in the form of visible light, which correlates to the accumulated absorbed dose. These luminate detectors have been used for UHDR beam dosimetry for a long time, due to their excellent dose rate independence and tissue equivalence. Additionally, TLD can be manufactured to be small, or in powder form, which allows for high spatially-resolved measurements. Although the detectors are passive, they have been used in several dosimetric validation studies for FLASH RT [27,28]. The Cherenkov dosimeter is an active luminescent detector. When charged particles travel at a phase velocity that exceeds the phase velocity of light in the irradiated medium, Cherenkov irradiation is emitted, where the intensity of light is proportional to absorbed dose [29]. Investigated Cherenkov based detector systems show characteristics such as high time resolution (prompt, ps), linear to absorbed dose, and dose rate independent [20], which suggest the detector to be an excellent candidate for clinical 2D real-time UHDR beam dosimetry [30]. However, Cherenkov based detector systems are expected to be energy dependent and determining the relation between light yield and absorbed dose can be complicated under experimental conditions. Intensive research is ongoing [20,31].

2.3 Chemical dosimeters
Any system that produces radicals, change colour, or undergo structural changes when irradiated, and where the measured transformation correlates to the absorbed dose, belongs to the group of chemical dosimeters. Typically, the chemical dosimeter forms stable radiation-induced changes which can be

read-out using optical laser, magnetic resonance imaging, or electron paramagnetic resonance spectroscopy [18]. Two frequently used chemical dosimeters in published pre-clinical FLASH studies are radiochromic films and alanine. Also, these two detectors were used for dose validation for the first human treated with FLASH-RT [32]. Alanine is an amino acid which forms a stable free radical upon irradiation. Electron paramagnetic resonance spectroscopy is then used to measure the amount of stable free radicals, which is proportional to the absorbed dose [33]. Alanine dosimeters exhibit linear dose response over a large range (2-150,000 Gy) and shows dose rate independence up to the order of magnitude 10^10 Gy/s [20]. These characteristics make the alanine dosimeter very useful in UHDR beam dosimetry [33], although the detector is a passive system, and the resolution is limited to the size of the pellets [20]. Radiochromic film is close to the passive ideal detector system with very high resolution in 2D, both energy- and dose rate-independent, as well as tissue equivalence. The disadvantage is that the read-out is recommended to be carried out 16-24 h post irradiation due to continued post-exposure polymerisation [18,20,28,34].

True 3D chemical dosimeters, such as polymer gel detectors, might be suitable detectors for UHDR beam dosimetry due to the high spatial resolution, energy independence, and tissue equivalence. The drawback is a known dose rate dependence [35,36]. The polymer gel dosimeter has been considered as a suitable candidate [20], however, a polymer gel FLASH study has not yet been reported.

Right dose in right place

Suppose we now have established standardised protocols for UHDR beam dosimetry and have access to active detector systems with ideal characteristics such as dose rate- and energy independence, tissue-equivalence, and high spatial- and time resolution. Then, another dosimetric challenge emerges, which is to validate that the absorbed dose is delivered to the correct position.

Motion management in radiotherapy is crucial for an accurate dose delivery [37]. The intra-fractional motion of tumour and organs at risks are the most addressed challenges during conventional motion management radiotherapy. With FLASH-RT the intra-fractional motion is nearly eliminated as the dose is delivered in a fraction of a second. However, there is a new demand on motion management, since during this short delivery time a real-time validation of the alignment between the beam and the target is crucial. Image guidance is standard in radiotherapy to verify the patient setup and adapt the treatment plan in the event of any anatomical changes. For FLASH-RT, the ultrafast delivery demands state-of-the art image guidance. For superior soft tissue contrast, MR-images would be ideal, however this is not yet available in combination with FLASH-RT. El Naqa et al. 2022 argues that future multiparametric on-board MRI protocols also could include functional information that could be explored in combination with FLASH-RT [31]. As for today, the widely accessible on-board imaging system is cone beam computed tomography (CBCT), which provides 3D volumetric kilovoltage imaging of the patient and has shown to reduce localization errors in radiotherapy by approximately 50% [38]. Recent developments in radiotherapy combine volumetric imaging with surface guided radiotherapy (SGRT) [39].

Real-time motion monitoring

Motion management systems using optical surface imaging, so called SGRT, are used clinically for patient positioning and real time motion monitoring and have been suggested to be a potential candidate for motion management system at FLASH-RT [31,40].

SGRT provides the largest available field-of-view (FOV) in radiotherapy [41] with approximately 1 m^3 scanning volume [42]. The capability to capture surface images of the patient in real-time makes SGRT systems excellent for motion management. The Catalyst^{TM} system is an optical surface scanning system that consists of a high-power LED projector, which projects a near-visible violet light (λ= 405 nm) for surface reconstruction purposes and a green (λ= 528 nm) and red (λ= 624 nm) projection light for live feedback of the patient posture position. The Catalyst^{TM} is a structured light system, which means that the near-visible violet light is projected as sequenced lines onto the object to be scanned. The irregularity of the object scanned disperses the sequenced lines, which are detected by a CCD camera.

Due to fixed geometry between the projector and the CCD, the principle of optical triangulation can be used to reconstruct a 3D surface of the scanned object. During treatment delivery the surface imaging system monitors the patient motion. The motion of the patient surface is recorded and tolerances for the detected motion are set prior to treatment. The optical surface scanning system is interfacing with the delivery control system to interrupt the beam delivery if the patient motion exceeds any pre-defined tolerance. According to the American Association of Physics in Medicine (AAPM) Task Group Report 142, the latency of beam-on/-off should be within 100 ms for a linac with synchronization of the radiation beam [43]. However, in the case of motion monitoring during FLASH-RT the latency must be much quicker, within a few ms, to be able to turn the beam off during a sudden motion.

3. Two ongoing studies at our FLASH linac

3.1 UHDR beam dosimetry using a polymer gel detector

Three-dimensional (3D) gel dosimetry has played an important role when introducing new treatments techniques into the radiotherapy clinic, such as IMRT/VMAT and breathing adapted conformal treatments [44,45,46]. All polymer gel dosimetry systems are characterized by polymerization, induced locally by free radical products of water radiolysis. The polymerization alters due to various physical and chemical properties of the polymer gel system. The 3D absorbed dose map can, for example, be read-out using MRI, optical- or x-ray computed tomography (CT).

The 3D polymer gel dosimeter is a linac-independent stand-alone detector system, tissue-equivalent, has high resolution in 3D, can be mould into various forms, and shows no energy- or beam-angle dependence [45]. However, a LET-dependence and a dose rate dependence have been identified [35,36]. Another disadvantage is that the read-out is recommended to be carried out about 24 h post irradiation due to continued post exposure polymerisation, and that each batch needs to be calibrated. We plan to investigate the possibilities of using electron-range modulating bolus in FLASH-RT and our ambition is to use 3D dosimetry for verification of the dose distribution. The initial aim was to investigate the feasibility of the NIPAM polymer gel detector for UHDR beam dosimetry.

NIPAM gel dosimeters were manufactured in a fume cupboard under normal levels of oxygen using 3 % w/w NIPAM (97 % Sigma-Aldrich), 3 % w/w BIS (99 % Sigma-Aldrich) [36]. Gelatine was used as the matrix substance and tetrakis(hydroxymethyl)-phosphonium chloride as an oxygen scavenger. The remaining constituent was water. The gel mixture was poured into vials of 15 ml (Ø 1.5 cm, length 6 cm) and stored in the dark at room temperature for about 24 h before irradiation. The vials from the same batch were irradiated to the same total doses at the convertible clinical Elekta Precise linac with a 10 MeV electron beam (Figure 1), either using conventional dose rate of 0.15 Gy/s or using UHDR (2 Gy/pulse, pulse dose rate = $0.6 \cdot 10^6$ Gy/s, average dose rates $\geq$75 Gy/s, total treatment times $\leq$0.22 s) [47]. All vials were irradiated at a 65 cm source to surface distance (SSD). The experiment was repeated once [48]. Radiochromic film (GafChromic EBT-XD) and a flatbed scanner for read-out (Epson Expression 12000XL) was used for validation of the dose to the gel dosimeters. Small pieces of film were placed into separate water-filled vials. The vials were irradiated with the 10 MeV electron beam, delivering 3, 7, 11 or 17 pulses at ultra-high dose rate or with 291, 679, 1068 or 1650 MU at conventional dose rate.

After irradiation the vials were stored in the dark at room temperature for about 24 h before MRI read-out. The vials were scanned at a General Electric (GE) Architect 3T camera using FSE pulse sequence with a TR 4000 ms and in plane pixel size 0.4 x 0.4 mm^2 with a slice thickness of 10 mm [39]. The data processing was carried out using MICE Toolkit 2021.2.1 (NonPi, Sweden). The average pixel intensity as a function of TE_{eff} was plotted for all vials. A mono-exponential curve fit was made to obtain the R2 values.

Journal of Physics: Conference Series **2630** (2023) 012010 doi:10.1088/1742-6596/2630/1/012010

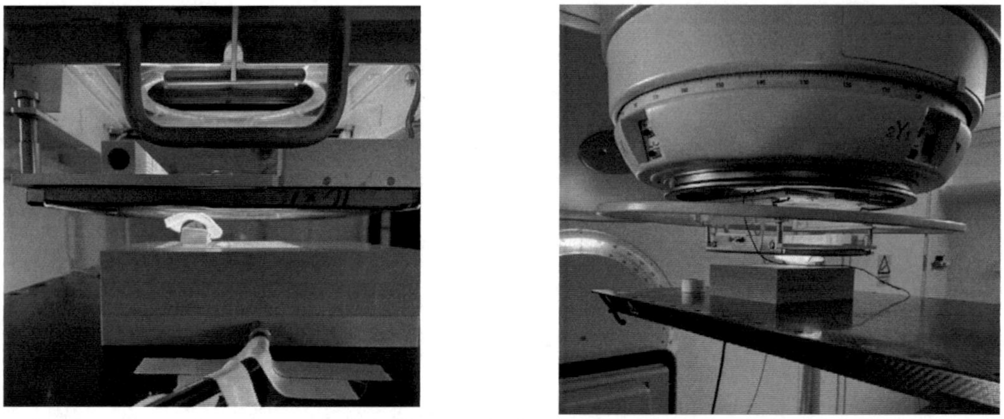

Figure 1. The setup of a polymer gel dosimeter during the electron beam at the convertible clinical Elekta Precise linac. The same setup was used for FLASH as for conventional irradiation.

The measurements carried out using Radiochromic film showed a linear relationship between absorbed dose and number of pulses for the ultra-high dose rate irradiation, and number of MU for conventional dose rate irradiation, respectively (Figure 2). The result enabled us to present the NIPAM gel MR readout signal R2 as a function of absorbed dose (Figure 3).

Gel measurements showed lower response for the ultra-high dose rate irradiation than conventional dose rate irradiation, regardless dose interval investigated, which reveals the detector dose rate dependence. However, the dosimeters did not seem to saturate even at the highest doses investigated (>45 Gy) (Figure 3), which was promising. These data follow a quadratic polynomial trend with R2>0.99 for all measurements presented (Figure 3). For absorbed doses up to 25 Gy a linear fit gives R2>0.98, which makes it possible to use a NIPAM gel dosimeter to measure relative absorbed dose distributions in that dose range, by using only background subtraction.

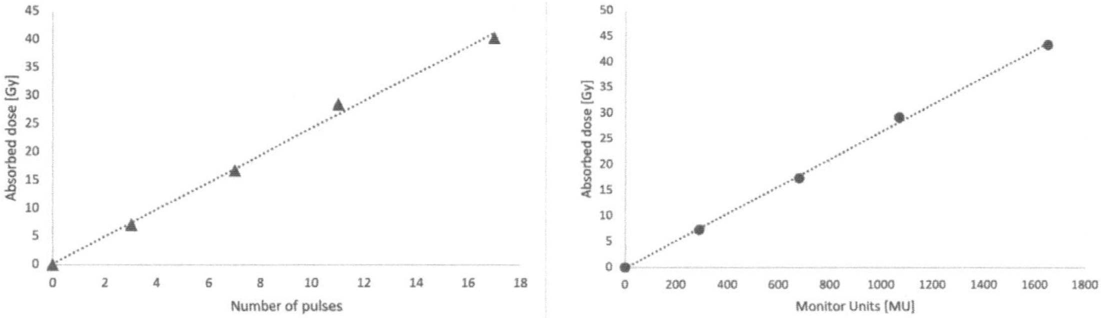

Figure 2. Radiochromic film dosimetry results for 10 MeV electron beam. Measured absorbed dose show a linear relationship to the number of pulses for ultra-high dose rate irradiation, R2>0.99 (left panel) and to the number of MU for conventional dose rate irradiation, R2>0.99 (right panel).

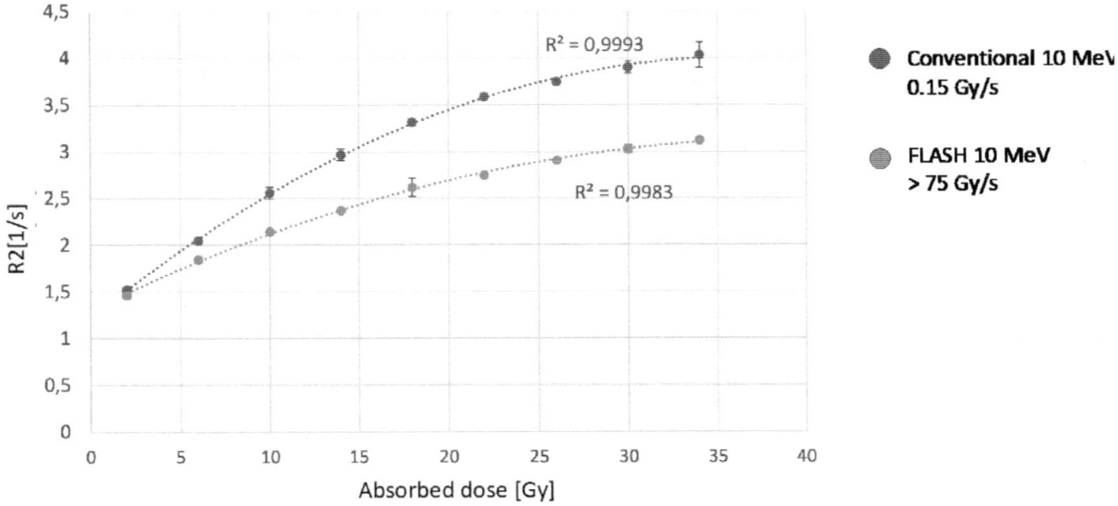

Figure 3. NIPAM polymer gel MR-redout signal R2 as a function of absorbed dose for conventional dose rate (blue) and ultra-high dose rate (orange) 10 MeV electron beam, respectively.

Our preliminary results showed a dose rate dependence, which speaks against the NIPAM gel dosimeter to be an ideal candidate for UHDR beam dosimetry. Nevertheless, relative dosimetry might be possible with a suitable calibration method. The investigation of using NIPAM polymer gel for 3D dose verification of electron-range modulated FLASH-RT is ongoing.

3.2 Surface guided FLASH radiotherapy

The fast FLASH-RT delivery almost eliminates any intra-fractional motion, nevertheless, it is crucial to have proper motion management to ensure a precise and accurate delivery of the potentially very high fraction dose. To enable a successful clinical translation of FLASH-RT, we need the ability to monitor the patient and target position in real-time and verify that the radiation is delivered to the target. Surface guided radiotherapy (SGRT) is a motion management alternative that provides real-time motion monitoring, sub-millimeter accuracy and a large off-isocentric field of view (FoV) [41,42]. Our research group has developed SGRT workflows and improved patient setup and breathing adapted treatment for several cancer patient groups [49-53], and we aim to use SGRT as part of our facility's canine patient research program in electron beam FLASH-RT [15]. As a critical component to the research, the surface imaging system will verify that the dog maintains stable in the correct position when the beam is triggered. However, since the surface imaging system is developed to detect light reflected from a human patient surface at a conventional SSD during conventional treatment times, our aim was to carry out feasibility studies to investigate the performance of the SGRT system for furry patients in a FLASH-RT setting [40].

Phantom studies were carried out using fur-like blankets of various colours placed on a platform which simulated breathing patterns (QUASAR Phantom, Modus Medical Devices). A breathing motion of 15 breaths/min and 10 mm amplitude was simulated and the capability of the surface imaging Catalyst™ system to correctly monitor the motion was investigated in SSD 100 cm, and in our off-isocentric electron FLASH setup at SSD 70 cm. Furthermore, to investigate how accurately the surface imaging system can detect a potential sudden patient related motion during FLASH-RT setup, an in-house moving phantom was utilized. The phantom uses a platform with a programmable vertical motion controlled by a solenoid, which enables the platform to move in a binary pattern. The same fur-like blankets were used and the preset vertical motion of 10 mm of the platform was set to various frequencies between 1-5 Hz, simulating sudden motion of a patient. The vertical motion was repeated up and down 10 times for each frequency investigated. Additionally, the surface imaging system was used for motion monitoring during FLASH-RT for 11 canine patients of various breeds with shallow situated tumors treated at the Elekta Precise linac. The canine patients were anesthetized prior to treatment and prescribed doses of 28-35 Gy. The total treatment times were ≤0.3 s. The treatment of the canine patients was approved by the Local Ethical and Administrative Committee at Department of Veterinary Clinical Sciences, University of Copenhagen, the Danish Experimental Animals Inspectorate (2020−15−0201−00429), the Animal Experiments Committee in Lund, Malmö (5.8.18-14316/2021), and the Swedish Board of Agriculture (reference number 5.2.18-02830/2020). An informed written consent form was also signed by all owners. The surface imaging system camera parameters for integration time and gain were individually optimized to enable as large detectable surface as possible for various fur colors and structures. The motion of the surface, well representing the shallow treatment volume, was monitored during the time from the confirmed setup at the treatment couch until after the treatment fraction was delivered. All motion data was analyzed by retrieving data from the Catalyst™ system log files.

Initial data showed that the simulated phantom breathing motion could be measured and verified by the surface imaging Catalyst™ system within 0.6 mm using all the fur-like materials at SSD 70 and 100 cm, respectively. The result from the detection of a simulated sudden motion of a patient at SSD 70 cm is presented for the various fur-like blankets in Figure 4. For the lowest frequency investigated, the surface imaging system detected the motion pattern very well regardless of fur-like material. However, with increased frequency, the surface imaging system ability to detect the motion pattern deteriorated. Nevertheless, the system managed to detect a motion, which might be enough for the system to indicate a "beam-hold" of the FLASH-RT delivery. For 10 out of the 11 canine patients receiving surface guided FLASH-RT, the total vector deviation in lateral, longitudinal and vertical directions ($\sqrt{\Delta lat^2 + \Delta long^2 + \Delta vert^2}$) was within 2 mm from planned setup position, measured from the point at which the setup position was approved in the treatment room until approximately 10 s after beam delivery (Figure 5). Patient 5 had the largest measured motion, but also the poorest surface coverage, thus the interpretation of the motion data was uncertain for this case. It was generally a challenge to obtain an adequate surface image for the canine patients due to different fur structures, lengths, and colors. For the patient cases with very dark fur a long integration time was necessary, which decreased the speed of the surface imaging system. Shorter integration time in those cases resulted in decreased reconstructed surface regions. For one of the patients motion up to 3 cm was recorded prior to beam on and the treatment was postponed until sufficient anesthesia had been administered. The FLASH-RT beam was delivered once the total vector deviation had been below the tolerance level of 10 mm for approximately one minute (Figure 6). The total vector deviation reported by the surface imaging system was 0.10 mm 1.72 s before beam-on and 1.63 s afterwards.

We have shown preliminary data demonstrating that motion management during FLASH-RT is important and feasible using the SGRT system Catalyst™. The surface imaging system showed to be feasible for motion monitoring of canine patients with various types of fur in off-isocentric electron FLASH-RT treatments. Presented phantom-studies indicated that the system also could detect sudden vertical motion, which could potentially result in a beam-hold until the patient is in the planned position. Furthermore, our surface guided FLASH-RT study for 11 canine patients showed that the motion monitoring system could be used to estimate the motion prior to treatment and to assure that the patient remains in the correct position during beam delivery. It should be noted that only shallow targets were treated with the electron FLASH beam in this study. For deep seated targets the additional complexity of internal target and organs-at-risk motion need further considerations.

4. Overall summary & concluding remarks

The technical development within radiotherapy during the last decades has allowed us to treat cancer patients successfully with decreased acute and long-term radiation-induced side effects. For further improvement, we need to increase the therapeutic window even more. A novel proposal is to use ultra-high dose rate irradiation, i.e., FLASH, since several preclinical studies recently showed less normal tissue toxicity than conventional dose rate radiotherapy and with maintained tumour effect. However, since the magnitude of the so-called FLASH effect varies between studies there is a need for further investigations. Also, standardised protocols for ultra-high dose rate (UHDR) beam dosimetry need to be established. Once the dosimetry can be completely controlled, the probability to fully understand the mechanism of the FLASH effect would increase, facilitating a clinical translation. The existing standard detector systems within the field of conventional radiotherapy meet severe challenges in FLASH-RT due to the ultra-high dose rate. Recently, published reviews and other novel studies present both suggested correction strategies and new promising detector candidates for UHDR beam dosimetry.

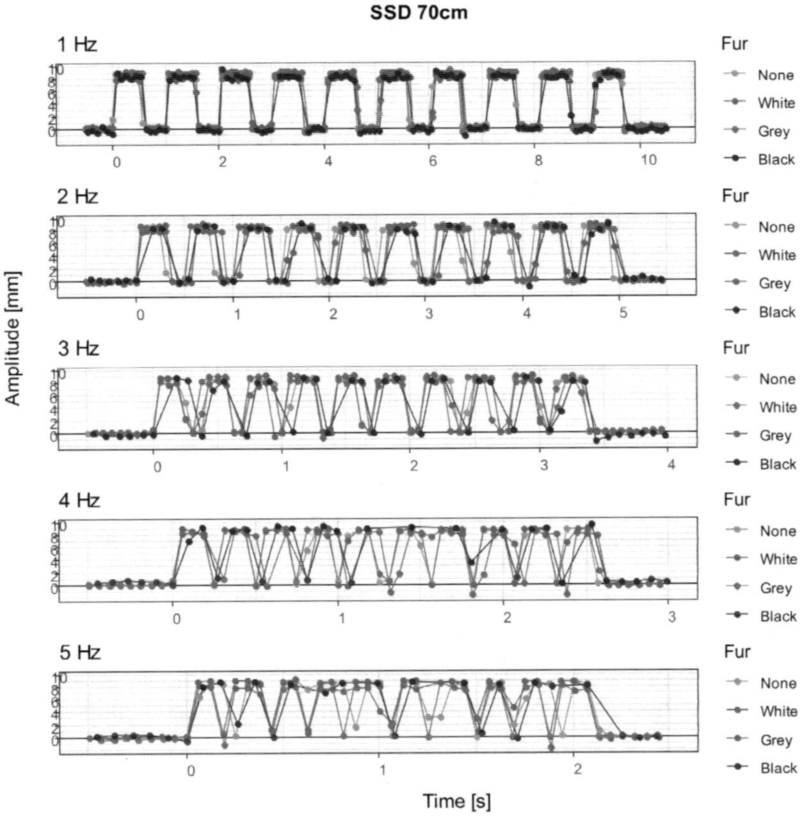

Figure 4. Surface imaging system Catalyst^{TM} motion monitoring result presented as vertical motion as a function of time for various fur-like blankets on a platform that moves vertically at various frequencies (1-5 Hz). The measurements were carried out at our electron FLASH-RT setup, i.e., at SSD 70 cm.

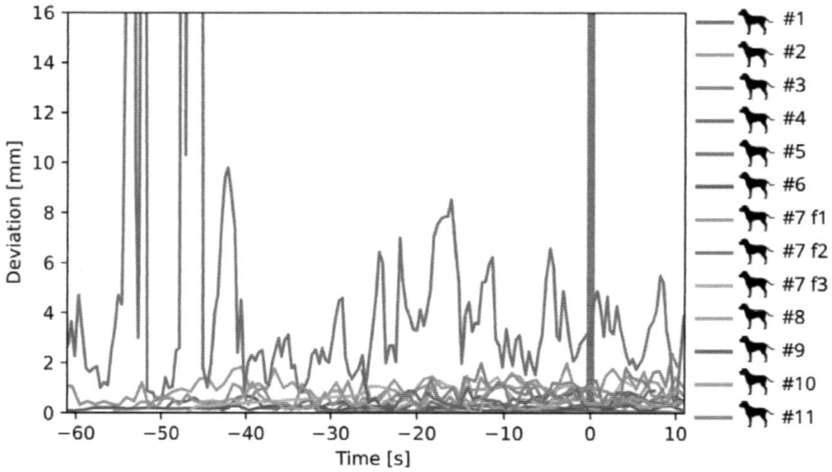

Figure 5. Motion data for 11 canine patients; the total vector of deviations in lateral, longitudinal and vertical directions as a function of time. The red line denotes beam-on normalized to time = 0 s to enable data visualization and comparison.

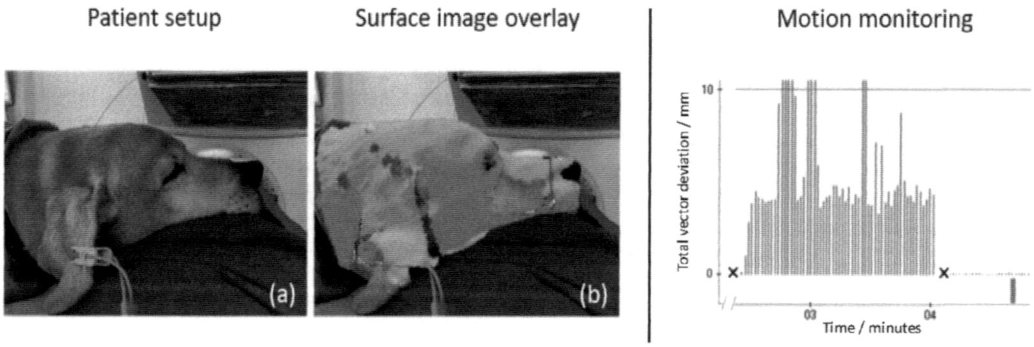

Figure 6. Left panel: The setup of a canine cancer patient (a), overlayed with the resulting surface image (b). Right panel: Real-time position (mm) relative to planned position over time (min); red bars indicate motion exceeding tolerance of 10 mm and the grey bar indicates the time for beam on [40].

It is not trivial to investigate the suitability of detector systems for UHDR beam dosimetry. Having the unique beam characteristics in mind, the dose rate dependence, as well as the spatial- and time-resolution are three important parameters of the system that need to be highlighted. All the luminescence-based dosimeters, such as the TLD, OSLD and Cherenkov-based detectors, as well as some of the chemical-based ones, such as radiochromic film and alanine, show outstanding dose rate independence. Although some of these (all but Cherenkov dosimetry) have rather long read-out times, they may still be suitable for passive FLASH measurements. Radiochromic film provides the best spatial resolution and is also tissue equivalent. Cherenkov dosimetry, with a ps time resolution and high 2D spatial resolution, was recently discussed as an ideal candidate for online FLASH-RT dosimetry. Charge-based detectors, such as diodes and diamonds, also have high time resolution, which makes them suitable for real-time dosimetry, although they are limited to point-dose measurements. Other unfavourable characteristics for charge-based detectors are their dose rate- and energy dependence and non-tissue equivalence. However, novel studies show that tuning of the physical properties of the diamonds can result in response linearity for UHDR irradiation. To conclude, the search for the ideal detector system for dosimetry in FLASH beams is still ongoing to meet the need for standardised protocols for UHDR beam dosimetry.

In the meantime, we have started initial FLASH measurements using the 3D polymer gel dosimeter NIPAM. We plan to use electron-range modulating bolus in FLASH-RT, requiring a detector system with high spatial resolution for 3D dose verification. Our first feasibility tests showed the expected dose rate dependence. However, the response can be fitted by a quadratic polynomial (linear in the typical FLASH fraction dose range up to 25 Gy) and together with a suitable calibration method, the NIPAM gel dosimeter might work for FLASH dose verification in 3D. This study is ongoing.
Regardless of the dosimetry system, once standardized procedures are established, the next challenge is to validate that the high fraction dose is delivered to the intended volume, given that the duration of the treatment is just a fraction of a second. We have utilized real-time motion monitoring using the in-room clinical optical surface imaging system Catalyst™ at our FLASH-modified Elekta linear accelerator. Our preliminary data showed that surface guided FLASH radiotherapy was feasible for canine patients with various types of fur in an off-isocentric FLASH radiotherapy setup. The surface imaging system ability to control the FLASH beam-hold is subject of an ongoing study.

Once an accurate dosimetry system for ultra-high dose rate beams is recognized, standardized dosimetry protocols are established, and a reliable motion management system for fast real-time monitoring is accessible, this would facilitate the clinical translation of accurate and safe FLASH radiotherapy.

5. References

[1] van de Lindt TN *et al* 2022 *Radiother. Oncol.* **167** 285-291
[2] Whitaker J *et al* 2022 *Trends Cardiovasc. Med.* **32** 287-296
[3] Favaudon V *et al* 2014 *Sci. Trans. Med.* **6** 245ra93
[4] Schüler E *et al* 2022 *Med. Phys.* **49** 2082-95
[5] Omyan G *et al* 2020 *J. Cancer Res. Ther.* **16** 1203-9
[6] Wilson J *et al* 2020 *Front. Oncol.* **9** 1563
[7] Vozenin MC *et al* 2019 *Clin. Oncol.* **31** 407-15
[8] Montay-Gruel P *et al* 2017 *Radiother. Oncol.* **124** 365-9
[9] Levy K *et al* 2020 *Sci. Rep.* **10** 1-35
[10] Fouillade C *et al* 2020 *Clin. Cancer Res.* **26** 1497-506
[11] Bourhis J *et al* 2019 *Radiother. Oncol.* **139** 11-7
[12] Simmons DA *et al* 2019 *Radiother. Oncol.* **139** 4-10
[13] Montay-Gruel P *et al* 2021 *Clin. Cancer. Res.* **27** 775-84
[14] Vozenin MC *et al* 2019 *Clin. Can. Res.* **25** 35-42
[15] Konradsson E *et al* 2021 *Front. Oncol.* **11** 658004
[16] Diffenderfer ES *et al* 2022 *Med. Phys.* **49** 2039-54
[17] Schüler E *et al* 2022 *Med. Phys.* **49** 2082-95
[18] Romano F *et al* 2022 *Med. Phys.* **49** 4912-32
[19] Schüller A *et al* 2020 *Physica Medica* **80** 134-50
[20] Ashraf MR *et al* 2020 *Front. Phys.* **8** 328
[21] Konradsson E *et al* 2020 *Radiat. Res.* **194** 580-586
[22] Petersson K *et al* 2017 *Med. Phys.* **44** 1157-67
[23] Cavallone M *et al* 2022 *Med. Phys.* **49** 4731-42
[24] Gómez F *et al* 2022 *Med. Phys.* **49** 4705-14
[25] Kranzer R *et al* 2022 *Phys. Med. Biol.* **67** 1361-6560/ac73d1
[26] Marinelli M *et al* 2022 *Med. Phys.* **49** 1902-10
[27] Konradsson E *et al* 2022 *J. Phys.: Conf. Ser.* **2167** 012003
[28] Jorge PG *et al* 2019 *Radiother. Oncol.* **139** 34-9
[29] Hachadorian RL *et al* 2022 *Med. Phys.* **49** 4018-25
[30] Rahman M *et al* 2021 *Phys. Med. Biol.* **66** 135009

[31] El Naqa I *et al* 2022 *Med. Phys.* **49** 4109-22
[32] Bourhis J *et al* 2019 *Radiother. Oncol.* **139** 18-22
[33] Gondré M *et al* 2020 *Radiat. Res.* **194** 573-9
[34] Jaccard M *et al* 2017 *Med. Phys.* **44** 725-35
[35] Jirasek A *et al* 2015 *Phys. Med. Biol.* **60** 4399-411
[36] Waldenberg C *et al* 2017 *J. Phys.: Conf. Ser.* **847** 012063
[37] Li H *et al* 2022 *Med. Phys.* 49 e50-81
[38] Bissonnette JP *et al* 2010 *Radiother. Oncol.* **96** 139-44
[39] Batista V *et al* 2022 *Tech. Innov. Patient Support Radiat. Oncol.* **22** 1-8
[40] Mannerberg A *et al* 2021 *Radiother. Oncol.* **161** *(suppl 1)* PO-1767
[41] Batista V *et al* 2020 *Radiother. Oncol.* **153** 34-42
[42] Al-Hallaq HA *et al* 2022 *Med. Phys.* **49** e82-112
[43] Klein EE *et al* 2009 *Med. Phys.* **36** 4197-212
[44] Karlsson A *et al* 2004 *J. Phys.: Conf. Ser.* **3** 284
[45] Ceberg S *et al* 2010 *Phys. Med. Biol.* **55** 4885-98
[46] Ceberg S *et al* 2010 *J. Phys.: Conf. Ser.* **250** 012051
[47] Lempart M *et al* 2019 *Radiother. Oncol.* **139** 40-45
[48] Hill B *et al* 2002 *Phys. Med. Biol.* **47** 4233-246
[49] Mannerberg A *et al* 2021 *Tech. Innov. Pat. Supp. Radiat. Oncol.* **19** 41-5
[50] Haraldsson A *et al* 2020 *J. Appl. Clin. Med. Phys.* **21** 139-48
[51] Kügele M *et al* 2019 *J. Appl. Clin. Med. Phys.* **20** 61-8
[52] Kügele M *et al* 2018 *J. Appl. Clin. Med. Phys.* **19** 25-38
[53] Pallotta S *et al* 2020 *Med. Phys.* **47** 6310-18

Contemporary Applications of Cherenkov Imaging in Radiation Therapy

R Zhang[1,2,3], S Decker[3], D Alexander[4], R Ashraf[5], M Rahman[6], R Hachadorian[7], G Gill[2], M Rose[2], B Hunt[2], P J Hoopes[2,3,8], L Jarvis[2,3], P Bruza[3], D Gladstone[2,3], B Pogue[2,3,7,8]

[1]Department of Radiation Medicine, New York Medical College, Vahalla, New York
[2]Department of Medicine, Geisel School of Medicine, Dartmouth College, Hanover, New Hampshire
[3]Thayer School of Engineering, Dartmouth College, Hanover, New Hampshire
[4]Department of Radiation Oncology, University of Pennsylvania, Philadelphia, Pennsylvania
[5]Department of Radiation Oncology, Stanford University, Stanford, California
[6]Department of Radiation Oncology, Southwestern University, Georgetown, Texas
[7]Department of Radiation Oncology, Harvard University, Cambriage, Massachusetts
[8]Department of Surgery, Geisel School of Medicine, Dartmouth College, Hanover, New Hampshire
[9]Department of Medical Physics, University of Wisconsin Madison, Madison, Wisconsin

E-mail: rongxiao.zhang@wmchealth.org

Abstract. Cherenkov radiation (CR) is produced from all high energy radiation sources and is part of the dose delivery process in tissue. As such, CR is a direct indicator of the dose delivery process and in recent years the ability to image and measure CR has provided a number of ways to help with radiotherapy dosimetry and delivery tracking. This review provides an overview of the fundamental physical principles of CR production and the radiation transport in tissue, along with applications of imaging CR that have seen significant development in the past few years.

1. Introduction

CR has a broad spectrum of light emission that is from the UV region down to near the infrared region and is generated when a charged particle travels through a dielectric medium at a phase speed greater than the speed of light in that medium. The requirement on the phase speed of the charged particle sets the threshold condition as $\frac{v}{c} \geq 1$, and the emission angle between the CR wavefront and moving vector of the charged particle determined as $cos\theta = \frac{c}{vn}$, where θ is the emission angle, c and v are the speed-of-light in a vacuum and the phase speed of the charged particle in the medium respectively, and n is the refractive index of the medium. In liquid water, the threshold energy of Cherenkov radiation is around 0.26 MeV and emission angle about 41 ° for electrons. The intensity and spectroscopic characteristics of CR are described by the Frank-Tamm formula $\frac{dN}{dl} = \frac{2\pi\alpha}{\lambda^2} sin^2\theta d\lambda = -\frac{\alpha}{c} sin^2\theta d\omega$, where N is the number of Cherenkov photons, l is the pathlength of the charged particle, α is the fine-structure constant, λ is the wavelength and the angular frequency $\omega = \frac{2\pi}{\lambda}$. Therefore, the spectrum of CR is proportional to $\frac{1}{\lambda^2}$ in the wavelength domain and flat in the energy domain. The polarization properties of CR are described by Jelley et al [9] based on classically electro-magnetic theory.

Theoretical analysis has demonstrated that, under transient charged particle equilibrium, the local intensity of CR is strongly proportional to the dose deposited, for both x-ray and electron beams (Fig. 1, adapted from [7]) [7, 10-12]. However, that proportionality in spatial domain can be complicated by many factors including the beam quality, spectral changes such as beam hardening and phase space properties of secondary charged particle [7], particularly the anisotropic distribution [13]. Corrections or mitigation methodologies are therefore generally warranted to account for those factors in order to use the local CR intensity distribution as surrogate for dose. In transparent medium such as water, the anisotropy of Cherenkov emission can be addressed by applying correction factors based on Monte Carlo simulations of specific experimental setups [7, 14]. Additionally, Cherenkov emission can be converted to be more isotropic by doping the solution (water) with fluorophores that absorb CR and emit fluorescence [15]. By choosing fluorophores with strong absorbance in the relative short wavelength range, large spectral separation between the absorption and emission peaks (to minimize self-absorption and emission), and high quantum yield, complex corrections due to the anisotropy of CR can be greatly mitigated to simplify measurements [15, 16]. However, note that even though the CR is largely absorbed, a mix of CR and fluorescence will be detected [16]. A recent study also showed that extracting polarized signal in combination with Monte Carlo simulated corrections can help address challenges posed by the anisotropic angular distribution of CR for generalized measurements [17]. In turbid medium, such as biological tissues or equivalent phantoms, the transport of CR depends on the absorption and scattering coefficients. Studies have shown that the sampling of CR in biological tissue is generally limited to the first several millimeters underneath the surface with the sampling sensitivity decreasing approximately exponentially as the depth increases [18-21]. The angular distribution of CR emitted from the surface is close to the Lambertian distribution with minor deviations affected by the g factor in Mie scattering [21] due to the strong scattering in turbid medium,. Unlike the spectrum of native CR, the emission spectrum is typically peaked in the red wavelength region from biological tissue surface due to the strong absorption in the short wavelength range [19-21].

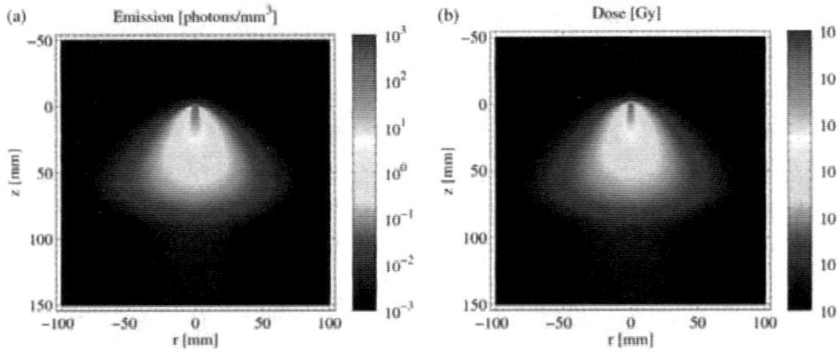

Figure 1. The proportionality between Cherenkov emission (a) and dose depositon (b) can be visually seen by comparing Monte Carlo point kernels for an 18 MeV x-ray beam in water. Adapted from [7].

In the context of radiation therapy (RT), the observation of Cherenkov radiation has been reported by multiple groups utilizing various modalities, including megavoltage electron [22], x-ray [15, 22, 23], proton/particle beams [24] and brachytherapy isotope seeds with the emission energy above the threshold energy [13, 25-27]. Due the relatively low intensity (on the order of 0.01–1 nW cm^{-2} per MBq g^{-1} for radionuclides, and 1–100 μW cm^{-2} per Gy s^{-1} for conventional beams used for external radiotherapy) [7, 28], CR was initially considered as noise in fiberoptics based scintillation dosimetry [29, 30] which needed to be removed via spectral or temporal filtering. Technological advances in sensitive optical detectors such as PMT, avalanche photodiode and intensified cameras have enabled direct measurements of CR in RT [26, 27, 31-34]. Fiber optic dosimeters, as point detectors or arrays, have shown promises for CR based dosimetry in phantom for various RT modalities [13, 25-27]. Sensitive camera systems including intensified CCD and CMOS, have enabled real-time 2D capture of CR and spectroscopy when integrated with a spectrometer [34]. To further increase the detectability and signal to noise ratio (SNR) with ambient light, gating in the time domain has been explored, particularly by synchronizing the intensifier to radiation pulses with a low duty cycle [35]. Spectral filtering and the spectral optimization of ambient light have also been investigated to boost the detection of CR [36].

With the theory correlating CR to dose being laid out and advances in detection technologies , the imaging of CR has emerged in RT over the past decade. Various applications have been explored, which can be roughly categorized as 1) dosimetry/QA in phantom, 2) chemical sensing, and 3) *in-vivo* imaging.

2. Dosimetry/QA in phantom

By using camera-based systems, CR 2D projection imaging has been demonstrated to profile electron and x-ray beams in water [14, 22]. Tomography of CR has demonstrated the reconstruction of volumetric dose distributions [37, 38] and been validated for quality assurance (QA) of conventional IMRT and VMAT plans [39, 40]. Given the versatile and non-contact nature of optical imaging, Cherenkov imaging by camera-based systems has been applied for routine machine QA [19, 20, 41], the optimization of total skin electron therapy (TSET) [42], commissioning, QA and the investigation of electron return effect on an MR-Linac [43-47].

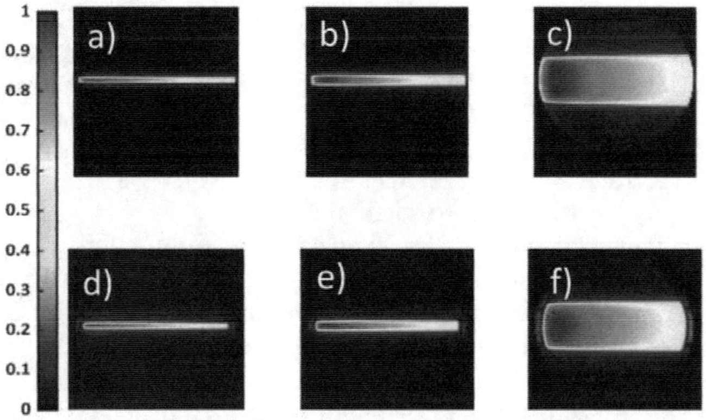

Figure 2. Projected dose distriburions from the TPS for 5, 10, and 50 mm square 6 MV photon beams are shown in (a), (b), and (c), respectively. Their corresponding Cherenkov images taken with an intensified CMOS camera are shown in (d), (e), and (f) [8].

Perhaps the greatest advantage of Cherenkov imaging is being able to achieve superb spatiotemporal resolution. In modern RT, advances in image-guidance and radiation delivery technologies allow the utilization of small beamlets to achieve optimized dose distributions with steep dose gradients, such as stereotactic radiation surgery (SRS), micro- and mini-beam RT and stereotactic body radiation therapy (SBRT) leading to challenges in dosimetry, beam characterization and QA [39, 48, 49]. Optical imaging such as Cherenkov imaging provides viable solutions to characterizing small fields with sub-millimeter spatial resolution [50] in 2D via projection imaging (Fig. 2) and in 3D via tomography [38, 51].

Recently, the temporal modulation of dose was explored to increase treatment efficacy in radiation oncology. RT with ultra-high dose rates (UHDR) termed as FLASH-RT, which uses average dose-rates upwards of 40 Gy/sec as opposed to dose rates of 0.01 Gy/s in conventional delivery, has shown improved therapeutic ratio via enhanced normal tissue sparing while retaining comparable tumor killing in translational studies. However, UHDR conditions cast unmet challenges to conventional dosimeters, particularly the independency on dose rate over a range of several orders of magnitude, high spatial and ultra-high temporal resolution [52] that are desirable for FLASH-RT. Cherenkov imaging in combination with fast electronics and feedback circuits has overcome those challenges under UHDR conditions and has been explored for dosimetry, monitoring, image-guidance, and control in FLASH-RT [53-57] in complement to traditional technologies.

3. Chemical sensing

Cherenkov spectroscopy and excited luminescence imaging have been explored in sensing chemical/physiological factors that are of interest in RT, particularly the tissue oxygenation which has a critical role to play in the efficacy and sensitivity of tissues to RT. Because of the invasiveness and complexity of conventional oxygen measurement techniques, the sensing of oxygenation *in-vivo* has remained difficult to find in routine use. Cherenkov emission spectroscopy, with spectral characteristics correlated to the tissue optical properties at various oxygenation levels, has been demonstrated to quantify and track tissue oxygen during RT non-invasively [58, 59].

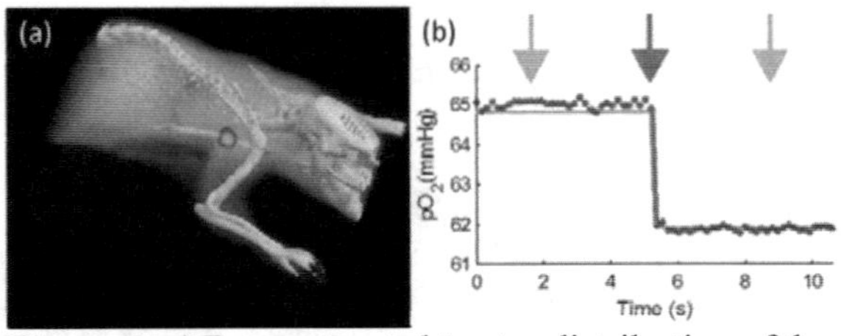

Figure 3. a) Reconstructed *in vivo* distribution of the pO_2 sensitive optical probe via CELSI [1-3]; b) The instant sensing of oxygen comsupiton during the FLASH irrdaition (red) [5, 6].

It has been shown that CR generated during RT can stimulate various optical (fluorescence or phosphorescence) probes [60, 61] that are biologically compatible for chemical sensing [62-64] and photodynamic therapy [65-67]. By introducing optical molecular probes stimulable by CR in the medium, spectroscopy and lifetime measurements of the excited luminescence which is sensitive to the local oxygen partial pressure (pO_2) has been applied for oximetry both *in-vitro* and *in vivo* [34, 68]. Diffuse optical tomography with the excitation pattern optimized by radiation beam has allowed 3D reconstruction of pO_2 to a couple of cm in depth [51, 69, 70]. This technique is suitable for preclinical studies with small animal and translational sites near the surface. The sampling depth and resolution can be further improved by Cherenkov-excited luminescence scanned imaging (CELSI) utilizing 2-dimensional sheets of radiation to produce Cherenkov photons, which then excite luminescence of probes distributed in biological tissues (Fig. 3a). By measuring total luminescence signal and then taking into account prior information about position of the scanning beam, the distribution of optical signal in the direction of scanning may be recovered and reconstructed in 3D with high spatial resolution [1-3, 71].

While current FLASH-RT studies focused on phenomenological observations have shown enhanced normal tissues sparing, termed the 'FLASH effect', the underlying mechanism is unknown. One of the initial most widespread hypotheses is that the effect is related to substantial oxygen consumption and depletion upon FLASH-RT. Like what has been demonstrated in conventional RT, most recently, Cherenkov excited luminescence lifetime imaging has been utilized for direct sensing of pO_2 dynamics during FLASH-RT (Fig. 3b), suggesting that radiologically relevant levels of hypoxia due to oxygen depletion is unlikely to occur in bulk tissue due to FLASH irradiation [6].

4. *In-vivo* imaging

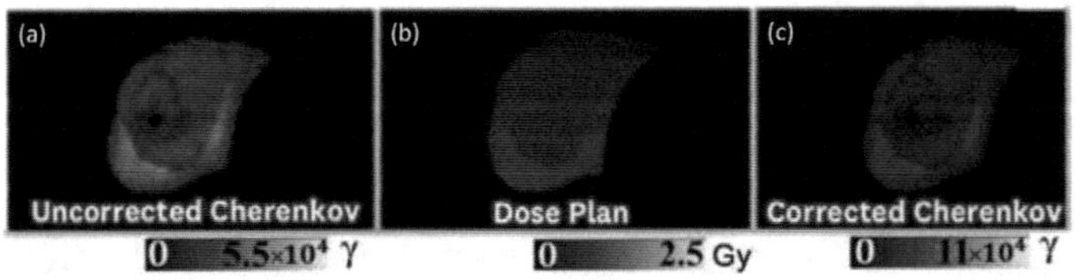

Figure 4. Patient Cherenkov images (a) can be corrected for subsurface tissue optical properties using routine CT scans. Correct images (c) are better surrogates for the dose distribution (b).

The imaging of CR in humans was first carried out in a trial for breast cancer patients receiving external beam RT [72]. By synchronizing the camera to the short radiation pulses, ambient, room light was rejected and real-time, background-subtracted CR images have been acquired at the speeds > 10 frames per second. The shapes of imaged CR revealed field segments projected on the patient surface [72, 73] while the intensity showed a good correlation with erythema observed on the skin. Soon after the proof-of-concept studies, CR imaging was explored as a method of image guidance to provide real-time assurance, motion tracking, post-treatment recording as well as treatment verification [74]. The human trial has been expanded to other sites including TSET via the Stanford technique [75-82], head and neck treatment with VMAT [83], and frame-based intracranial SRS [84], where the purposes of CR imaging have been feasibility-driven and exploratory with qualitative analysis such as motion monitoring and the validation of coverage.

What remains challenging is the quantitative correlation between CR emitted from patient surface and the deposited dose, as the relationship is sensitive to multiple factors including tissue optical properties, beam energy and geometry, imaging angle as well as treatment modality [21]. Progress has been made by correcting the *in-vivo* Cherenkov image with patient specific skin optical properties measured by Spatial

Frequency Domain Imaging (SFDI) [85]. X-ray attenuation extracted from patients' simulation CT scans has shown good correlation with the optical absorption and thus has been explored as another promising correction factor correlating CR emission to the superficial dose distribution (Fig. 4) [86-88]. In future work, by combining geometry related corrections such as the patient surface profile and imaging angle with additional optical corrections, such as that for skin pigmentation, the correlation between the Cherenkov intensity and superficial dose as well as the responses to radiation in the imaged regions can be improved. Despite the challenge of imaging superficial dose quantitatively, CR is a free signal that exists in the presence of any megavoltage RT, and comes at no additional dose or time to the patient. It has been exploited for many unique applications in the domain of quality assurance and quality improvement. In addition to the real-time motion monitoring and coverage validation, it has been shown that CR imaging can be used for the verification of field matching [89]. The concept of utilizing an always-on CR imaging system for uninterrupted RT recording to support offline review has been realized by permanently mounting cameras in the therapy bunker. Such a prototype system is now commercialized, providing software and hardware in one solution to complement existing technologies such as surface guided radiation therapy [90]. In a yearlong review of over 600 patients imaged with this system, study members identified 9 treatment anomalies across different stages of RT including simulation, planning and treatment, that would be have gone otherwise unnoticed without CR imaging [91]. While none of these incidents were deemed clinically significant by the treating physician upon review, visualizing the beam on a patient provides the unique ability to detect these types of cases. Future work is focused on developing an automated detection algorithm to catch treatment anomalies in real-time through Cherenkov image analysis [92]. The increased scale of data which can be made available through always-on CR imaging should enable several important machine learning (ML) applications. Several ML applications which have already been demonstrated in other imaging domains and have high potential in CR imaging include: deep image denoising [93, 94], image registration and motion estimation [95], anomaly detection [96, 97], and *in vivo* landmark segmentation (vasculature, for example) for automated patient alignment vertification [98]. A novel three-channel camera that was time-gated was used to image color Cherenkov emissions from patients during treatment. Color shades of Cherenkov emission in radiotherapy may be used to interpret oxygen saturation, tissue blood volume and major vessels within the body [99]. Given the nature of instant emission, CR imaging via fast electronics is suitable for ultrafast monitoring and recording of FLASH-RT delivery which typically takes less than tenth of a second. Figure 5 demonstrates imaging CR in real time of a UHDR electron FLASH delivery to a minipig's skin with optically clear bolus resolving the treated field [4, 5].

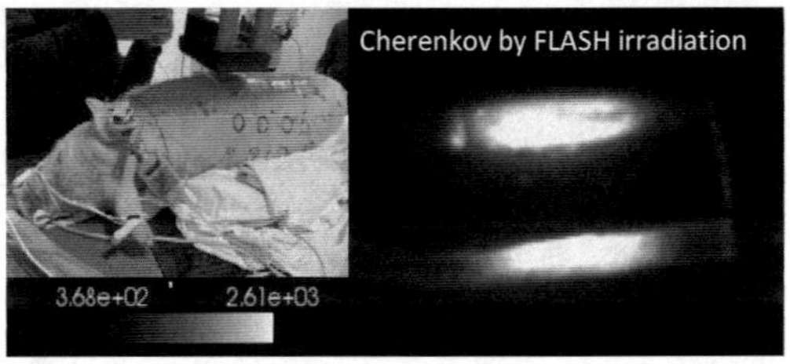

Figure 5. Cherenkov imaging during large animal FLASH-RT as quality assurance and delivery control [4, 5].

5. Conclusion

In conclusion, through the review of contemporary applications of CR imaging in RT, the technology has matured over the past decade and is now readily transferable to clinics for scaled and diversified investigations. Despite the challenge in correlating CR intensity to absolute dose, new avenues are being created for CR imaging in emerging and advanced RT regimens, including FLASH-RT.

6. Acknowledgements

This work has been partially supported by NIH grant R01 EB023909 and the Dartmouth Cancer Center.

7. References

[1] Zhang R *et al* 2015 *Opt Lett* **40** 827-30
[2] Pogue B W *et al* 2018 *Nat Biomed Eng* **2** 254-64
[3] Jia M J *et al* 2019 *Med Phys* **46** 3067-77

[4] El Naqa I *et al* 2022 *Med Phys* **49** 4109-22
[5] Rahman M R 2022 Development of a Flash Radiotherapy Platform: Characterization, Safety, and Delivery Validation. Dartmouth College)
[6] Cao X *et al* 2021 *Int J Radiat Oncol Biol Phys* **111** 240-8
[7] Glaser A K *et al* 2014 *Physics in Medicine & Biology* **59** 3789
[8] Ashraf M R *et al* 2019 *Medical physics* **46** 5227-37
[9] Jelley J V 1955 *British Journal of Applied Physics* **6** 227
[10] Zlateva Y *et al* 2019 *Medical Physics* **46** 2383-93
[11] Zlateva Y *et al* 2019 *Medical Physics* **46** 2370-82
[12] Glaser A K *et al* 2013 *Biomedical optics express* **4** 741-59
[13] Darafsheh A *et al* 2015 *Design and Quality for Biomedical Technologies VIII*: SPIE pp 135-42
[14] Glaser A K *et al* 2013 *Med Phys* **40** 012101
[15] Glaser A K *et al* 2013 *Physics in Medicine & Biology* **58** 601
[16] Jean E *et al* 2019 *Physics in Medicine & Biology* **64** 155019
[17] Cloutier E *et al* 2021 *arXiv preprint arXiv:2106.08218*
[18] Decker S *et al* 2021 *J Biomed Opt* **26**
[19] Zhang R *et al* 2013 *Phys Med Biol* **58** 5477-93
[20] Zhang R *et al* 2013 *Med Phys* **40** 101914
[21] Zhang R *et al* 2017 *J Biophotonics* **10** 645-56
[22] Axelsson J *et al* 2011 *Medical physics* **38** 4127-32
[23] Blumenthal D T *et al* 2015 *Radiotherapy and Oncology* **116** 331-3
[24] Helo Y *et al* 2014 *Physics in Medicine & Biology* **59** 7107
[25] Maraghechi B *et al* 2020 *Optics and Ionizing Radiation*: International Society for Optics and Photonics p 112240E
[26] Jang K W *et al* 2012 *Opt. Express* **20** 13907-14
[27] Jang K W *et al* 2013 *J Biomed Opt* **18** 27001
[28] Glaser A K *et al* 2015 *Physics in Medicine & Biology* **60** 6701
[29] Beddar A S *et al* 2004 *Phys Med Biol* **49** 783-90
[30] Archambault L *et al* 2006 *Med Phys* **33** 128-35
[31] Andreozzi J M *et al* 2015 *Medical physics* **42** 994-1004
[32] LaRochelle E P M *et al* 2018 *Physics in medicine and biology* **63** 085019-
[33] Alexander D A *et al* 2020 *Physics in Medicine & Biology* **65** 225013
[34] Glaser A K *et al* 2012 *Opt Lett* **37** 1193-5
[35] Ashraf M R *et al* 2019 *Med Phys* **46** 1044-8
[36] Rahman M *et al* 2021 *J Biomed Opt* **26**
[37] Glaser A K *et al* 2013 *Opt Lett* **38** 634-6
[38] Glaser A K *et al* 2015 *Medical physics* **42** 4127-36
[39] Ashraf M R *et al* 2022 *Journal of Physics: Conference Series* **2167** 012027
[40] Bruza P *et al* 2017 *IEEE Trans Med Imaging* **36** 2099-103
[41] Miao T *et al* 2019 *Medical physics* **46** 811-21
[42] Andreozzi J M *et al* 2016 *Med Phys* **43** 993-1002
[43] Andreozzi J M *et al* 2018 *Med Phys* **45** 2647-59
[44] Andreozzi J M *et al* 2020 *Med Phys* **47** 1258-67
[45] Alexander D A *et al* 2021 *Med Phys* **48** 2750-9
[46] Alexander D A *et al* 2020 *Med Phys* **47** 3861-9
[47] Zhang R *et al* 2020 *MEDICAL PHYSICS* pp E436-E
[48] Das I J *et al* 2008 *Medical physics* **35** 206-15
[49] Roussakis Y *et al Phys Med Biol* **60** N419-25
[50] Ashraf M R *et al* 2019 *Med Phys* **46** 5227-37
[51] Demers J L *et al* 2013 *Opt Lett* **38** 1364-6
[52] Ashraf M R *et al* 2020 *Frontiers in Physics* **8**

[53] Rahman M *et al* 2021 *Int J Radiat Oncol Biol Phys* **110** 872-82
[54] Ashraf M R *et al* 2021 *Med Phys* **48** 2673-81
[55] Rahman M *et al* 2021 *Phys Med Biol* **66**
[56] Ashraf M R *et al* 2022 *Phys Med Biol* **67**
[57] El Naqa I *et al* 2022 *Medical physics, 49(6), pp.4109-4122.*
[58] Axelsson J *et al* 2011 *Med Phys* **38** 4127-32
[59] Zhang X *et al* 2017 *Opt. Lett.* **42** 3101-4
[60] Dunphy I *et al* 2002 *Anal Biochem* **310** 191-8
[61] Esipova T V *et al* 2019 *Cell Metab* **29** 736-44 e7
[62] Lin H *et al* 2016 *Phys Med Biol* **61** 3955-68
[63] Pogue B W *et al* 2021 *J Biomed Opt* **26**
[64] Soter J A *et al* 2020 *Phys Med Biol* **65** 095004
[65] Darafsheh A *et al* 2015 vol 9308: SPIE
[66] Yoon S W *et al* 2018 *International Journal Of Radiation Oncology* Biology* Physics* **100** 794-801
[67] Oldham M *et al* 2016 *PloS one* **11** e0162078
[68] Zhang R *et al* 2012 *Biomed Opt Express* **3** 2381-94
[69] Zhang R *et al* 2013 *J Biomed Opt* **18** 50503
[70] Holt R W *et al* 2014 *Phys Med Biol* **59** 5317-28
[71] Jia M J *et al* 2019 *Opt Lett* **44** 1552-5
[72] Jarvis L A *et al* 2014 *Int J Radiat Oncol Biol Phys* **89** 615-22
[73] Zhang R *et al* 2013 *Journal of biomedical optics* **18** 110504
[74] Zhang R *et al* 2015 *Phys Med Biol* **60** L1-14
[75] Xie Y *et al* 2020 *Med Phys* **47** 201-12
[76] Zhong W *et al* 2021 *Proc SPIE Int Soc Opt Eng* **11628**
[77] Andreozzi J M *et al* 2018 *Med Phys* **45** 2639-46
[78] Zhu T C *et al* 2021 *Proc SPIE Int Soc Opt Eng* **11628**
[79] Zhu T C 2019 *Journal of Physics: Conference Series* **1305** 012016
[80] Pogue B W *et al* 2017 *Journal of Physics: Conference Series* **847** 012046
[81] Miao T *et al* 2020 *Journal of Medical Imaging* **7** 034002
[82] Petroccia H *et al* 2019 vol 10860: SPIE
[83] Alexander D A *et al* 2019 *Physics in Medicine & Biology* **64** 145021
[84] Tendler I I *et al* 2020 *International Journal of Radiation Oncology* Biology* Physics* **106** 422-9
[85] Hachadorian R *et al* 2019 vol 10874: SPIE
[86] Hachadorian R *et al* 2020 *Nature communications* **11** 1-9
[87] Hachadorian R *et al* 2018 *Journal of biomedical optics* **24** 071609
[88] Hachadorian R L *et al* 2022 *Medical Physics, 49(6), pp.4018-4025.*
[89] Hachadorian R *et al* 2021 *Radiotherapy and Oncology* **160** 90-6
[90] Jarvis L A *et al* 2021 *International Journal of Radiation Oncology*Biology*Physics* **109** 1627-37
[91] Alexander D A *et al* 2021 *arXiv preprint arXiv:2110.07494*
[92] Decker S M *et al* 2022 *The British Journal of Radiology, 95(1137), p.20211346.*
[93] Geng M *et al* 2022 *IEEE Trans Med Imaging* **41** 407-19
[94] Rai S *et al* 2021 *arXiv preprint arXiv:2103.06575*
[95] Fu Y *et al* 2020 *Physics in Medicine & Biology* **65** 20TR01
[96] Nayak R *et al* 2021 *Image and Vision Computing* **106** 104078
[97] Ali R *et al* 2020 *arXiv preprint arXiv:2006.09654*
[98] Mou L *et al* 2021*Medical Image Analysis* **67** 101874
[99] Alexander D A *et al* 2021 *Light Sci Appl* **10** 226

Journal of Physics: Conference Series **2630** (2023) 012012 doi:10.1088/1742-6596/2630/1/012012

The role of standards laboratories in reducing uncertainty in clinical dosimetry: A Canadian perspective

J Renaud[1], B Muir[1] and M McEwen[1]
[1]Metrology Research Centre, National Research Council Canada, Ottawa, ON, Canada

E-mail: james.renaud@nrc-cnrc.gc.ca

Abstract. As a National Metrology Institute, the National Research Council Canada (NRC) provides confidence in measurement results, and data traceable to SI units. This paper outlines some of the ways that the NRC contributes to reducing measurement uncertainty in clinical medical physics dosimetry. These activities include, (i) the improvement of existing primary standards and traceability for established beam modalities, such as MV photon and electron beams; (ii) improving measurement accuracy for new beam modalities through the development of transportable systems which permit operation at the user's facility; (iii) contributing to new dosimetry protocols, best practice reports and educational outreach; and (iv) supporting the verification of clinical implementation by offering dosimetry auditing capabilities through coordination with the clinical medical physics community.

1. Introduction

The National Research Council Canada (NRC) was created though an Act of Parliament in 1916 as a Federal applied research and technology organization. A clause in the original act states that the NRC shall act as the National Metrology Institute (NMI) for Canada, and NRC continues to fulfil that role today. In fact, Canada had already become a signatory to the Metre Convention (Convention du Mètre) in 1907 - this is the international treaty that forms the basis for international agreement on units of measurement - and the NRC is responsible for maintaining that international participation.

For some, the NRC is just "that place in Ottawa that calibrates our instruments", and is perhaps also known for the longest-running radio show on the CBC (1 pm every day!) where you can check your wristwatch or other timepiece. However, an NMI plays a much larger role in the application of measurements and measurement science, and in this paper, we will outline some of the ways that, primarily for Canada but ultimately internationally, the NRC contributes to improving accuracy in measurements in medical physics.

Before discussing improvements, it is important to note that the concept of maintenance is essential in the field of measurement standards. A calibration certificate is issued with the underlying assumption that the measurements underpinning the calibration are consistent with those made the previous day/week/year and can be reproduced in the equivalent times in the future. That requires the active maintenance of equipment and procedures that is supported by i) a recognized quality system, and ii) an international framework of comparisons, thus demonstrating competence and equivalence [1].

2. Improving measurement accuracy in medical physics

2.1. Improved primary standards with reduced uncertainty

In parallel with the essential maintenance activities noted above, a core activity of an NMI is to improve current primary standards to achieve increased accuracy and reduced uncertainty. For most standards, accuracy is limited by the measurement uncertainty and therefore reducing uncertainty is often the primary goal for established standards. An example of this at the NRC is the recent review carried out by Mainegra-Hing of the Co-60 air kerma standard [2]. In that work, initiated by the requirement to implement the recommendations of ICRU Report 90 [3], an analysis was carried out to determine each component that is required to determine air kerma using a cavity ionization chamber, resulting in an improved realization (even though no change was made to the physical standard).

2.2. Improved traceability for existing beam modalities

The development of primary standards is not always in phase with user applications, and this is often seen in radiation therapy. Ideally, one would have the necessary measurement infrastructure in place ready for any new treatment technique or beam modality, but it can be challenging to develop measurement standards without access to those radiation beams, and therefore standards development tends to lag clinical practice. Interim methodologies can be developed to provide adequate traceability but the aim should be to develop as direct a measurement standard/method as possible for each application or beam modality. User take-up and the required uncertainty for treatment delivery will impact the timescale for development of a new standard (and, of course, resourcing is always a constraint). In 1999, the AAPM TG-51 dosimetry protocol was published, detailing a procedure to realize absorbed dose to water in clinical photon and electron beams produced by linear accelerators [4]. The basis for traceability was an ionization chamber calibrated in a Co-60 beam and calculated conversion factors were required to transfer this calibration to the user beam. This basic formalism was also used by the IAEA TRS-398 Code of Practice, and therefore projects were initiated at a number of NMIs to provide direct traceability for MV photon and electron beams [5]. In 2009, the first comparison organized by the BIPM of MV photon beams was carried out at the NRC and currently around ten countries have the capability of providing direct traceability, and reduced uncertainty, for such beams [6]. The improvement in uncertainty is significant; The TG-51 route, as detailed in its Addendum, yields a standard uncertainty of around 1%, while the direct calibration route reduces that to less than 0.5% [7]. Similar initiatives have been completed, or are ongoing, by NMIs for protons, heavy ions and Ir-192.

2.3. Improved traceability for new beam modalities

As a more direct means of establishing traceability in modern beam modalities, new dose to water standards, most often calorimetry-based, are being developed as transportable systems which permit operation at the user's facility. This trend towards developing portable standards purpose-built for clinical use, potentially coordinated within the framework of a wider consortium [8], is likely to become a default approach, as it is unfeasible for NMIs to possess, or even have local access to, every specialized radiotherapy modality brought to market. In the long run, it will be more practical to perform at least some dose calibrations in the clinic using transfer standards, reducing the uncertainty in the beam qualities used by NMIs and those by the end user. Recent examples include water and graphite calorimeters adapted by NMIs for direct use in MR-guided linac [9] and proton beams [10]. In this approach, calorimeter measurements can form the basis of a direct dose calibration of an ionization chamber in the clinically-relevant field as a more accurate alternative to determining correction factors for those beams. ICRU Report 78 and 91, which provide prescribing, recording, and reporting guidelines for protons and small photon beams recommend the direct use of calorimetry in the user's beam where possible [11-12].

At the NRC, an easy-to-use, but accurate aluminum calorimeter system has been commissioned, successfully shipped (at the height of the pandemic!) and operated remotely at a BC Cancer Agency radiotherapy clinic. By leveraging a proven design [13], simplifying the data acquisition process, and

engineering the system to be lightweight and robust, it was possible to construct a truly mail-able calorimeter that does not require support staff on-site for operation.
In the same vein, a bespoke calorimeter was designed at the NRC and was recently used to measure the dose rate of a 2-mm wide 140 keV synchrotron beam at the Canadian Light Source in Saskatoon [14]. Preliminary data indicate that the overall uncertainty on dose determination can be significantly improved from the air kerma-based methods currently in use.

Perhaps the best current example of coordinated dosimetry research in response to an emerging beam modality is the joint research project, UHDpulse, where NMI collaborators are trialing several new calorimeter systems for direct use in ultra-high dose rate (i.e., FLASH) and laser-driven particle beams [15-17]. At the NRC, work is underway to characterize pyroelectric-based calorimeters capable of FLASH dosimetry that can be read out with only an electrometer [18].

2.4. Improved protocols to make use of new understandings
There is a long history of NRC involvement in the development of reference dosimetry protocols. One of the lead authors of the AAPM TG-51, and an advocate for this transition to dose standards [19], was David Rogers, long-time NRC researcher and group leader at the time. Also, many of the key advances that allowed dissemination of dose standards based on water calorimetry were made by NRC scientists [20].

An Addendum to TG-51 for photon beam reference dosimetry was published in 2014, and the lead author, Malcolm McEwen, is based at NRC. The Addendum provides new k_Q data, recommendations to improve the accuracy and consistency of the implementation of TG-51 (including the extension to flattening-filter free beams), uncertainties and how the clinical physicist can influence uncertainties, and specifications for a reference-class ionization chamber [8]. The new k_Q data recommended in the Addendum is from the work of Muir and Rogers [21], and the lead author of that publication, Bryan Muir, is also now based at NRC. Muir is also the lead on an Addendum to TG-51 for electron beam reference dosimetry, which is under review in the AAPM committee structure. The electron beam Addendum provides new data and recommendations based on a review of the literature on electron beam reference dosimetry published in the decades since TG-51 was released.

The AAPM TG-351 on clinical reference dosimetry in MR-guided radiotherapy is extending the guidelines of TG-51 for use in MR-linacs. A draft is in the final stages of preparation. Canadian researchers are leading this effort to address the growing user base for MR-linacs.

NRC is also contributing to AAPM TG-359, on FLASH dosimetry. The goal of this TG is to publish a guidance report on determining the dose and reviewing the need for standardization in dosimetry for UHDR beams to be used in research experiments and in pre-clinical applications. General guidelines will be provided on calibration, dosimetry, operation, and monitoring of beams in FLASH-type modes.

2.5. Improved guidance to reduce errors in clinical implementation
Although the TG-51 protocol and its Addendum clearly specify the procedure for reference measurements of external beams, the protocol lacks implementation guidance. Muir is the lead author and Renaud, also based at NRC, is a contributing author on an AAPM report that provides procedural guidance to achieve the most accurate reference dosimetry measurements. Topics include: (i) measurement of depth-ionization curves to obtain beam quality specifiers for the selection of beam quality conversion factors, (ii) considerations for the dosimetry system and specifications of a reference-class ionization chamber, (iii) commissioning a dosimetry system and frequency of measurements, (iv) positioning the water tank and ionization chamber, (v) requirements for ancillary equipment needed to measure charge and to correct for environmental conditions, and (vi) translation from dose at the reference depth to that at the depth required by the treatment planning system. Commonly-used simplified procedures are described and the impact on the uncertainty quantified to aid the physicist on where to allocate resources. This report is in the final stages of Medical Physics Journal review.

In addition to contributing to protocols and reports, researchers at NMIs provide guidance through educational outreach. There have been several educational sessions presented by NRC researchers that focus on improving clinical understanding of reference dosimetry protocols with the goal of achieving better accuracy and reducing errors in clinical reference dosimetry measurements [22-30].

2.6. Development of audit capabilities to verify clinical implementation

Alanine dosimetry capabilities have been established at the NRC since the early 2000s [31-32]. The system had been inactive for around 10 years but was re-established for postal dosimetry audits by Mansour in 2018 [33]. Alanine dosimeters are used to measure dose to water at therapy dose levels (10 Gy – 100 Gy) in Co-60 and MV photon beams with an uncertainty in the calibration of a batch of alanine dosimeters irradiated in Co-60 and in a clinical MV beam of 0.5 % and 0.6 %, respectively. To date, six Canadian cancer clinics have participated in postal dosimetry audits for MV photon beams. The agreement between stated and measured doses were typically within 1 % (k=1).

In addition to TG-51 verifications for conventional linac photon beams, two audits were performed for Canadian clinics with Elekta Unity MR- linacs [34]. The results show that combining the uncertainty budget for NRC measurements of alanine dosimeters irradiated in the MR-linac beam in terms of dose to water with the data given in the TG-51 Addendum and removing the common component of the Canadian primary standard of absorbed dose to water, then the level of agreement between stated and measured doses is within the overall k=1 uncertainty.

The alanine dosimetry system is also being used for reference dose measurements for FLASH beams from the IntraOp Mobetron system at CHUM, along with a portable beam current monitor [35], since conventional real-time dosimeters (i.e., ionization chambers) perform poorly in beam with ultra-high doses per pulse.

3. References

[1] BIPM 2021 CIPM MRA-P-11
[2] Mainegra-Hing E 2022 *Metrologia* **59** 045003
[3] Seltzer S *et al* 2016 ICRU Report 90
[4] Almond PR *et al* 1999 *Med Phys* **26** 1847-70
[5] Andreo P *et al* 2002 *Phys. Med. Biol.* **47** 3033
[6] Picard S *et al* 2010 *Metrologia* **47** 06025
[7] McEwen M *et al* 2014 *Med. Phys.* **41** 041501
[8] Kerkmeijer LG *et al* 2016 *Frontiers in Oncology* **6** 215
[9] De Prez L *et al* 2019 *Phys. Med. Biol.* **64** 035013
[10] Romano F *et al* 2019 IAEA-CN-273 International Atomic Energy Agency (IAEA)
[11] Newhauser W 2009 ICRU Report 78
[12] Wilke L *et al* 2019 *Strahlentherapie und Onkologie* **195** 193-8
[13] Bourgouin A 2020 *Determination of Wair value in high energy electron beam.* Carleton University
[14] El Gamal I 2022 Annual CIRMS Meeting http://cirms.org/wp-content/uploads/cirms2022/junior-investigator-essays/CIRMS2022_JI_Essay-Islam_El_Gamal.pdf
[15] Bourgouin A *et al* 2021 http://uhdpulse-empir.eu/wp-content/uploads/2021-04_CIRMS2021_ABourgouin_Abstract.pdf
[16] McCallum S 2021 FRTP Conference http://uhdpulse-empir.eu/wp-content/uploads/FRPT_McCallum-presentation.pdf
[17] Bourgouin A *et al* 2022 *Med. Phys.* **49** 6635-45
[18] Renaud J *et al* 2021 *Med. Phys.* **48** 4667
[19] Rogers DWO 1992 *Med. Phys.* **19** 1227-39
[20] Seuntjens J *et al* 1999 *NRC Report PIRS* **584**
[21] Muir BR *et al* 2010 *Med. Phys.* **37** 5939-50
[22] McEwen M *et al* 2009 In *Clinical Dosimetry Measurements in Radiotherapy* 1027-1058. Madison,

Wisconsin: Medical Physics Publishing
[23] McEwen M R 2009 *In Clinical Dosimetry Measurements in Radiotherapy* 501-547. Madison, Wisconsin: Medical Physics Publishing
[24] McEwen M 2011*Med Phys* **38** 3752
[25] McEwen M 2012 *Med. Phys.* **39** 4006
[26] McEwen M 2015 *Med. Phys.* **42** 3739
[27] McEwen M 2017 AAPM Annual. aapm.org/education/vl/vl.asp?id=12223
[28] Muir B 2017 AAPM Annual. https://www.aapm.org/education/vl/vl.asp?id=12222
[29] Muir B 2019 AAPM Annual. https://aapm.org/education/VL/vl.asp?id=13809
[30] McEwen M 2020 Joint AAPM COMP Virtual Meeting. amos3.aapm.org/abstracts/pdf/155-50581-1531640-159703-1978106521.pdf
[31] Zeng GG *et al* 2004 *Phys. Med. Biol.* **49** 257
[32] Zeng GG *et al* 2005 *Phys. Med. Biol.* **50** 1119
[33] Mansour I 2018 Doctoral dissertation, Carleton University
[34] McEwen M *et al* 2020 *Med. Phys.* **47** E281
[35] Renaud J *et al* 2022 *Physica Medica* **94** S105-6

Journal of Physics: Conference Series **2630** (2023) 012013 doi:10.1088/1742-6596/2630/1/012013

A review of small animal dosimetry techniques: image-guided and spatially fractionated therapy

Chris D. Johnstone[1] and *Magdalena Bazalova-Carter[2]
[1]Princess Margaret Cancer Centre, Department of Radiation Oncology, University of Toronto, Toronto, ON, Canada
[2]Department of Physics and Astronomy, University of Victoria, Victoria, BC, Canada

E-mail:bazalova@uvic.ca

Abstract. Research in small animal radiotherapy is a crucial step in clinical translation of novel radiotherapy techniques, either delivered as stand-alone treatment or in combination with other treatments, such as chemotherapy and immunotherapy. In order to efficiently translate preclinical findings to the clinical setting, preclinical radiotherapy must replicate clinical therapy in terms of mode of delivery as well as dose delivery accuracy as closely as possible. In this review article, we focused on the description of dosimetry tools for radiotherapy of small animals delivered with kilovoltage x-ray beams on image-guided irradiators and in a spatially-fractionated manner by means of microbeam therapy. The specifics of dosimetry of kilovoltage x-ray beam deliveries with small, often sub-millimeter, beams are highlighted, and suitable dosimeters, phantoms, and dose measurement and calculation techniques are reviewed. Future directions for accurate real-time high spatial resolution dosimetry of small animal irradiations are also discussed.

1. Introduction

Cancer radiotherapy research performed with small animals in a preclinical setting is crucial in advancing patient cancer care. It is used to not only evaluate novel radiotherapy techniques, alone or in combination with chemotherapy or immunotherapy, but also to evaluate imaging techniques to assess radiotherapy treatment outcomes.

Compared to clinical irradiations, small animals are typically irradiated with much smaller fields of 1-10 mm in size. Small field dosimetry has been in the frontline of dosimetry research for a about a decade. While preclinical radiotherapy research has been investigated with proton beams and other particles, preclinical radiotherapy is most often delivered with kilovoltage (kV) x-ray beams, i.e., with beams of lower energies compared to clinical radiotherapy performed with megavoltage photon beams [1]. Such difference in beam energy also poses challenges to small animal irradiation dosimetry. The focus of the review article will be on dosimetry of radiotherapy of small animals performed with kilovoltage x-ray beams, delivered either with image-guided or single-field irradiators, or in a spatially-fractionated manner.

1.1. Small animal radiotherapy

Since the 1970s cabinet irradiators have been used as a small scale, cost-effective tool in preclinical research [2, 3]. Current commercial irradiators for the radiation therapy of small animals includes the PXi X-RAD 320 (Precision X-Ray Inc., North Brandford, CT), Xstrahl RS225 and RS320 (Xstrahl

Inc, Suwanee, GA), Faxitron MultiRad 160, 225, and 350 (PharmaMedSci, St. Laurent, Quebec), Gammacell 40 (Best Theratronics, Ottawa, Ontario), Pantak Therapax-150 and DXT-300 (Pantak Inc., Branford, CT). These irradiators usually treat with only a single stationary field with very simple collimation. In addition, they inherently lack the ability for precision radiation therapy as they do not contain on-board imaging technology to accurately plan and deliver conformal treatments, and are mainly used for general flank tumor, total body irradiation (TBI), and cell irradiations [4, 5]. Irradiation times are usually approximated from simple in-phantom measurements where custom made jigs and lead blocks must be created to provide more conformal and specialized treatments [6-9]. A higher degree of treatment accuracy can be achieved with medical linear accelerators (LINACs), currently commercially produced by Varian and Elekta as the two largest LINAC manufacturers. While LINACs are excellent at performing conformal irradiations, they can cost upwards of a million dollars and are expensive to maintain, require specialized training to operate, and are specifically calibrated to treat humans.

A shift in preclinical research away from standard cabinet irradiators began in the mid 2000's when the first image-guided small animal irradiator was created, creating for the first time, conformal preclinical radiation therapy on a practical and small scale [10-15]. This shift allows for more precise and accurate treatment delivery that includes treatment planning systems (TPS) and onboard image guidance as a small scale analogue to a LINAC. This introduces the ability to accurately treat tumours implanted within the animal, and increases the geometric dose delivery accuracy to within 0.2 mm.[11,16]

1.2. Spatially-fractionated radiotherapy

Spatially-fractionated radiotherapy (SFRT) delivers very small beams of high doses interleaved with areas of low doses (Figure 1). It was first demonstrated to protect normal tissues in brain following irradiations with 22 MeV deuterons [17]. Curtis *et al.* observed no tissue necrosis developed in the normal mouse brain after focal administration of >1 kGy along the peak dose areas of microbeam arrays [17]. SFRT can be delivered with kV x-rays as microbeam radiation therapy (MRT) with beams of ~25-100 µm in size with centre-to-centre (c-t-c) distances ~200-400 µm gaps [18, 19] or as minibeam radiation therapy (MBRT) with 500-700 µm beams and ~mm-long c-t-c distances [20]. X-ray MRT delivered with synchrotron beams has shown to reduce normal tissue toxicity in brain tumors in rats [21] and mammary carcinoma in mice [22]. MRT and MBRT are characterized by a number of dosimetry parameters, such as the peak doses, the valley doses, and the peak-to-valley ratios (PVDRs). A recent review paper summarized the history, current status, and future prospects of SFRT [23]. To this date, however, it is not clear what are the dosimetric parameters driving the reduced normal tissue toxicity in SFRT and it is therefore important to evaluate these parameters in radiobiological studies accurately.

2. Radiation Sources

2.1. Image guided irradiators

Two commercial image-guided small animal irradiators are currently available, the PXi X-RAD SmART and the Xstrahl Small Animal Radiation Research Platform (SARRP). Both units are entirely self-shielded, contain a TPS, and include on-board image guidance. The PXi X-Rad SmART operates at a 220 kVp tube voltage, 13 mA tube current, 0.8 mm Be internal and 0.30 mm Cu external filtration, 20° tungsten anode, and a 3.0 mm focal spot for therapy beam delivery. The maximum uncollimated dose rate at a 30 cm isocenter distance in 2 cm of water is approximately 4 Gy/min. Treatment collimators range from 1 to 25 mm circular and 5 × 5 to 40 × 40 mm^2 square field sizes at isocenter. CBCT images are obtained using a stationary couch and 360° rotating source and flat-panel detector, operating at a common imaging protocol of 60 kVp tube voltage (40 to 100 kVp range), 0.200 mm pixel size, 1 mA tube current, 5 frames per second (fps), and 60 s scan time, using a 2.0 mm Al external filtration with a 0.4 mm focal spot [16].

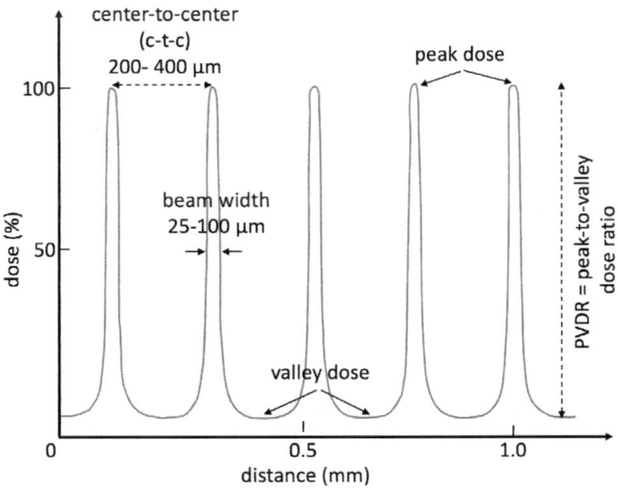

Figure 1. A typical dose profile in microbeam therapy (MRT). Important beam parameters are denoted.

The Xstrahl SARRP operates at a 220 kVp tube voltage with a 13 mA tube current, 0.8 mm Be internal and 0.15 mm Cu external filtration, 20° tungsten anode, and a 3.0 mm focal spot for therapy beam delivery. The maximum uncollimated dose rate at a 35 cm isocenter distance in 2 cm of water is approximately 3.6 Gy/min. Treatment collimators range from circular 0.5 to 40 mm and 3 × 3 to 10 × 0 mm^2 squared field sizes at isocenter. CBCT images are obtained using a stationary source and flat-panel detector with a 360° rotating couch, operating at a common imaging protocol of 60 kVp tube voltage (40 to 80 kVp range), 0.200 to 0.350 mm pixel sizes, 0.8 mA tube current, 6 fps, and 60 s scan time, using a 1.0 mm Al external filtration with a 0.4 mm focal spot [10-12]. Imaging dose using the common imaging protocol is approximately 3.5 cGy to the entire body of a 20 g mouse [24, 25].

2.2. *Synchrotron beams*

A synchrotron is a type of cyclical particle accelerator where a beam of charged particles are accelerated through a changing electromagnetic field in circular or oval paths that can be used to generate x-ray beams with dose-rates up to 5,000 times higher than clinical radiotherapy [26]. Synchrotrons deliver highly-collimated quazicontinuous polychromatic x-rays with mean energies of 30-500 keV, therefore of higher energies compared to image-guided small animal irradiators. For example, the synchrotron beam at the European Synchrotron Research Facility (ESRF) in Grenoble is 50-μm thin, has a mean energy of 102 keV, and delivers extremely high dose rates of 16 kGy/s [27]. An overview of available synchrotron beamlines and their properties can be found in a recent publication by Montay-Gruel *et al* [28]. The high dose-rate available in synchrotron beams can be used in combination with MRT allowing for spatially fractionated dose using an array of parallel beams of 25-100 μm widths to deliver tumoricidal dose while providing normal-tissue sparing effects [29-31].

Synchrotrons lend themselves to small animal applications for therapeutic radiation therapy such as microbeam dosimetry, FLASH, establishing tumour and normal tissue affects in small animals, generating genetic models such as epilepsy and human tumour models in small animals, and the study of brain function with microbeam irradiation [30-35], including synchrotrons used for small animal imaging [36, 37]. These synchrotron beams can be generated by either large-scale beam-lines [30, 34, 35, 38], or more recently, through compact laboratory-sized synchrotrons for lower kilovoltage x-ray beam energies [32, 36].

3. Dosimeters

3.1. Ionization chambers

Ionization chambers are the gold standard in reference dosimetry for kV beams for delivering a high level of accuracy, reproducibility, sensitivity, and real-time output. This is especially true for cylindrical ionization chambers as these chambers have a nearly constant energy response between the 40 and 300 kV tube voltages used in image-guided small animal irradiators [39]. In North America, the American Association of Physicists in Medicine Task Group 61 (AAPM TG-61) is one of the most used dosimetry protocols and recommends that absolute dosimetry and beam quality be exclusively obtained with air-filled ionization chambers that are calibrated by an Accredited Dosimetry Calibration Laboratory [39]. For relative dosimetry, parallel plate ionization chambers and smaller volume cylindrical ionization chambers may also be used.

Since image-guided small animal irradiators require small fields ($\leq$ 4 cm), small volume cylindrical ionization chambers are utilized to prevent volume averaging effects. PinPoint ionization chambers can be 0.015 cm^3 in volume (2 mm diameter, 5 mm length) with a 0.20 nC/cGy response and are well suited for field sizes as small as 2×2 cm^2 (as specified by the manufacturer, PTW) to avoid volume averaging effects. Multiple studies show an uncertainty greater than 6% when measuring 5 mm diameter field sizes, and an uncertainty within 3% when measuring field sizes 6 mm and larger when using PinPoint ionization chambers [40, 41]. In a study by Charles *et al* [42], it was determined that a 0.015 cm^3 volume PinPoint ionization chamber had an uncertainty in output factor measurements for small field sizes of approximately 1.0 and 3.5% for 1.5 and 1.0 cm diameter field sizes, respectively, however this analysis was completed with a 6 MV beam. Hill *et. Al.* recommends using a PinPoint chamber with an aluminum electrode, as a PinPoint chamber with a steel electrode reported over responding by as high as 8% due to the photoelectric effect from the high atomic number of steel [43]. Compared to the larger volume ionization chambers for absolute dosimetry, a 0.6 cm^3 volume (6 mm diameter, 23 mm length) with a 20 nC/cGy sensitivity can be used for field sizes down to 4×4 cm^2. If smaller field sizes are to be measured, diamond detectors, small scintillators, or radiochromic film should be used.

3.2. Radiochromic films

One of the greatest challenges in small animal dosimetry is the high spatial resolution required to measure small fields, especially when employing spatially fractionated microbeams, and this is where radiochromic films are second to none. The most widely used and characterized radiochromic film for kilovoltage dosimetry currently is GAFChromic EBT3 film (Ashland Advanced Materials, Bridgewater, NJ) [44-55]. EBT3 film has a nominal 28 µm thick active layer (Li-PCDA with dominant peaks at 633 and 595) [56] sandwiched between 125 µm thick matte layers of polyester-substrate, with an optimal dose range between 0.20 and 10 Gy and $\leq$ 25 µm measurement resolution [57, 58]. This makes EBT3 film optimal for measuring small field output and profiles down to a 25 µm field size. For doses up to 40 Gy, EBT-XD should be used [59]. For FLASH and/or microbeam dosimetry, GAFchromic HD-V2 (Li-PDCA active layer with dominant absorption peaks of 670 and 635 nm) can be used as this film has a dose range between 50 to 1000 Gy and an extremely high spatial resolution of $\leq$ 5 µm [56, 60, 61]. Vaiano *et al* [53] developed a method to use EBT3 film up to 100 Gy with a 4% reading uncertainty for 250 kVp x-rays.

For an increase in dosimetric accuracy, it is recommended that a triple-channel calibration (red, green, and blue channels) be used over single-channel (red channel) and double channel calibration methods, including creating a separate calibration for each film batch and given beam quality [56, 62]. A number of triple-channel calibration techniques are available [62-64]. Using a triple channel method can extend the dose range of EBT3 film to 40 Gy [44]. The red channel is most sensitive for the lower energy range and the green channel is most sensitive for the higher energy range of radiochromic film, and the blue channel provides a response signal to automatically correct for nonuniformities in the radiochromic film [44]. For the reading of radiochromic film, it is recommended to use a professional colour flatbed scanner (such as the Epson 10000XL), circular ROIs, 600 dpi for small field dosimetry, RGB colour mode, ≥ 16 bit depth, and saved in an uncompressed .TIFF format [56]. Great care must be exercised in the calibration, measurement, handling, and reading of radiochromic film to produce accurate dosimetry, especially kilovoltage dosimetry (due to energy sensitivities), and readers are directed to the recently updated AAPM TG-235 report on radiochromic film dosimetry [56]. However, for SFRT, flatbed scanners might not be sufficient to resolve the peak and valley doses due to microscopic spatial inhomogeneities that arise due to the granular structure, defects, and compositional variations in the film. This can be mitigated by using a microscope for improved resolution scans in microbeam dosimetry [61].

3.3. Plastic scintillator detectors

Plastic scintillators have been used in megavoltage therapy, with a growing interest in examining their properties for kilovoltage energies [65-67]. Recently, plastic scintillators have been characterized and successfully validated for use in the 40 to 220 kVp tube voltage energy range within multiple small animal irradiators [49, 68-72]. Due to an energy dependence that is present for energies below 250 kVp, output must be corrected for by using a combination of mass-energy absorption coefficients, Monte Carlo (MC) obtained beam spectra information, and quenching corrections [66, 71].

The greatest advantage of plastic scintillators lies in their high special resolution to measure very small fields and profiles down to a 3 x 3 mm^2 field size [49]. This is possible due to a very small 1 mm^3 active volume of polystyrene. While radiochromic film can measure even smaller field sizes in 2D, the advantage of plastic scintillators over film is that they can measure output in real-time. This includes the ability for real-time *in-vivo* mouse measurements with an average accuracy within 4.2% when compared to TPS calculations as shown by Le Deroff *et al* [68]. Esplen *et al* [72] acquired point dose measurements inside a heterogeneous mouse phantom with field sizes between 3 x 3 and 10 x 10 mm^2 and reported accuracy within 3.2% when compared to MC simulations. Johnstone *et al* [49] acquired in-air profiles using 3 × 3 and 5 × 5 mm^2 field sizes and reported a mean 2.4 % difference in-between the penumbra region when compared to radiochromic film measurement and MC simulations, but found that measuring a 1 mm diameter profile with the scintillator under responded by 20% due to volume averaging effects. More recently, plastic scintillators have been demonstrated to be a suitable dosimeter for 120-kVp ultrahigh dose-rate experiments performed with a modified x-ray tube [73, 74].

3.4. Diamond detectors

A real-time commercially available detector with the ability to measure field sizes smaller than PinPoint ionization chambers are diamond detectors. The properties of an ideal detector system are high reproducibility, sensitivity, linearity, spatial resolution, and tissue equivalence as well as low dose rate, angular, and energy dependence, and diamond detectors are excellent for these categories [75]. Natural and synthetic diamond detectors have approximate tissue equivalence, with a push towards synthetic chemical vapor deposition (CVD) due to the relatively availability, low cost and dose-rate linearity compared to natural diamond detectors [76].

A common commercial synthetic diamond detector is the PTW 60019 microDiamond (PTW, Freiburg, Germany) that has a small nominal 0.004 mm^2 sensitive volume, 1.1 mm radius, 0.001 mm thickness, waterproof, contains a nominal 1 nC/Gy response, and conveniently has a 0 V bias [75]. This detector has been characterized for kilovoltage energies between 40 and 250 kVp, compared to other commonly used detectors, and achieves good dose-rate linearity, but requires pre-irradiation and has an energy dependence below 250 kVp, especially below 60 kVp [77].

Diamond detectors are excellent for small field dosimetry of kilovoltage x-ray beams when energy dependence is accounted for, including measuring profiles in spatially fractionated x-ray beams of 30-120 kVp, up to 700 Gy/s dose rates, using 600 μm minibeams using the PTW 60019 microDiamond detector [26]. Other studies have also used microDiamond detectors to measure small field size output and profiles from 2 cm down to 2 mm for < 300 kVp x-ray beams [78, 79]. Small fields have also been measured with diamond detectors in image-guided small animal irradiators [80, 81].

3.5. Other detectors

Other small-volume detectors are also suitable for dosimetry of small animal radiotherapy. Their sizes must be smaller than the irradiation field and their response to kV x-ray beams must be carefully characterized. Thermoluminescent detectors (TLDs) are popular detectors for small animal radiotherapy due to their small size down to $1\times1\times1$ mm^3 microcubes [82, 83]. They can be conveniently used for *in vivo* dosimetry [84]. TLDs can also be inserted into cadavers for accurate dose measurements mimicking experimental conditions for mouse brain [85] and bone marrow irradiations [86]. TLDs were used for a mail audit to verify the accuracy of small animal dosimetry by the University of Wisconsin Medical Radiation Research Center [87] and MD Anderson Cancer Center [88]. TLDs overrespond to low-energy x-rays (< 3 mm Cu HVL) by as much as 13% and need to be carefully characterized in order to achieve dosemetric accuracy [89]. Injectable 0.6-mm long TLD rods have also been developed for small animal radiotherapy. They were injected subcutaneously in the dorsal neck of mice irradiated in a Cs-137 irradiator [90].

Optically stimulated luminescent detectors (OSLDs) would be suitable due to their small size, unfortunately their high energy dependence limits their use for pre-clinical applications. For example, nanoDot OSLDs are 0.2-mm thick disk-shaped detectors with a diameter of approximately 5 mm encased in a light-tight $10 \times 10 \times 2$ mm^3 plastic carrier overrespond by 85% in a 2-mm Al HVL beam compared to a 9.5-mm Al HVL beam [91]. The nanoDot OSLDs have been shown by Poirier *et. al.* to have a strong variation in light output in the 40-300 keV range that over responds by as much as a factor of 4 compared to a 6 MV photon beam, including a 15% variation in dose between different x-ray units of equivalent effective energies [92]. For accurate dosimetry, OSLDs must be well characterized for a given beam quality and irradiation geometry in order to account for their energy dependence. Similarly, microMOSFET dosimeters are of a small physical size of 1×1 mm^2 and 3.5 mm long but exhibit large energy dependence. Their overresponse to a 2 mm Al HVL beam compared to a 6 MV photon beam is 400% [93].

4. Phantoms

4.1. Homogeneous phantoms

Solid water is the phantom type of choice for kilovoltage x-ray reference dosimetry in small animal irradiators for its practicality and water equivalence. The phantom must be water equivalent or the kilovoltage absorbed dose measurements acquired will not adequately represent dose to water and can resulting in incorrect dose calculations of as much as 6% [94]. The ICRU Report 44 recommends for solid water to be considered water equivalent; it must not introduce uncertainties to the absorbed dose greater than 1%, where a given thickness of phantom material should have the same radiation and scattering properties as the same thickness of water [95]. This is important as there are a wide range of phantom materials being used without the exact material types specified within image-guided small animal irradiators [7, 10-12, 14, 15, 20, 24, 69, 96-111]. Hill *et al* [112] examined depth doses and profiles for 10 commonly used solid water materials compared to water under low energy kilovoltage dosimetry (40 to 300 kVp) and recommends the most accurate materials (dose percentage differences less than 2% to water) to be RMI-457 (Gammex Inc., Wisconsin, USA), PWDT (CIRS, Virginia, USA), and Virtual Water (CMNC, Tennessee, USA). The worst performing solid water were plastics, namely PMMA, Polystyrene, and Plastic Water (CIRS), with percentage differences between 3-6% compared to water. To bypass this water equivalence issue, Arango *et.al* [40]. used a small $334 \times 336 \times 422.5$ mm^3 (L × W × H) 3D scanning water tank MP3-XS (PTW, Freiburg, Germany), which fit within the small confines of the self-shielded Xstrahl SARRP.

Most users of small animal irradiators don't have access to a small scanning water tank, thus solid water phantoms are almost exclusively used. The venders of the image-guided small animal irradiators provide small 6 x 6 x 0.5 cm^3 plastic solid water slabs that fit into a custom jig that is attached to the head of the irradiator to be used with radiochromic film [24, 109, 113]. Large solid water is provided for LINACs (i.e., large 20 × 20 cm^2 slabs) must be specially machined smaller to fit on the small animal treatment couch, or else the user will have to create a jig to raise up and support the larger solid water slabs from the bottom of the irradiator [25]. Solid water slabs can be used with radiochromic film or machined for the insertion of ionization chambers for dosimetry measurements. Simple cylindrical homogenous mouse phantoms (i.e. ~6 cm long and 2.5 cm diameter) that dosimeters such as PinPoint ionization chambers, OSLDs, and TLDs have been used for pre-irradiation dose verification [87, 114, 115].

4.2. Heterogeneous phantoms

While the material of solid water must be carefully chosen to match water equivalence, the dosimetric properties of heterogeneous materials for low kilovoltage photon beams such as bone in small animals are not well established, with material densities and elemental compositions shown to produce a significant absorbed dose difference compared to using ICRU44 human tissues [95, 116, 117]. To accurately match the physical densities and elemental composition of mouse tissues to create mouse-equivalent materials, it can be accomplished through elemental composition analysis methods such as x-ray fluorescence (XRF), atomic absorption spectroscopy (AAS), and inductively coupled plasma mass spectrometry (ICP-MS) [118-122]. However, most small animal research using heterogeneous phantoms do not create custom mouse-equivalent materials and instead use human-equivalent materials (or non-tissue equivalent materials such as cork [8] and PMMA [87]) that may not be validated at low kilovoltage energies. The most accurate way to validate the dosimetric properties of heterogeneous materials is to compare experimental measurement with TPS or Monte Carlo Simulations [24, 117, 123-129]. Validation measurements are usually carried out by placing radiochromic film adjacent to solid water slabs aside heterogeneous slabs such as bone and lung [24, 123, 124, 130]. Commercial heterogeneous materials can be obtained from companies such as Gammex and CIRS.

Recently, a number of studies have replicated the physical geometry and material composition of mice by creating custom realistic heterogeneous and anthropomorphic mouse phantoms (Figure 2) [72, 131-138]. These phantoms are useful in better replicating the actual radiation scattering events, geometry, and tissue attenuation of actual mice for treatment dose verification. These phantoms have been created using realistic mouse geometry that includes tissue materials such as bone, lung, brain, liver, heart, and kidney. Inserts for dosimeters within these phantoms include radiochromic film, TLDs, scintillators, and PinPoint ionization chambers.

5. Dose measurements

Many protocols exist for kilovoltage reference dosimetry such as the AAPM TG-61 [39], the Code of Practice from the Institution of Physics and Engineering in Medicine and Biology (IPEMB) [139], the International Atomic Energy Agency (IAEA) TRS-398 [140], the Netherlands Commission on Radiation Dosimetry NCS-10 [141], and the most recent German Institute for Standardization DIN 6809-4:2020-04 [142]. Dosimetric comparisons of the different protocols have been examined [143-145], however, as of this writing, no official protocol for reference dosimetry exists for standard cabinet and image-guided small animal irradiators. The limited geometry within the self-shielded irradiators do not allow for the scatter-free geometry recommended by the aforementioned protocols. Thus, a dosimetry protocol specifically for cabinet irradiators is needed.

In North America, the most common approach is following a modified version of the AAPM TG-61 protocol [14, 24, 25, 41, 49, 98-100, 103, 107, 110, 146-148]. For absorbed dose to water measurements, the AAPM recommends the "in-air method" or the "in-phantom method" using an ADCL accredited cylindrical ionization chamber. The setup geometry and correction factors for each method are described in detail in the AAPM TG-61 protocol [39]. As a result of setup and reproducibility difficulties using water phantoms within image-guided small animal irradiators, many studies used the in-air method [6, 14, 25, 41, 49, 96, 99, 103, 107]. In this method, a large volume ionization chamber (i.e., 0.6 cm^3 volume) can be secured to the treatment couch and leveled at isocenter with the active volume extended off the edge of the couch to avoid couch backscatter, while using a collimated field no less than 4 cm in diameter for large ionization chambers. To establish the most accurate positioning possible, centering the active volume of the ionization chamber with the CBCT and not with the lasers has been done [99]. An EPID detector can also be used to position the chamber in the x- and y-directions using a collimator for centering guidance, and a custom-built front-pointer can be used for positioning in the z-direction [25, 149]. The output may be acquired at the four cardinal angles to verify that the chamber is correctly positioned at isocenter, and to verify output with changing gantry angle. In the event that a small water phantom is available (i.e., $6 \times 6 \times 10$ cm^3 to $30 \times 30 \times 30$ cm^2) and can be positioned in a reproducible way, the in-phantom method can be used with a cylindrical ionization chamber placed at isocenter at a depth of 2 cm in water [40].

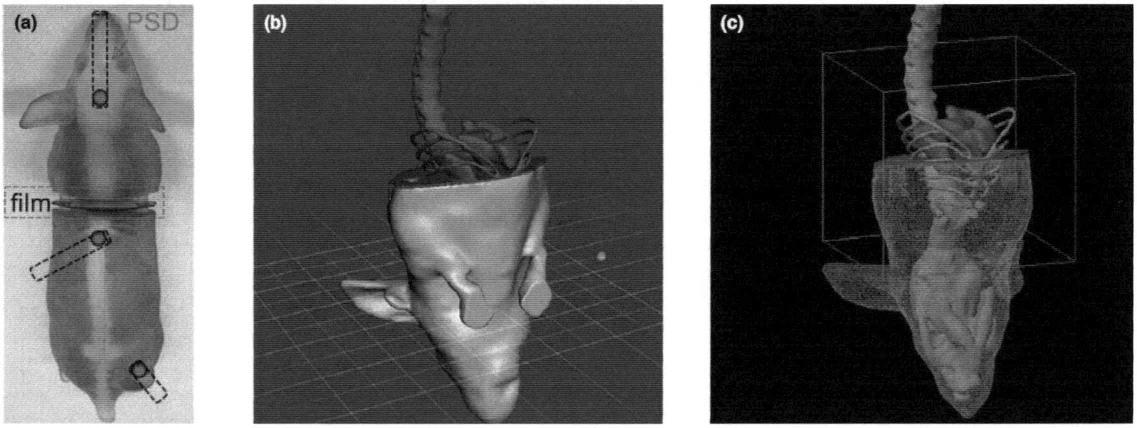

Figure 2. Photograph (a) and 3D volume renderings (b, c) of a 3D printed heterogeneous mouse phantom. The plastic scintillator and film dosimeter (1 mm active element in orange, not to scale) sites are outlined in a). Reprinted with permission from Esplen *et al* [72].

For daily pre-treatment output verification and monthly output measurements, it is recommended to measure air-kerma at isocenter to verify any changes in machine output over time, as this method allows for a quick, simple, and reproducible setup compared to in-phantom measurements with less risk for setup error [25]. Ease of setup is of high importance as low kilovoltage x-rays have a high dose gradient with distance from the source, and a setup reproducibility on the order of 1 mm is required for accurate and reproducible dosimetry [25, 139]. For commissioning and annual dosimetry, the in-phantom method for output measurements should be used, with great care taken into the accuracy of the setup [96, 103, 110]. Recently, the use of a built-in EPID imager has been examined as a means of simplified dosimetry within an image-guided small animal irradiator [149-151].

It is recommended to not use the reference dosimetry dose rate as a means to establish the irradiation time for therapy treatments for mice or cell irradiations if the irradiation geometry and mouse/cell-thickness is not identical to the reference dosimetry. The different scattering conditions can misrepresent the dose in mice by as much as 15% due to large differences in the lateral scatter, backscatter, and field size differences within the full-scattering conditions recommended by the AAPM TG-61 compared to the relatively small thicknesses of mice or irradiated cells [6, 152, 153]. Instead, it is recommended to use a cylindrical solid water mouse phantom that houses a PinPoint ionization chamber to better model the realistic scattering conditions of mice treatments [114, 115]. If a mouse phantom is not available, the backscatter factors from Chen *et al* [153] that report scatter factors with varying phantom thickness can be used with reference dosimetry measurements, and likewise, could be used for quantifying the mm-thick backscatter conditions for cell irradiations.

Beam quality for low kilovoltage x-rays is quantified by using both the first half-value layer (HVL) and kVp and are required properties for reference dosimetry [39, 124, 154]. According to the AAPM TG-61, HVL should be measured under a scatter free and narrow-beam geometry by placing the ionization chamber approximately 100 cm from the source with collimation and attenuating material (Cu or Al) approximately 50 cm from the chamber. The collimation of the beam at 50 cm for narrow-beam geometry should create a field size just large enough to provide a uniform exposure across the sensitive volume of the chamber [39]. This 100 cm geometry cannot be achieved within the small self-shielded confines of image-guided small animal irradiator cabinets, so modifications must be made. These modifications include placing a large cylindrical ionization chamber approximately 50 cm from the source, with the attenuating material and beam collimation placed as far from the detector as possibly [24, 99, 103, 106]. The distance of the chamber may be varied to allow at least 20 cm between the walls of the irradiator and the chamber to reduced beam scatter to the measurements [24, 103]. A three-point semi logarithmic interpolation technique has been recommended to determine HVL for kilovoltage energies as larger amounts of measurement points may inherently change the beam quality during HVL analysis [124].

Percentage depth dose (PDD), off-axis profile measurements, and relative output factors can also be obtained. These measurements are used for commissioning of the irradiator, quality assurance testing, and dosimetry for treatment planning calculations. Measurements can be performed with radiochromic film wedged in-between small solid water slabs placed on the treatment couch [110], or placed within a jig attached to the head of the irradiator [24, 109, 113]. When placing solid water on the couch, care must be taken to prevent any bowing of the couch that will skew output readings [25]. If a small scanning water tank can be setup in a reproducible way, a waterproof high resolution PinPoint ionization chamber can be used for PDD, profile, and output factor measurements [40, 97, 103].

5.1. Dosimetry accuracy and reproducibility

The accuracy and reproducibility of kV beam dosimetry is heavily dependent on quantifying the irradiation geometry. This includes the source-to-surface distance (SSD), the attenuation and material above the point of measurement, the backscatter thickness and material below the point of measurement, field size, and beam quality (kVp and HVL). In conformal small animal irradiators, the dose can change by 10% with the following: 2 cm change in SSD, 10% per cm of attenuating material [48, 155], 0.05 mm change in Cu external filtration [154], or 1 cm of backscatter (field sizes > 1 cm) [153, 156]. Changes in dose from backscatter thickness become negligible at backscatter thicknesses of approximately 5 cm and thicker [153]. A 5% change in dose can occur with a change in field size from 5×5 mm^2 to 3×3 mm^2 and to 10×10 mm^2 [24]. An increase in field size from 1 to 4 cm increases the backscatter factor for mice irradiations by 10% [153]. Changes in field size or beam quality affect the attenuation, backscatter, and overall dose [39, 48, 124, 153, 155, 156], and changes in beam quality can affect the relative biological effectiveness (RBE) [154]. The quantification of dose changes with irradiation geometry in a small animal irradiator are illustrated in Figure 3. An absorbed dose variation of only ±5% can affect tumour tissue control probability by 10-20% [1, 157, 158]. This is of course valid for both clinical and preclinical radiotherapy and accurate dosimetry is of highly importance during the commissioning of clinical linear accelerators as well as for small animal irradiators [159].

6. Dose calculations

6.1. Monte Carlo simulations

Monte Carlo (MC) based dose calculations are gold standard for dose simulation algorithms within the field of clinical radiotherapy as well as for preclinical radiotherapy. A number of MC codes for x-ray beam dose calculations exist, such as the EGSnrc code [160], GEANT4 [161], and its wrapper TOPAS [162], VMC++ [163], MCNPX [164], and FLUKA [165]. MC dose calculations with kV x-ray beams are highly sensitive to the accuracy of modelling of the incident beam [166] as well as the irradiated object. For example, Bazalova *et. al.* demonstrated that segmenting at least 39 tissues is required to calculate dose with a 2% accuracy for a 250 kVp beam [116]. Muhavava *et al* presented a MC study that can be used to convert measured dose at a point to dose at a different depth in single-field irradiators [167]. In addition, it is crucial that appropriate interaction cross-sections and particle energy or distance cutoffs are used and therefore an appropriate application of MC dose calculations requires a good knowledge of the MC code as well as the irradiation system.

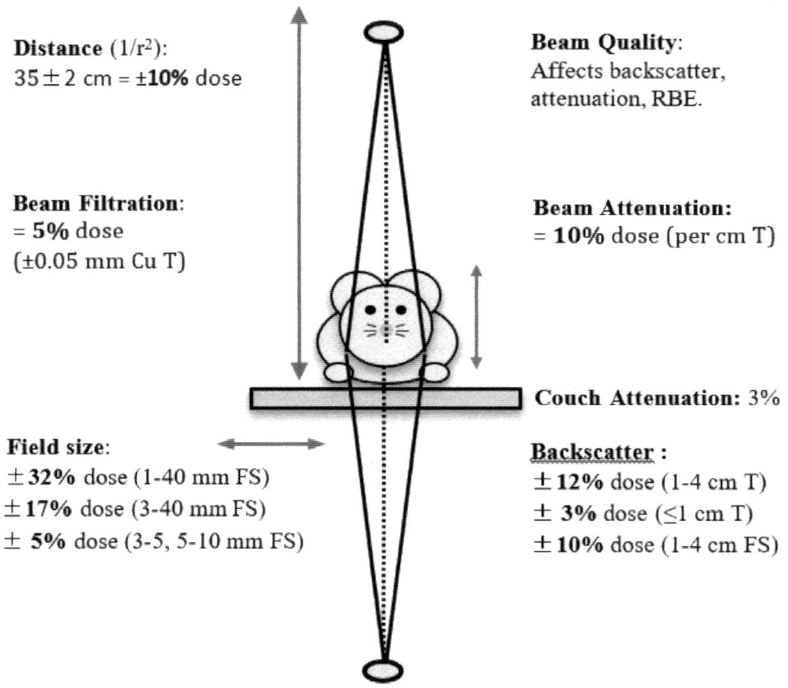

Figure 3. Effects of changing geometry on delivered dose in an image-guided small animal irradiator as a function of field size (FS), distance from the source, backscatter, material thickness (T), and attenuation. Adapted from Draeger et al.1 with values specific to conformal small animal irradiators [153, 154, 156].

In order to enable access to MC dose calculations for small animal radiotherapy, an MC dose calculation engine based on an analytical source model and VMC++ was recently implemented in commercial software for small animal treatment planning μ-RayStation [126, 168]. Full plan dose calculation speeds were quoted in the order of 3-5 minutes, which was up to 680 times faster than full GATE MC dose calculation, rendering μ-RayStation online dose calculations feasible.

Non-commercial MC dose calculations, while accurate, are still time consuming and might not be feasible for small animal radiotherapy treatment planning tasks. More time-efficient dose calculations are therefore of high importance in small animal radiotherapy.

6.2. Other techniques

A number of analytical models that result in faster dose calculations in small animal radiotherapy exist. In general, the analytical approximation is performed in the source description, such as in μ-RayStation described above. Granton and Verhaegen also built an analytical source model of the 225Cx image-guided small animal irradiator and reported a high dose calculation accuracy for field sizes >1 mm and speed ups of a factor of 19-1200 depending on beam size [146]. In addition, GPU-accelerated superposition-convolution dose calculation algorithm was first implemented for SARRP [169]. Dose differences between the analytical dose calculations and EBT film measurements and MC of <5% were found for a 5×10 mm^2 beam with ultrafast dose calculation times of < 10 s. In order to fully evaluate analytical dose calculation algorithms, smaller fields sizes of <1 mm should be studied. It is likely that analytical dose calculation algorithms will not be feasible to implement in SFRT employing submillimeter x-ray beams.

7. Future prospects

In the last decade there has been great progress within the field of small animal radiotherapy including the associated dosimetry. There is a clear need for online dosimetry with energy-independent detectors of sub-millimeter sizes capable of accurate dose measurements of millimeter-size kilovoltage x-ray beams. Small-volume ionization chambers, plastic scintillator detectors, and diamond detectors are currently the most suitable detectors for this task.

In terms of spatially-fractionated radiation therapy, however, no such online dosimeter currently exists. Dose measurements are currently done with radiochromic films that require offline readout and therefore cannot be used for real-time measurements. One future possibility to perform online SFRT dosimetry could be the combination of a scintillating film and a high-resolution CCD camera [169]. FLASH radiation therapy is moving into the forefront of small animal radiation therapy research, and as such, suitable detectors need to be characterized for the novel high dose rates and small field sizes of small animal FLASH irradiations [171-173].

8. Conclusions

We provide an overview of small animal radiotherapy dosimetry with a focus on small field and spatially fractionated radiation therapy. Preclinical research is crucial in ultimately making advancements in patient care. We have discussed the challenges with properly measuring the small fields encountered in small animal dosimetry, which is critical in realizing the clinical transition from preclinical research. The main radiation sources that deliver kilovoltage energies in modern small animal studies are presented, including the most optimal dosimeters and phantom materials required to accurately measure kilovoltage x-rays and small field sizes. Current standards and best practices in kilovoltage dosimetry are discussed including the methods used to apply the gold standard in small animal dose calculations. This review will help serve as a compact reference compilation for current standards in small animal dosimetry techniques and provides a foundation for implementing new dosimetry techniques while moving forward within this growing field.

9. Acknowledgements

We would like to thank Dr. Yannick Poirier for allowing us to reproduce an adapted version of Figure 3 from the original publication and for his very helpful and valuable discussions.

10. References

[1] Draeger E *et al* 2020 *Int. J. Radiat. Oncol. Biol. Phys.* **106** 243-52
[2] Hranitzky E *et al* 1973 *Radiology* **107** 641-4
[3] Tillner F *et al* 2014 *Zeitschrift ffβr Medizinische Phys.* 24 335-51
[4] Eke I *et al* 2018 *Mol. Cancer Res.* **16** 1855-64
[5] Belley MD *et al* 2014 *Med. Phys.* **41** 034101
[6] Noblet C *et al* 2014 *Physica Medica* **30** 63-68
[7] Jensen MD *et al* 2013 *Med. Phys.* **40** 081706
[8] Arndt CD *et al* 2011 *Radiat. Res.* **175** 784-9
[9] McGurk R *et al* 2012 *Health Phys.* **103** 454-62
[10] Wong J *et al* 2008 *Int. J. Radiat. Oncol. Biol. Phys.* **71** 1591-9
[11] Matinfar M *et al* 2008 *Bone* **23** 1-7
[12] Deng H *et al* 2007 *Phys. Med. Biol.* **52** 2729-40
[13] Verhaegen F *et al* 2011 *Phys. Med. Biol.* **56** R55-83

[14] Rodriguez M *et al* 2009 *Phys. Med. Biol.* **54** 3727-40
[15] Stojadinovic S *et al* 2007 *Med. Phys.* **34** 4706-16
[16] Clarkson R *et al* 2011 *Med. Phys.* **38** 845-56
[17] Curtis H 1967 *Radiat. Res.* **7** 250-7
[18] Slatkin D *et al* 1992 *Med. Phys.* **19** 1395-400
[19] Esplen NM *et al* 2018 *Phys. Med. Biol.* **63** 175004
[20] Prezado Y *et al* 2017 *Sci. Rep.* **7** 1-10
[21] Dilmanian F *et al* 2002 *Neuro. Oncol.* **4** 26-38
[22] Dilmanian F *et al* 2003 *Radiat. Res.* **159** 632-641
[23] Yan W *et al* 2020 *Clin. Transl. Radiat. Oncol.* **20** 30-38
[24] Johnstone CD *et al* 2018 *Phys. Med. Biol.* **63** 115012
[25] Johnstone CD 2019 University of Victoria, Dissertation
[26] Livingstone J *et al* 2016 *Med. Phys.* **43** 4283-93
[27] Montay-Gruel P *et al* 2018 *Radiother. Oncol.* **129** 582-588
[28] Montay-Gruel P *et al* 2022 *Med. Phys.* **49** 2055-67
[29] Trappetti V *et al* 2021 *Int. J. Mol. Sci.* **22** 7755
[30] Engels E *et al* 2020 *Sci. Rep.* **10** 1-13
[31] Eling L *et al* 2019 *Radiother. Oncol.* **139** 56-61
[32] Burger K *et al* 2020 *Med. Phys.* **47** 5183-93
[33] Titt U *et al* 2022 *Med. Phys.* **49** 497-509
[34] Studer F *et al* 2015 *Physica Medica* **31** 607-14
[35] Romanelli P *et al* 2011 *Neurol. Res.* **33** 825-31
[36] Bayat S *et al* 2022 *Front. Physiol.* **13** 1-7
[37] Donnelley M *et al* 2010 *AIP Conf. Proc.* **1266** 30-34
[38] Livingstone J *et al* 2017 *J. Synchrotron Radiat.* **24** 854-65
[39] Ma C-M *et al* 2001 *Med. Phys.* **28** 868-93
[40] Arango EM *et al* 2020 *Phys. Med. Biol.* **65** 015012
[41] Pidikiti R *et al* 2011 *Phys. Med. Biol.* **56** 2585-99
[42] Charles PH *et al* 2014 *Med. Phys.*41 1-8
[43] Hill R *et al* 2009 *Med. Phys.* **36** 3971-81
[44] Eduardo Villarreal-Barajas J *et al* 2014 *J. Appl. Clin. Med. Phys.* **15** 331-8
[45] Robinson SM *et al* 2020 *J. Appl. Clin. Med. Phys.* **21** 314-24
[46] Brown TAD *et al* 2012 *Med. Phys.* **39** 7412-17
[47] Morales JE *et al* 2016 *Med. Phys.* **43** 4687-92
[48] Johnstone CD *et al* 2018 *Phys. Med. Biol.* **63** 115012
[49] Johnstone CD *et al* 2019 *Med. Phys.* **46** 394-404
[50] Mourya A *et al* 2018 *Health Technol.* **8** 171-7
[51] Massillon-JL G *et al* 2019 *Physica Medica* **61** 8-17
[52] Wang YF *et al* 2018 *Phys. Med. Biol.* **63** 145001
[53] Vaiano P *et al* 2019 *Physica Medica* **61** 77-84
[54] Lindsay P *et al* 2010 *Med. Phys.* **37** 571-6
[55] Hill B *et al* 2002 Phys. Med. Biol. **47** 4233-46
[56] Niroomand-Rad A *et al* 2020 *Med. Phys.* **47** 5986-6025
[57] Bazyar S *et al* 2017 *Phys. Med. Biol.* **62** 8924-42
[58] Kwon NH *et al* 2019 *J. Korean Phys. Soc.* **74** 399-404
[59] Palmer AL *et al* 2015 *Phys. Med. Biol.* **60** 8741-752
[60] Martorana F *et al* 2019 *Nucl. Technol. Radiat. Prot.* **34** 79-84
[61] Bartzsch S *et al* 2015 *Med. Phys.* **42** 4069-79
[62] Micke A *et al* 2011 *Med. Phys.* **38** 2523-34
[63] Pérez Azorín JF *et al* 2014 *Med. Phys.* **41** 1-10
[64] Méndez I *et al* 2014 *Med. Phys.* **41** 1-10

[65] Williamson JF *et al* 1999 *Phys. Med. Biol.* **44** 857-71
[66] Boivin J *et al* 2016 *Phys. Med. Biol.* **61** 5569-86
[67] Beaulieu L *et al* 2016 *Phys. Med. Biol.* **61** R305-43
[68] Deroff C Le *et al* 2017 *Br. J. Radiol.* **90** 20160454
[69] Le Deroff C *et al* 2019 *Phys. Med. Biol.* **64** 115015
[70] Le Deroff C *et al* 2020 *Med. Phys.* **47** 234-41
[71] Williamson J 1999 *Phys. Med. Biol.* **44** 857-71
[72] Esplen N *et al* 2019 *Med. Phys.* **46** 5294-303
[73] Cecchi DD *et al* 2021*Med. Phys.* **48** 7399-409
[74] Hart A *et al* 2022 *Phys. Med. Biol.* **67** 105007
[75] Lárraga-Gutiérrez JM *et al* 2015 *Phys. Med. Biol.* **60** 905-24
[76] Di Venanzio C *et al* 2013 *Med. Phys.* **40** 1-9
[77] Khan AU *et al* 2020 *Med. Phys.* **47** 4553-62
[78] Butler DJ *et al* 2018 *Med. Phys.* **45** 943-52
[79] Damodar J *et al* 2018 *Appl. Radiat. Isot.* **135** 104-9
[80] Muñoz Arango E *et al* 2021*Med. Phys.* **48** 4038-52
[81] Kampfer S *et al* 2018 *Z. Med. Phys.* **28** 303-9
[82] Da Rosa LAR *et al* 1999 *Appl. Radiat. Isot.* **50** 573-577
[83] Stasko J *et al* 2022 *Radiat. Res.* **197** 650-4
[84] Kuess P *et al* 2014 *Z. Med. Phys.* **24** 363-72
[85] Li MD *et al* 2015 *Glia.* **63** 754-67
[86] Gott KM *et al* 2020 *Dose-Response* **18** 1559325820916572
[87] Pedersen KH *et al* 2016 *Radiat. Res.* **185** 163-8
[88] Gronberg MP *et al* 2020 *Radiat. Res.* **193** 341-50
[89] Nunn AA *et al* 2008 *Med. Phys.* **35** 1859-69
[90] Garty G *et al* 2019 *Phys. Med. Biol.* **64** 3-10
[91] Al-Senan RM *et al* 2011 *Med. Phys.* **38** 4396-4
[92] Poirier Y *et al* 2018 *Med. Phys.* **45** 402-13
[93] Ramaseshan R *et al* 2004 *Phys. Med. Biol.* **49** 4031-48
[94] Hill R *et al* 2014 *Phys. Med. Biol.* **59** R183-231
[95] White D *et al* 1989 *ICRU Report 44*
[96] Feddersen TV *et al* 2019 *Australas. Phys. Eng. Sci. Med.* **42** 541-51
[97] Ford E *et al* 2017 *Phys. Med. Biol.* **62** 43-58
[98] Ghita M *et al* 2017 *Radiat. Oncol.* **12** 1-10
[99] Granton PV *et al* 2012 *Med. Phys.* **39** 4155-66
[100] Van Hoof SJ *et al* 2013 *Radiother. Oncol.* **109** 361-6
[101] Jermoumi M *et al* 2015 *Phys. Medica.* **31** 529-35
[102] Kim MM *et al* 2019 *Phys. Med. Biol.* **64** 135013
[103] Lindsay PE *et al* 2014 *Med. Phys.* **41** 1-12
[104] Na YH *et al* 2018 *Med. Phys.* **45** 3330-9
[105] Newton J *et al* 2011 *Med. Phys.* **38** 6754-62
[106] Noblet C *et al* 2016 *Phys. Med. Biol.* **61** 3521-35
[107] Sharma S *et al* 2017 *Technol. Cancer Res. Treat.* **16** 45-56
[108] Tillner F *et al* 2016 *Phys. Med. Biol.* **61** 3084-108
[109] Tryggestad E *et al* 2009 *Phys. Med. Biol.* **54** 5341-57
[110] Wang YF *et al* 2018 *Phys. Med. Biol.* **63** 145001
[111] Zhou H *et al* 2010 *Int. J. Radiat. Oncol. Biol. Phys.* **78** 297-305
[112] Hill R *et al* 2010 *Med. Phys.* **37** 4355-63
[113] Biglin ER *et al* 2019 *Radiat. Oncol.* 14 1-10
[114] Lu L *et al* 2013 *Int. J. Radiat. Biol.* **89** 1094-9
[115] Kazi AM *et al* 2014 *Health Phys.* **106** 97-105

[116] Bazalova M *et al* 2011 *Med. Phys.* **38** 3039-49
[117] Verhaegen F *et al* 2014 *Z. Med. Phys.* **24** 323-34
[118] Cheremisinoff N *et al* 1996 In *Polymer Characterization* 43-81
[119] IAEA 1980 *Elemental Analysis of Biological Materials - Report 197*
[120] Mavromichalis I *et al* 2000 *J. Food Compos. Anal.* 13 799-807
[121] Golovanova O *et al* 2011 *J. Appl. Spectrosc.* **78** 145-8
[122] Cho J *et al* 2013 *Phys. Med. Biol.* **58** 3815-35
[123] Poirier Y *et al* 2014 *Med. Phys.* **41** 1-11
[124] Poirier Y *et al* 2012 *Med. Phys.* **39** 3041-50
[125] Motomura AR *et al* 2010 *Med. Phys.* **37** 590-9
[126] Chiavassa S *et al* 2020 *Phys. Med. Biol.* **65** 035006
[127] Chow JCL 2010 *Radiat. Phys. Chem.* **79** 567-74
[128] Chow JCL *et al* 2010 *Med. Phys.* **37** 5322-9
[129] Bazalova M *et al* 2014 *Med. Phys.* **41** 5-7
[130] Wack L *et al* 2014 *Physica Medica* **30** 36-46
[131] Lee CL *et al* 2016 *JINST* **11** T01003
[132] Esplen N *et al* 2019 *Med. Phys.* **46** 1030-6
[133] Soultanidis G *et al* 2019 *Phys. Med. Biol.* **64** 12NT02
[134] Zhang H *et al* 2018 *Phys. Med. Biol.* **63** 195011
[135] Welch D *et al* 2017 *Radiat. Res.* **187** 433-42
[136] Welch D *et al* 2015 *Phys. Med. Biol.* **60** 3589-98
[137] Wu L *et al* 2008 *Med. Phys.* **35** 3759-64
[138] Woods K *et al* 2019 *Med. Phys.* **46** 5733-47
[139] Klevenhagen S *et al* 1996 *Phys. Med. Biol.* **41** 2605
[140] Andreo P *et al* 2000 *IAEA* TRS-398
[141] Grimbergen T *et al* 1997 *Report 10 NRS10*
[142] German Institute for Standardization (DIN). *Clinical Dosimetry - Part 4* DIN 6809-4
[143] Munck Af Rosenschöld P *et al* 2008 *Phys. Med. Biol.* **53** 4431-42
[144] Jhala E *et al* 2009 *Australas. Phys. Eng. Sci. Med.* **32** 11-15
[145] Peixoto JGP *et al* 2000 *Phys. Med. Biol.* **45** 563-75
[146] Granton PV *et al* 2013 *Phys. Med. Biol.* **58** 3377-95
[147] Johnstone CD *et al* 2015 *J. Appl. Clin. Med. Phys.* **16** 118-130
[148] Poirier Y *et al* 2020 *Med. Phys.* **47** 2013-22
[149] Anvari A *et al* 2018 *Med. Phys.* **45** 3246-57
[150] Anvari A *et al* 2018 *Med. Phys.* **45** 4642-51
[151] Anvari A *et al* 2020 *Med. Phys.* **47** 1871-9
[152] Eaton DJ *et al* 2013 *J. Appl. Clin. Med. Phys.* **14** 5-17
[153] Chen Q *et al* 2019 *Phys. Med. Biol.* **64** 055001
[154] Poirier Y *et al* 2019 *Br. J. Radiol.* **92** 1-8
[155] British Institute of Radiology 1996 *Br. J. Radiol.* Suppl. 25 London (UK)
[156] Subiel A *et al* 2020 *Phys. Med. Biol.* **65** 085016.
[157] Brahme A 1984 *Acta Oncol.* **23** 379-391.
[158] Brahme A *et al* 1988 *Acta Oncol. (Madr)* Suppl. 1 1-76
[159] Chen Q *et al* 2021 *Med. Phys.* **48** 881-889
[160] Rogers DWO *et al* 2017 *Source* **509** 1-260
[161] Agostinelli S *et al* 2003 *Nucl. Instruments Methods Phys. Res. Sect. A* **506** 250-303
[162] Perl J *et al* 2012 *Med. Phys.* **39** 6818-37
[163] Kawrakow I *et al* 2000 In *The Use of Computers in Radiation Therapy*. Springer
[164] Pelowitz D 2005. MCNPX user's manual version 2.5.0. *Los Alamos Natl Lab.*
[165] Battistoni G *et al* 2016 *Front. Oncol.* **6** 116
[166] Bazalova M *et al* 2009 *Med. Phys.* **36** 4991-9

[167] Mahuvava C *et al* 2022 *Med. Phys.* **49** 1911-23
[168] Poirier Y *et al* 2020 *Med. Phys.* **47** E722
[169] Cho N *et al* 2018 *Med. Phys.* **45** 2252-65
[170] Jenkins CH *et al* 2015 *Med. Phys.* **42** 5-13
[171] Schüler E *et al* 2017 *Int. J. Radiat. Oncol. Biol. Phys.* **97** 195-203
[172] Esplen N *et al* 2020 *Phys. Med. Biol.* **65** 23TR03
[173] Poirier Y *et al* 2022 *Med. Phys.* **49** 4682-92

Optimization of MRI pulse sequences and gadobutrol-doped polymer gel for real time 4D radiation dosimetry on the MRI-Linac

Yves De Deene[1,2] and Daniel Mason[3]
[1]School of Science, University of Western Sydney, Sydney, Australia
[2]Physikum, Stockholm University, Sweden
[3]Nepean Blue Mountains Local Health District, NSW Health, Penrith, Australia

Email: Y.DeDeene@westernsydney.edu.au

Abstract. Different MRI pulse sequences can be applied to acquire real time dose distributions on an MRI-Linac. The temporal resolution is affected by the MRI pulse sequence type, the MRI pulse sequence parameters and the NMR relaxation-dose response of the polymer gel. To obtain a sufficiently high temporal resolution, it is crucial to optimize the MRI pulse sequence parameters such as TE, TR and flip angle. The optimization goal is to obtain the highest dose resolution within the shortest time frame, i.e. the smallest temporal uncertainty. We here also demonstrate that it is possible to further improve the temporal uncertainty of the polymer gel dosimeter by adding an MRI contrast agent. The optimization of two MRI pulse sequences in combination with a gadobutrol-doped MAGAT polymer gel dosimeter is discussed.

1. Introduction

In previous work on real time 4D radiation dosimetry on an MRI-Linac [1] we showed that a RARE pulse sequence (Rapid Acquisition with Relaxation Enhancement, equivalent with FSE, TSE) can be used to acquire dose maps in short time intervals, allowing dosimetry in pseudo-real time. As with any imaging technique, there is a trade-off between temporal resolution and signal-to-noise ratio (SNR), which can be comprised in the concept of temporal uncertainty. We emphasize the importance of making a clear distinction between 'temporal or imaging resolution' and 'temporal uncertainty'. 'Imaging resolution' refers to the time period between two consecutive images while 'temporal uncertainty' refers to the minimum time span in which a valid dose readout can be made [1]. In this paper, we provide a theoretical framework for optimizing both the sequence parameters and the gel formulation to obtain the highest imaging rate for which a significant measurable dose change applies, while minimizing the temporal uncertainty. While in this paper, two different MRI pulse sequences are considered, the concept is generally applicable to other MRI pulse sequences.

2. Theory

The temporal uncertainty for the different studies can be calculated using the concept of dose resolution which was introduced as a quantitative metric to define the minimal separation between two absorbed doses that can be distinguished with a given level of confidence [2, 3]. The relation between temporal uncertainty $TP_{p\%}$ and dose resolution D_{Δ}^{p} is given by

$$\mathrm{TP_{p\%}} = \frac{\mathrm{D^p_\Delta}}{\dot{\mathrm{D}}} = \frac{k_p\sqrt{2}\sigma_D}{\dot{\mathrm{D}}} \tag{1}$$

where $\dot{\mathrm{D}}$ is the dose rate and the dose resolution is given by: $\mathrm{D^p_\Delta} = \mathrm{k_p}\sqrt{2}\sigma_\mathrm{D}$ with $\mathrm{k_p}$ the coverage factor for the level of confidence p. σ_D is the standard deviation in the measured dose map. The standard deviation of the estimated dose is determined by the sensitivity of the gel dosimeter and the measurement parameters. In MRI, measurement parameters are determined by the pulse sequence and pulse sequence parameters such as the repetition time (TR), the echo time (TE) and flip angle (α). It is clear that for a certain dose rate, the optimum is reached when the standard deviation (σ_D) is minimal. In order to make a fair comparison, the measurement time to acquire a volumetric data set with a certain number of slices (N_{sl}) needs to be taken into account. Indeed, allowing a longer allocated measurement time frame allows more averages to be taken, hence increasing the signal-to-noise ratio (SNR) in the base images. As the SNR increases with the square root of the measurement time $\mathrm{T_m}$ to acquire the volumetric data set, the optimization problem ***for the RARE pulse sequence*** can be written as

$$\min_{\mathrm{TE,TR}}\{\sigma_\mathrm{D}\sqrt{\mathrm{T_m}}\} = \min_{\mathrm{TE,TR}}\left\{\left(\frac{\partial \mathrm{D}}{\partial \mathrm{R}}\right)\left(\frac{\partial \mathrm{R}}{\partial \mathrm{S}}\right)\sigma_\mathrm{S}\sqrt{\mathrm{T_m}}\right\} \cong \min_{\mathrm{TE}}\left\{\frac{\sqrt{\mathrm{T_m}}}{\mathrm{r_2^D TE\ SNR}}\sqrt{1+\mathrm{e}^{2\mathrm{r_2^D D TE}}}\right\} \tag{2}$$

where: $\mathrm{R} = \frac{1}{\mathrm{r_2^D TE}}\ln\left(\frac{\mathrm{S(0)}}{\mathrm{S(D)}}\right)$, $\mathrm{r_2^D}$ is the dose relaxivity (i.e. $\mathrm{r_2^D} = \partial\mathrm{R_2}/\partial\mathrm{D}$).

The derivation of equation 2 can be found elsewhere [4]. The last equality on the right hand side of equation 2 is only approximately valid when the longitudinal relaxation can be neglected (that is, when $TR \gg T_1$). Note however that this condition is not necessarily satisfied. In our optimization, we do not use this assumption and use the general applicable equation.

For the balanced Steady State Free Precession (bSSFP) pulse sequence the optimization becomes

$$\min_{\alpha\in\left[0,\frac{\pi}{2}\right]}\{\sigma_\mathrm{D}\sqrt{\mathrm{T_m}}\} = \min\left\{-\left(\frac{\partial \mathrm{S(D)}}{\partial \mathrm{D}}\right)^{-1}\left(\sqrt{1+(1-\mathrm{R})^2}\right)\sigma_\mathrm{S}\sqrt{\mathrm{T_m}}\right\} \tag{12}$$

where $\mathrm{R} = \frac{\mathrm{S(0)-S(D)}}{\mathrm{S(0)}}$. The derivation and the expression for $\frac{\partial \mathrm{S(D)}}{\partial \mathrm{D}}$ can be found elsewhere.

The optimization problem of a pulse sequence thus results in a two stage process: 1) finding the expression of σ_D as a function of the imaging parameters and 2) finding the minimum objective function $\sigma_\mathrm{D}\sqrt{\mathrm{T_m}}$ or the minimum temporal uncertainty $\mathrm{TP_{p\%}}$ in the imaging parameter space.

Minimization of the objective function towards concentrations of contrast agent and different flip angles can be performed numerically [4].

3. Methods and Materials

3.1. Polymer gel fabrication and irradiation

Cylindrical vials (4 ml) were filled with MAGAT gel doped with various amounts of gadobutrol (GadoVist®, Bayer). The exact gel formulation and fabrication steps can be found elsewhere [1] with the exception of the addition of gadobutrol. The samples were irradiated on a 6 MV Linac (Elekta) at reference depth with a dose rate of 5.4 Gy/min.

3.2. MRI measurements

R_2 versus dose response curves are acquired with a multi-echo pulse sequence with 32 echoes (ΔTE = 9.5 ms; TR = 3 s), while R_1 versus dose response curves were obtained by use of an inversion recovery turbo spin echo sequence with different inversion times (TI = 100 ms – 1500 ms in steps of 100 ms). To validate the optimization approach experimentally, MRI measurements are conducted on a 3T MRI scanner (Verio, Siemens) with both a RARE sequence where the echo time and repetition time were varied and a bSSFP sequence where the flip angle was varied.

3.3. Sequence optimization

For the RARE sequence, the imaging parameters that are optimized are the echo train length (ETL) and the repetition time (TR). The echo time spacing (ΔTE) is taken as low as possible (8 ms) without compromising the receiver bandwidth (130 Hz/pixel). A current constraint in our RARE sequence is that $TE_{eff} = ETL \cdot \Delta TE$. For the bSSFP sequence, the flip angle (α) is optimized while TR is taken as small as possible and TE is fixed at half the repetition time (TE = TR/2). Note that the optimal set of parameters is dependent on the relaxation rates and thus on the absorbed dose. We here consider a dose interval of [0 Gy, 4 Gy], typical for a single fraction. For the minimization objective function we consider the maximum σ_D within the considered dose interval, i.e. $\max(\sigma_D(D))$. The minimization problem is implemented in a MATLAB® script. The optimal imaging parameters and minima are determined for different amounts of gadobutrol concentrations from which the ultimate optimum with respect to gadobutrol concentration is determined.

4. Results and discussion

4.1. Gadobutrol-doped MAGAT gel response

R_1 and R_2 dose response plots for gadobutrol-doped MAGAT are shown in figure 1. Only a very small dependence of R_1 on the dose is found (figure 1b). It is clear from figure 1c that the transverse gadobutrol relaxivity r_2^{Gd} increases with increasing dose for the first 2 Gy. The increase in r_2^{Gd} can be attributed to a decrease in the rotational correlation time of the gadobutrol contrast agent with the increased formation of the methacrylic interpenetrating network. It is not clear yet, if covalent bonds between gadolinium and methacrylic acid are also responsible for the change in relaxivity.

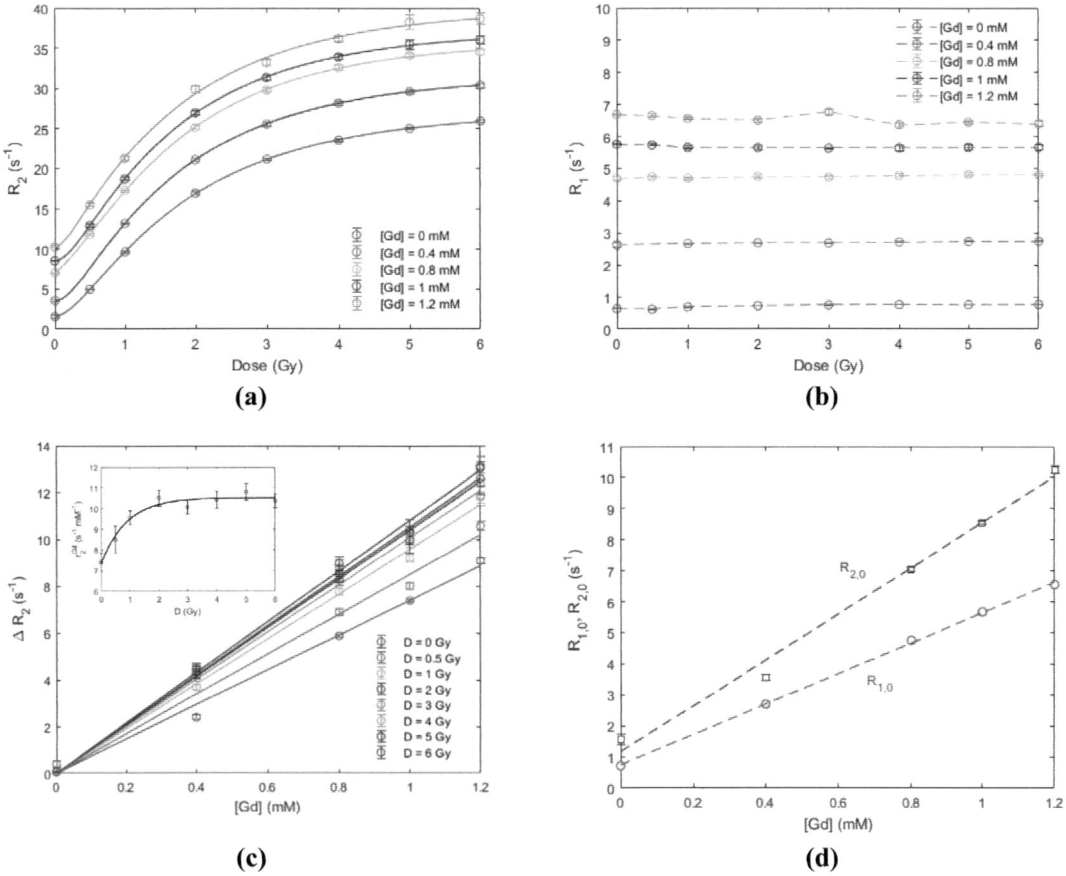

Figure 1. R_2 (a) and R_1 (b) dose response of gadobutrol doped MAGAT polymer gel, ΔR_2 as a function of gadobutrol concentration for various dose values (c) and the R_1 and R_2 relaxation rate at zero dose $R_{1,0}$ and $R_{2,0}$ (d) as a function of gadobutrol concentration. The R_1 and R_2 relaxivities for a non-irradiated MAGAT gel are respectively $r_1 = 4.908$ s^{-1} mM^{-1} and $r_2 = 7.3941$ s^{-1} mM^{-1}. The inset in (c) shows the r_2 relaxivity as a function of dose.

4.2. Optimization of the RARE pulse sequence

The signal intensity as a function of radiation dose for various echo times and repetition times is shown in figure 2. The good correspondence between measured (symbols) and theoretical (solid lines) derived signal intensity can be appreciated.

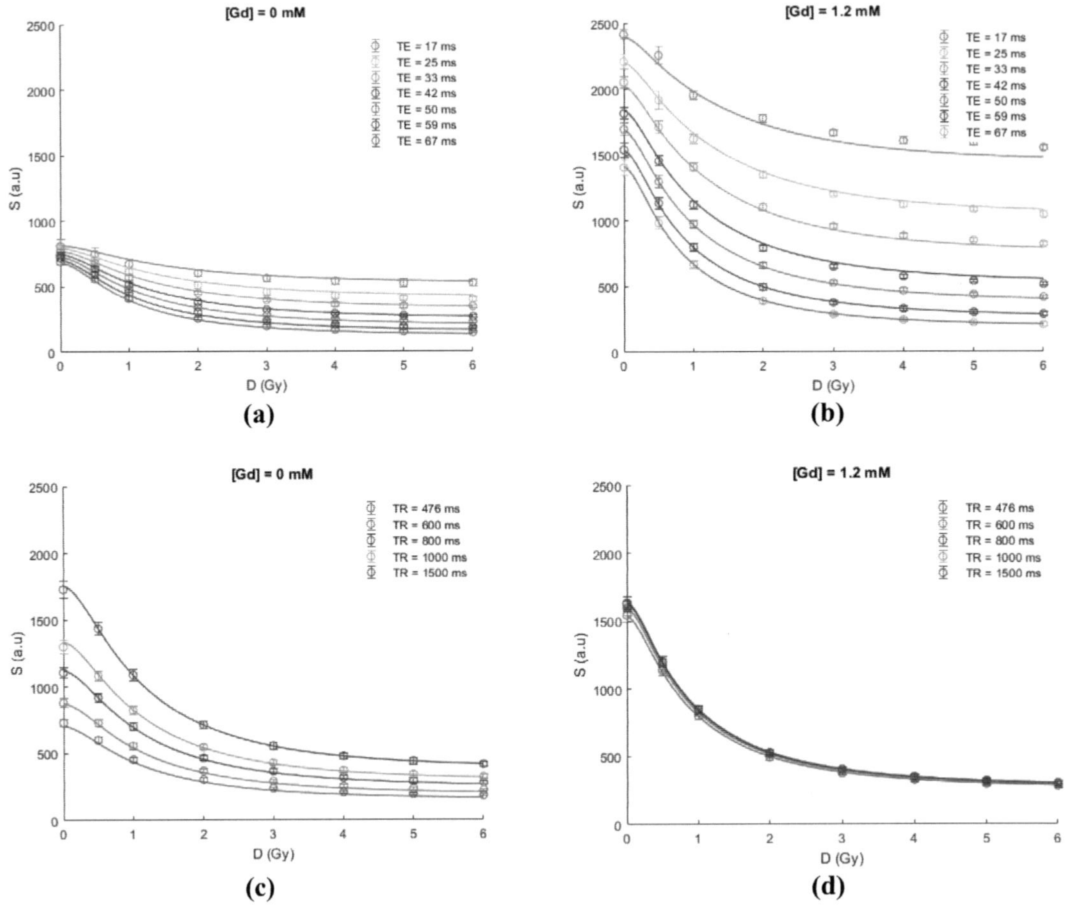

Figure 2. Signal intensity as a function of dose for various echo times (TE) with repetition time (TR) fixed at 476 ms (a-b), and for various repetition times with echo time fixed at 59 ms (c-d) for MAGAT gels doped with 0 mM gadobutrol (a,c) and 1.2 mM gadobutrol (b,d). The discrete data points are measured values while the solid lines are theoretically derived signal intensities on the basis of the experimentally derived R_1 and R_2 values.

The optimum echo train length (ETL_{opt}), repetition time (TR_{opt}), temporal uncertainty and number of experiments (averages) (NEX) for the RARE sequence are shown in figure 3 as a function of gadobutrol concentration. The optimal gadobutrol concentration is 0.58 mM which results in a temporal uncertainty $TP_{95\%}$ of 18 s for an imaging volume of 7 slices and a matrix size containing 128 phase encoding lines. The corresponding optimal TR and TE are 420 ms and 56 ms (ETL = 7, $\Delta TE = 8$ ms) respectively. A constraint in our particular RARE sequence is that the effective TE corresponds with the last recorded echo. A higher dose resolution can be obtained when this constraint is lifted but comes at the cost of a broader imaging point-spread-function (PSF).

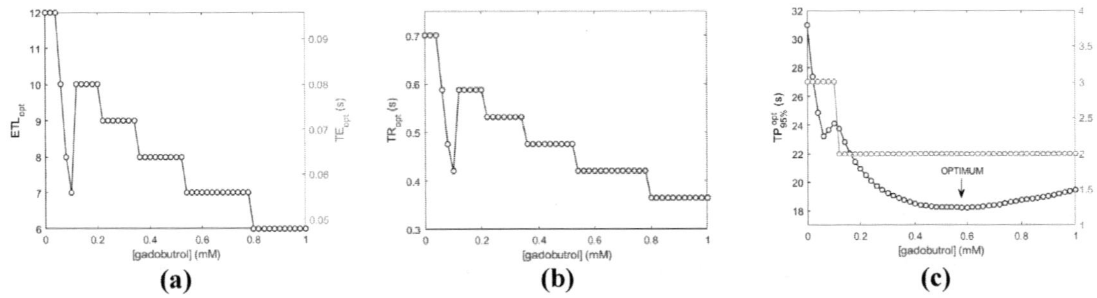

Figure 3. Optimal ETL (a), TR (b) and objective function $TP_{95\%}$ (c) for MAGAT gel doped with various amounts of gadobutrol. The optimal gadobutrol concentration is 0.58 mM. The number of experiments NEX is also shown.

4.3. Optimization of the balanced SSFP pulse sequence

For the bSSFP sequence, the total measurement time is fixed and is given by $T_m = TR \cdot N_{sl} \cdot N_{ph} \cdot NEX$. Only the number of experiments NEX is considered as variable in the total measurement time. The optimum gadobutrol concentration with the bSSFP-sequence is 1.2 mM with an optimal flip angle of 90 degrees. The corresponding temporal uncertainty is 31 s (NEX = 3).

5. Conclusions

We here present an optimization strategy of the pulse sequence parameters in combination with a contrast agent doped polymer gel dosimeter. Two pulse sequences are considered. The parameter optimization takes into account the number of slices, the longitudinal relaxation and the measurement time. It is found that the effect of longitudinal relaxation can not be neglected in the optimization of the imaging parameters. For the RARE sequence the optimal MAGAT gel formulation is obtained for 0.58 mM gadobutrol with an optimal TR/TE of 420 ms/56 ms. The addition of 0.58 mM gadobutrol results in a reduction of the temporal uncertainty of 42%. For the bSSFP sequence, the optimal gadobutrol concentration is 1.2 mM with an optimal flip angle of 90°. The achievable temporal uncertainty is significantly higher for the bSSFP sequence than for the RARE sequence (31 s for bSSFP versus 18 s for RARE). It is concluded that the optimal imaging parameters and gadobutrol concentration is highly dependent on the pulse sequence and experimental design (e.g. number of slices). The reader is advised to conduct the optimization for their specific experimental design. Further reduction in the temporal uncertainty can be achieved by acquiring more averages of the pre-radiation reference scan. Future work will focus on the implementation of sparse sampling and a spiral keyhole imaging strategy to further reduce the temporal uncertainty.

6. References

[1] De Deene Y *et al* 2020 *Phys. Med. Biol.* **65** 225031
[2] Baldock C *et al* 2001 *Phys. Med. Biol.* **46** 449-60
[3] De Deene Y *et al* 2002 *Phys. Med. Biol.* **47** 3117-41
[4] De Deene Y 2023 *J. Phys.: Conf. Series* (Review paper, this volume)

Journal of Physics: Conference Series **2630** (2023) 012015 doi:10.1088/1742-6596/2630/1/012015

Spatial and Dosimetric accuracy of 3D polymer gel with CBCT readout - Varian HyperArc® SRS implementation

Tenzin Kunkyab[1], Michelle Hilts[1,2], Andrew Jirasek[1], Derek Hyde[1,2]
[1]Department of Mathematics, Statistics, Physics & Computer Science, University of British Columbia – Okanagan campus, Kelowna, V1V 1V7 Canada
[2]BC Cancer - Kelowna, Kelowna, V1Y 5L3, Canada

Email: kunkyab.tenzin@bccancer.bc.ca

Abstract. 3D polymer gel dosimetry is a promising means to verify complex radiation treatments such as stereotactic radiosurgery (SRS), as it provides both 3D dosimetric and spatial information. The purpose of the study is to use a polymer gel read-out with cone-beam computed tomography (CBCT) to commission Varian's HyperArc®- treatment planning and delivery. Three targets (3 cm, 2 cm, 1 cm diameter respectively) were defined on a treatment plan with a maximum dose of 25 Gy, resembling a single isocentre, multiple-lesion SRS plan. Pre- and post-irradiation CBCT images of the gel were obtained for dosimetry analysis. One slice containing two large targets was used to self-calibrate the entire gel volume for dose comparison. We were able to achieve sub-millimeter spatial accuracy and all evaluated gamma criterion (5% 1mm, 3% 1mm, 2% 1 mm were > 95%). In summary, in this study we have demonstrated that CBCT polymer gel dosimetry can be a highly valuable tool for commissioning complex radiation treatment techniques such as SRS.

1. Introduction

3D polymer gel dosimeters (PGD) are a promising tool for treatment verification of complex radiotherapy techniques such as SRS.[1] The inherent three-dimensional nature of PGD's have the advantages of providing both 3D dosimetric and spatial information. Generally, gel dosimeters require a read-out method to extract dose information after the irradiation. This can be achieved with magnetic resonance imaging (MRI), optical computed tomography (OCT) and x-ray computed tomography (CT),[2] however, more recently, LINAC integrated cone-beam CT (CBCT) has become a promising alternative [3, 4]. The advantage of CBCT read-out is that it eliminates any positioning inaccuracies introduced when setting up the irradiated dosimeter for imaging off-line (for example with MRI or CT). Jirasek et al. have shown that acquiring CBCT images of the gel dosimeter 20-30 min after irradiation provides excellent polymerization yield (> 90%).[3] Due to this short wait-time after the irradiation, we can achieve a close to "real-time" image acquisition after the dose delivery with CBCT modality.

In this preliminary study, we use a polymer gel dosimeter imaged with CBCT to evaluate the dosimetric and spatial accuracy of a non-coplanar SRS technique, using a HyperArc® treatment plan (Varian Medical Systems, Palo Alto CA).

2. Material and Methods

2.1. Gel fabrication

All the gels in this study were produced using the established manufacturing protocol outlined by Maynard et al.[5] To summarize, the gel composition was 15% N-isopropylacrylamide (NIPAM, Sigma Aldrich, St. Louis, MO), 5% gelatin (Sigma), 5% N-N'-methylenebisacrylamide (Sigma), 10 mM tetrakis hydroxymethylphosphonium chloride (THPC, Sigma) and 75% deionized water. Following mixing under precisely controlled temperatures, the final mixture was transferred into a 1 L gel jar (Modus Medical, London, ON), and then allowed to cool in a water bath in the refrigerator for a minimum of 5 hours.

2.2. Treatment planning

The Eclipse treatment planning software (Varian Medical Systems, CA) was used to create a HyperArc® plan with 3 spherical targets (diameters 3 cm, 2 cm, and 1 cm), prescribed to a dose of 20 Gy (and maximum dose up to 25 Gy), mimicking an SRS treatment for multiple brain metastases. The Acuros dose calculation algorithm was used with a 1 mm dose grid. The "HyperArc®" treatment plan comprised of 3 arcs (2 non-coplanar): couch rotation of 0° with the co-planar arc, 315° and 45° with the non-coplanar arcs.

2.3. Irradiation

The gel dosimeter was allowed to equilibrate at room temperature prior to each irradiation. Similar to Johnston et al.[6], the gel was positioned in a custom-made gel holder affixed securely to the treatment couch to maintain alignment during each couch rotation of the treatment. The set-up is shown in figure 1.

Figure 1. Illustrates the experimental set-up for the PGD irradiation and imaging.

A Varian Truebeam LINAC utilizes a High-Density Multi-leaf collimator (HDMLC), with 6MV Flattening-filter-free (FFF) to deliver a volumetrically modulated arc therapy (VMAT) at up to 1400 MU/min. Couch rotations were manually set-up to the next position during the treatment since automated couch movement can only be enabled when delivering a true HyperArc® treatment technique with encompass immobilization. The overall irradiation time took < 7 minutes.

2.4. CBCT read-out
LINAC integrated CBCT was used to acquire 5 pre- and 5 post-irradiation scans of the gel. Pre-irradiation scans were used for background subtraction because setting up a separate background gel would introduce additional positioning set-up error. For all CBCT images, scan parameters were: tube voltage of 125 kVp, tube current of 1620 mAs, 5400 projections and a 1 mm reconstructed slice thickness. In order for the gel to polymerize, a 20 minutes wait time was allocated after the irradiation, prior to imaging.

2.5. Image processing
All image processing and analyses were completed using an in-house MATLAB (MathWorks, Natick MA) program. 5 pre- and 5 post-irradiation CBCT images were separately averaged and the averaged pre-irradiation images were subtracted from the averaged post-irradiation image. An Adaptive mean filter (3×3 kernel) and remnant artefact removal (RAR) (window span, 7) to reduce image noise and artefacts.[3], [7] Using the DICOM coordinate system of the CBCT image, the treatment plan was aligned to enable dose comparison.

2.5.1. Dose analysis
A central slice containing the two largest targets was used to obtain a calibration curve to convert the CT number (pixel values) to dose using the corresponding treatment plan slice. The entire volume of gel was "self-calibrated" using the calibration curve obtained from the central slice. 3D Gamma analysis (https://github.com/mwgeurts/gamma) was performed with dose criterion of 5% 1mm, 3% 1mm and 2% 1 mm with 30% threshold throughout all the slices that contained 3 targets (~ 49 slices).

2.5.2. Spatial accuracy analysis
A 3D center of mass calculation was performed on the 3 targets, both in the plan and measured image. The distance between the two center of mass coordinates was calculated on the same targets, which was then converted to mm in terms of spatial distance between the two.

3. Results

3.1. 3D gamma analysis
The gamma pass percentages achieved with thresholding the dose to 30% of the maximum dose (25 Gy) was 99.9%, 98.5%, 96.6% for 5% 1mm, 3% 1mm and 2% 1mm respectively. All gamma pass percent rates are above 95%. A dose profile along three directions (horizontal: x = 120, vertical: y = 130, and diagonal) of an example slice from the measured and planned dose images are shown in figure 2 a.). The diagonal profile was plotted to evaluate the profile across both the targets. Figure 2 b.) shows the dose profile along the image slice containing the third target (1 cm).

3.2. Spatial accuracy analysis
The spatial accuracy analysis shows that delivered target center of mass agrees with planned position within 0.7mm for all targets. The spatial accuracy of the 3 cm, 2 cm, and 1 cm targets were achieved within 0.58 mm, 0.62 mm, and 0.60 mm respectively.

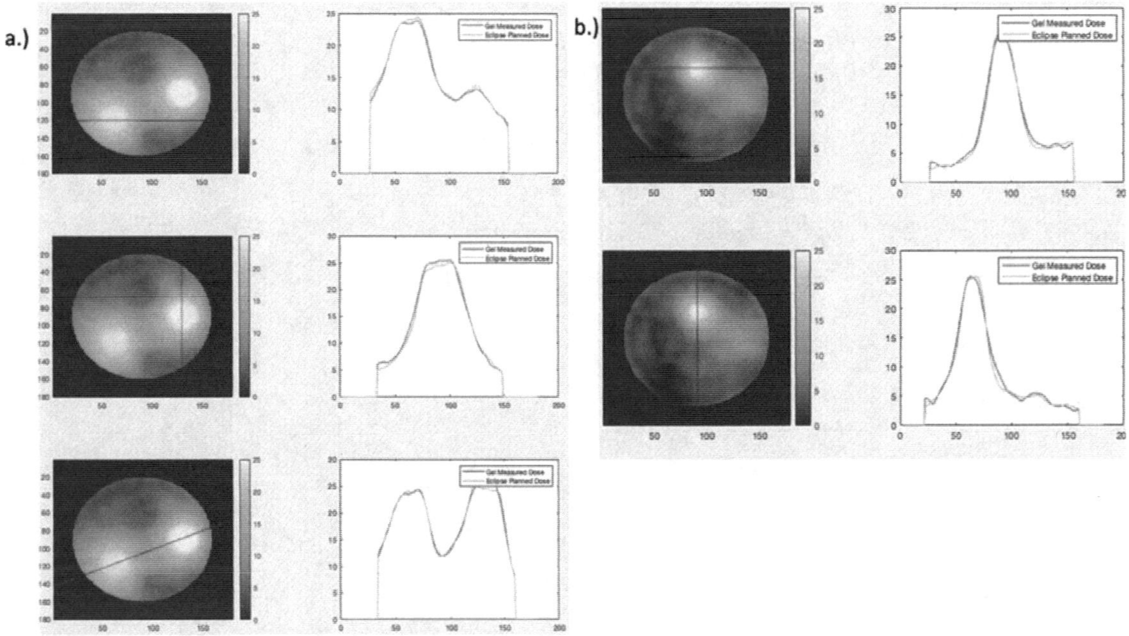

Figure 2. a.) Illustrates the dose profile along x = 120, y = 130, and diagonal direction through the 2 out of the 3 targets. 2.b.) The dose profile on the image slice containing the third target (1 cm), along x = 62, y = 91 horizontal and vertical profile.

4. Discussion

In the current study, we evaluated the dosimetric and spatial accuracy of a "HyperArc®" treatment plan comprising of 3 targets, using a novel CBCT polymer gel dosimetry technique. Both 3D spatial and dosimetric conditions were in satisfactory agreement between the treatment plan and the measurement, with gamma criterion of 5% 1mm, 3% 1mm and 2% 1mm all > 95%. Dose threshold set to 30% is used to evaluate for target coverage and conformality. 30% of maximum dose (25 Gy) delivered in this study is 7.5 Gy, which is well below the 12 Gy dose threshold for the occurrence of radiation necrosis. Furthermore, Acuros XB algorithm utilizes an adaptive mesh refinement, such that any doses below 15% would be calculated at a lower spatial resolution. For dosimetric comparison purposes, threshold is set above the 15% level.

This study forms an essential step towards fully commissioning the clinical workflow of the HyperArc® technique in our clinic. The results from the current study may also provide incentive for other institutions to utilize polymer gel dosimetry in their clinics, to verify treatment technique such as SRS, where spatial localization of the dose delivery is of utmost important. Geometric miss in an actual patient treatment delivery may cause a loss of target coverage, thereby, underdosing the PTV. Our study has shown that a sub-millimeter accuracy can be achieved with SRS treatment delivery, using the 3D spatial information extracted from the polymer gel dosimeter. Our study also highlighted the advantages of implementing CBCT for the read-out of the polymer gel dosimeter to minimize positioning inaccuracies, and essentially providing a "real-time" dose read-out. 3D reproducibility of the PGDs with multiple irradiations will be analysed in our future work.

5. Conclusion

The current study showed that CBCT polymer gel dosimetry can be a valuable clinical tool to commission highly complex radiation delivery techniques such as SRS. A 3D gamma analysis showed that all dose criterions were met and a sub-millimeter accuracy of the dose delivery was achieved.

Journal of Physics: Conference Series **2630** (2023) 012015 doi:10.1088/1742-6596/2630/1/012015

6. References

[1] Farhood B *et al* 2019 *Appl. Radiat. Isot.* **143** 47-59
[2] Baldock C *et al.* 2010 *Phys. Med. Biol.* **55** R1-63
[3] Jirasek A *et al* 2020 *Phys. Med. Biol.* **65** 225030
[4] Adamson J *et al* 2019 *Int. J. Radiat. Oncol. Biol. Phys.* 103 1271-1279
[5] Maynard E *et al* 2017 *Med. Phys.* **44** 736-746
[6] Johnston H *et al* 2012 *Phys. Med. Biol.* **57** 3155-3175
[7] Jirasek A *et al* 2012 *Phys. Med. Biol.* **57** 3137-3153

Journal of Physics: Conference Series **2630** (2023) 012016 doi:10.1088/1742-6596/2630/1/012016

A characterization of dye gel dosimeters for boron neutron capture therapy - dose response to gamma rays

Ryosuke Narita[1] and Yoshinori Sakurai[2]
[1]Graduate school of Engineering, Kyoto University, Kyoto 615-8540, Japan
[2]Institute for Integrated Radiation and Nuclear Science, Kyoto University, Osaka 590-0494, Japan

Email: sakurai.yoshinori.8n@kyoto-u.ac.jp

Abstract. A three-dimensional verification method using dye gel dosimeters is studied for the external exposure in boron neutron capture therapy (BNCT). The characterization experiments of the dye gel dosimeters were performed for the dose response to gamma rays. As the dye gel dosimeters, leucomalite-green, leucocrystal-violet and genipin gels, which have different coloration characteristics, were selected. The results for the dose response to gamma rays showed linearity up to 100 Gy. However, it was found that the correction for auto-oxidation and dose rate dependency were necessary.

1. Introduction

In boron neutron capture therapy (BNCT) as well as the other radiotherapies, it is important to estimate the patient's exposure dose accurately from the viewpoint of radiation protection. The dose distribution near the target volume for treatment is estimated by simulation calculation using a treatment planning system. The external exposure can also be estimated by simulation calculation in the same way. However, no effective method has been established for the experimental verification of three-dimensional (3D) dose distribution. We are considering a method using gel dosimeters [1].

Gel dosimetry has been attracting attention as a method of external exposure dose estimation. The gel dosimeter is composed mainly of water as same as the human body. It is flexible and has a high degree of freedom in shape, which makes it possible to create a human equivalent phantom from the dosimeter itself. The gel dosimeter phantom after irradiation can be read by diagnostic devices such as MRI and CT to estimate the 3D dose distribution.

The BNCT irradiation field is a mixed field of thermal neutrons, epi-thermal neutrons, fast neutrons, and gamma rays, and the biological effectivenesses (RBEs) of these radiation components are different. It is necessary to discriminate the radiation components in the dose estimation. It is possible to create gel dosimeters with enhanced sensitivity for each radiation component by adjusting the ingredients. Although there are various types of gel dosimeters, dye gel dosimeters are attracting attention because they can be read by compact devices such as optical CT. By using a combination of dye gel dosimeters with different coloration characteristics and different sensitivities to the radiation components, the component-discrimination estimation can be expected.

We are studying a 3D verification method using dye gel dosimeters for the external exposure in BNCT. The characterization experiments of the dye gel dosimeters were performed mainly for the dose response to gamma rays.

2. Material and methods

2.1. Dye gel dosimeters

As the dye gel dosimeters for BNCT, micellar gel dosimeters with two organic dyes such as leucomalite-green (LMG) and leucocrystal-violet (LCV), and genipin gel dosimeter with genipin dye were selected. These gel dosimeters have the different characteristics for coloration [2], [3], [4]. Using the combination of the coloration characteristics and the sensitization specific to the radiation components such as thermal neutrons, fast neutrons and gamma rays, the component-discrimination estimation by these gel dosimeters can be expected. To improve the sensitivity to thermal neutrons, boric acid was added to the LMG gel as a sensitizer. To decrease the sensitivity to fast neutrons, the genipin gel was prepared using heavy water instead of light water.

The materials and composition of the dye gel dosimeters are shown in Table 1. The prepared gels were poured into a PMMA cuvette (disposable cell) with 10 mm x 10 mm x 43 mm in inner volume and placed in an incubator at 4 degrees Centigrade for 1 hour for cooling and solidification. The light absorbance of the irradiated gel dosimeters was measured using a spectrophotometer (V730-Spectrophotometer, Japan Spectroscope Co., Ltd.).

Table 1. The materials and composition of the dye gel dosimeters.

Materials	LMG gel	LCV gel	Genipin gel
Water; H_2O [wt%] *	96	96	0
Water; D_2O [wt%]	0	0	96
Gelatin [wt%]	4	4	4
LMG [mM] *	0.3	0	0
LCV [mM]	0	1	0
Genipin [mM]	0	0	0.5
Tx-100 ** [mM]	7	4	0
TCAA *** [mM]	15	25	0
H_2SO_4 [mM]	0	0	100
n-$B(OH)_3$ [mM]	50	0	0

* wt%: weight percentage. mM: millimolar (mol/m^3). ** Since the leuco dye is insoluble in water, it is dispersed as micelles in an aqueous solution using a Tx-100: Triton X-100, surfactant. *** TCAA: Trichloroacetic acid is used for pH adjustment and it acts as a sensitizer.

2.2. Characterization for coloration by auto-oxidation

It has been reported that micellar gel colorates itself due to auto-oxidation [5]. The experiment was performed for the coloration of micellar gels and fading of genipin gel due to auto-oxidation. The unirradiated gels were stored in an incubator at 4 degrees Centigrade for 1 day, and the coloration was confirmed for the gels at 1, 2, 4, 8 and 24 hours after the preparation.

2.3. Characterization for dose response to gamma rays

To characterize the dose response to gamma rays, the gels were irradiated with doses up to 100 Gy using Co-60 Gamma-ray Irradiation Facility at Institute for Integrated Radiation and Nuclear Science, Kyoto University. The light absorbance of the gel dosimeter immediately after the irradiation was measured by the spectrophotometer. Since the main component such as light water or heavy water affects the dose response of genipin gels, the differences in dose response were compared between light water and heavy water. Similarly, since additive dependency was reported, the differences in dose response of LMG gels were compared between additive and non-additive, such as added boric acid and non-additive [6].

2.4. *Characterization for dose rate dependency*
It has been reported that micellar gel dosimeter has the dependency for dose rate [7]. The experiment was performed for the dose rate dependency. In the experiment, the dose rate was changed to 15, 30, 45, 60 and 75 Gy/h, and the irradiated dose was fixed to 60 Gy. The light absorbance of the irradiated gel dosimeters immediately after the irradiation was measured using the spectrophotometer.

2.5. *Standard error and correction*
Five measurement data of unirradiated gels, which were prepared and measured on the different days, were used for the estimation of the standard error after correcting for the effect of auto-oxidation. The changes of the light absorbance to gamma ray dose were re-estimated using the standard errors after correcting for the effects of auto-oxidation and dose rate dependence.

3. Results

3.1. *Characteristics for coloration by auto-oxidation*
Figure 1 (a) shows the changes for the light absorbance due to auto-oxidation. Measurement wavelength were 600 nm for LCV gels and genipin gels, and 625 nm for LMG gels. The light absorbance of micellar gels such as LCV and LMG increases slightly during almost 24 hours. For genipin gel, the light absorbance decreases over time during almost 24 hours.

3.2. *Characteristics for dose rate dependency*
The changes of the light absorbance for the dose rate of gamma ray irradiation are shown in Figure 1 (b). The dependency for dose rate was confirmed for all the dye gel dosimeters. The genipin gel showed a high correlation by linear approximation and the micellar gels showed a high correlation by logarithmic approximation.

3.3. *Characteristics for dose response to gamma rays*
The light absorbance for the dose response corrected for the auto-oxidation and dose rate dependency are shown in Figure 1 (c). In this figure, the standard error was shown. Some data of LMG gel did not follow the approximate straight line. The dose response of the dye gels to gamma rays showed linearity up to 100 Gy, with strong correlation coefficients of R = 0.99 or more. For genipin and LMG gels, the color change could be clearly observed visually, while for LCV gels, only a slight change could be observed visually.

3.4. *Influences for D_2O vs H_2O and additive in gel*
The light absorbance was almost two times larger for the genipin gel with light water than that with heavy water, as shown in Figure 1 (d). However, the linearity for the gamma ray dose was almost same for the both genipin gels. The coloration of the gel with the boric acid additive was smaller than non-additive when it was prepared, and the coloration rate was large.

4. Conclusion
It is considered that the correction for the effect of auto-oxidation and dose rate dependency using the approximate equation is necessary because the effect and dependency are large and not negligible.

It is difficult to confirm the color change visually and estimate the dose using the dye gel dosimeters under the same irradiation condition for BNCT clinical study, since the maximum dose is almost 10 Gy for normal tissue. The usability under the BNCT condition should be re-considered after the characterization for the response to neutrons and the dependency of linear energy transfer.

It was confirmed that there was no problem in the utilization of the genipin gel dosimeter with heavy water, and that the gel with the boric acid additive was more suitable for the measurement. The pH measurements for the gel dosimeters will be performed in future experiments.

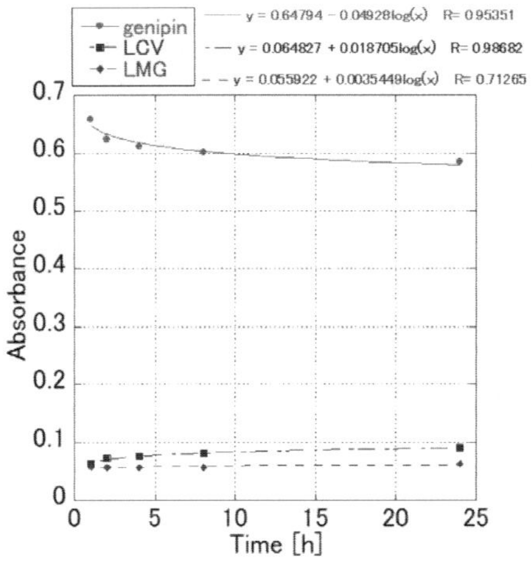

(a) due to auto-oxidation

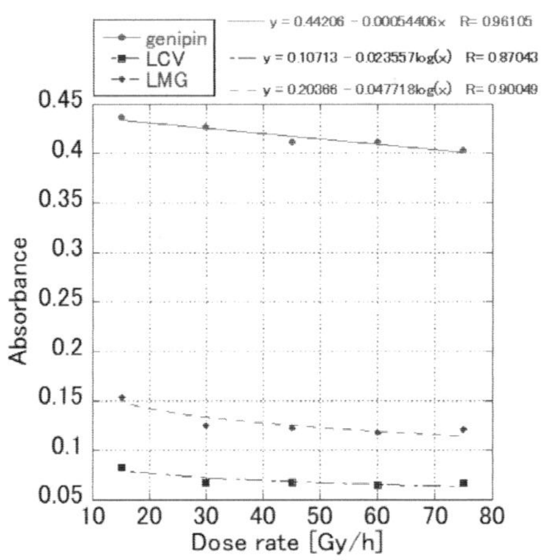

(b) for dose rate

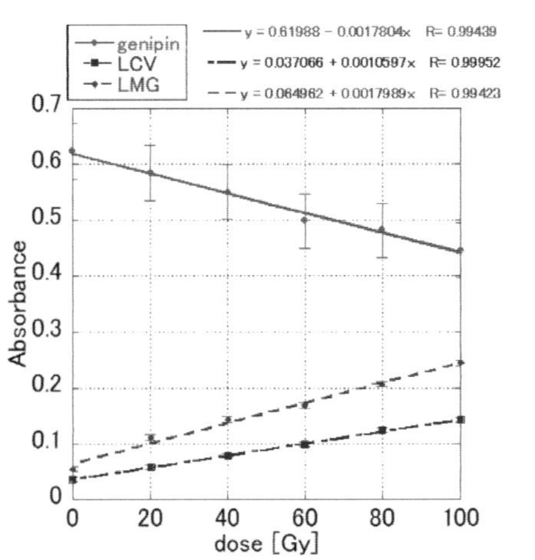

(c) for dose response

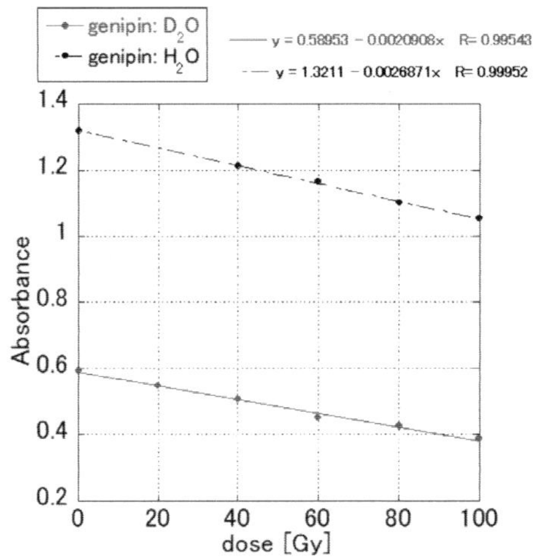

(d) for D_2O vs H_2O

Figure 1. The changes of the light absorbance

The standard error of genipin gel was larger than that of micellar gel. In addition to the weighting error during preparation, it is thought that the variation was larger due to the possibility that the dye was not uniform depending on the mixing conditions.

For the response characteristics corrected for the auto-oxidation and the dose rate dependency with standard error, some data of LMG gel did not follow the approximate straight line. Considering this fact, the sensitivity was slightly greater at doses as low as 40 Gy for LMG gel.

5. References

[1] Gambararini G *et al* 2007 *Radiat. Prot. Dosim.* **126** 604-607
[2] Jordan K,*et al* 2009 *Phys. Med. Biol.* **54** 6773-6789
[3] Babic S *et al* 2009 *Phys. Med. Biol.* **54** 6791-6808
[4] Davies J B *et al* 2010 *J. Phys: Conf. Ser.* **250** 012008
[5] Nasr A T *et al* 2015 *Phys. Med. Biol.* **60** 4665-4683
[6] Nasr A T *et al* 2015 *Phys. Med. Biol.* **60** 4685-4704
[7] Vandecasteele J *et al* 2011 *Phys. Med. Biol.* **56** 627-651

Journal of Physics: Conference Series **2630** (2023) 012017 doi:10.1088/1742-6596/2630/1/012017

Preliminary research on the dose response of polymer gel dosimeter with carbon beam irradiation

Libing Zhu[1], Xincheng Xiang[1], Xiangang Wang[1]
[1]Institute of Nuclear and New Energy Technology, Tsinghua University, Beijing, 100084, P.R. China

Email: wangxiangang@tsinghua.edu.cn

Abstract. This paper is to investigate dose response of polymer gel dosimeter after carbon irradiation. We employed polymer gel dosimeter, MAGAT-f to measure the dose response curve, single spot shape and Bragg peak range of carbon beam. Irradiation energy was 80.55 MeV/u. In order to characterize the dose response, R_2-dose curve and absorbance-dose curve were built up. An approximate linearity is found for R_2-dose relationship. Saturation is found for the absorbance of the gel after 10 Gy. The results illustrate the feasibility of utilizing polymer gel dosimeter for Bragg peak and spot shape measurement of carbon ion beam.

1. Introduction
Carbon ion beams have the advantage of dose conformity compared with X-rays radiotherapy. Polymer gel dosimeter is a promising tool for patient-specific QA due to its advantage of 3D dose measurement with high resolution. Kantemiris et al [1] investigated carbon beam dosimetry using VIP polymer gel. Maeyama et al [2] studied the radiological characteristics of VIP polymer gel under carbon beam irradiation. Mizukami et al [3] studied a nanocomposite Fricke gel dosimeter under carbon ion beams irradiation. Furuta et al [4] compared Monto Carlo simulation and measurement with PAGAT gel in biological samples for carbon ion beam. There is little exploration in the dose response of MAGAT-f polymer gel dosimeter with carbon beam irradiation. The purpose of this paper is to investigate the dose response of MAGAT-f gel and to utilize the gel to measure the Bragg peak and spot shape of carbon ion beam.

2. Material and methods

2.1. Gel preparation
The gel was fabricated according to the procedures in [5]. All the ingredients were manufactured by Sigma-Aldrich and mass concentration of each component is in Table 1. After preparation, the gel was poured into ten plastic cuvettes for dose response measurement. For Bragg peak and spot shape measurement, the gel was poured into a large volume container.

Table 1 Mass concentration of polymer gel dosimeter

	MAGAT-f
Distilled water	83%
Gelatin (Type A, 300 Bloom)	8%
Methacrylic acid 99.5%	6%
Formaldehyde solution with 37% minimum and stabilized with 10% methanol	3%
N, N'-methylene-bis acrylamide 99%	-
Acrylamide 99%	-
Tetrakis(hydroxymethyl) phosphonium Chloride (THPC) 80%	10 mM

2.2. *Irradiation*
The Heavy Ion Research Facility in Lanzhou (HIRFL) can accelerate ions from hydrogen to uranium to high energies. It consists of Electron Cyclotron Resonances (ECR) ion sources, the Sector Focusing Cyclotron (SFC), the Separated Sector Cyclotron (SSC) and the Radioactive Ion Beam Line in Lanzhou (RIBLL) etc. In our experiments, the cuvette was positioned at ISO-center and irradiated with carbon ion beam with dose ranging from 0 to 18 Gy. The irradiation energy was 80.55 MeV/u. For Bragg peak and spot shape measurement, the irradiation dose was 10 Gy and dose rate 2 Gy/min.

2.3. *Imaging*
The gel was scanned by a 3.0 T MRI (*Phillips Achieva TX*, Best, Netherlands) with a head coil. The spin-to-spin relaxation rate ($R_2=1/T_2$) was determined using a 16-echo multi-spin echo (MSE) sequence. The scanning condition was echo spacing =22.5 ms; maximum echo time (TE) =360 ms; repetition time (TR) =3000 ms; field of view (FOV) =180 mm; matrix size (MS) =512*512 pixels; slice thickness =3 mm; number of excitations (NEX) =1. In addition, the cuvette was scanned by a spectrometer (*GE*, Boston, the USA) to investigate the relationship between the absorbance and dose. The wavelength ranged from 380 nm to 900 nm.

3. Results

3.1. *Dose response*
The dose response curve (Fig.1) illustrates linear relationship for MAGAT-f gel dosimeter. The absorbance decreases with the scanned wavelength. The integration of absorbance with the wavelength increases with the dose up to 10 Gy and an obvious saturation can be seen in Fig.2 (b) after 10 Gy.

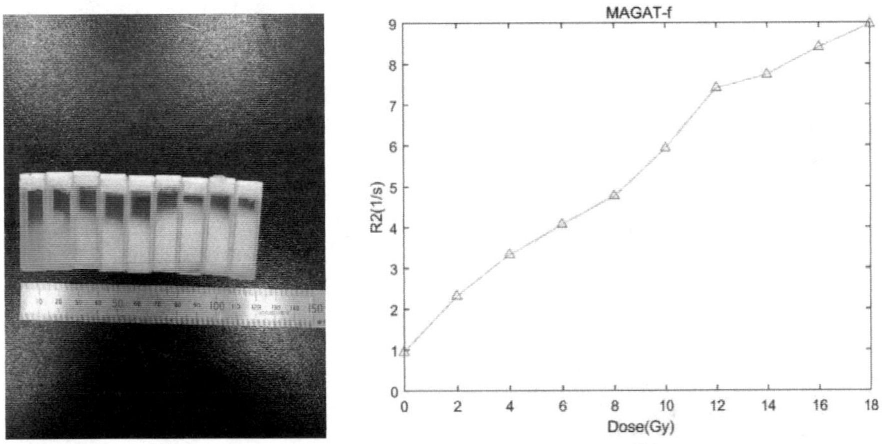

(a) (b)

Figure 1 MAGAT-f gel after carbon irradiation from left to right: 2, 4, 6, 8, 10, 12, 14, 16, 18 Gy and, (b) dose response curve

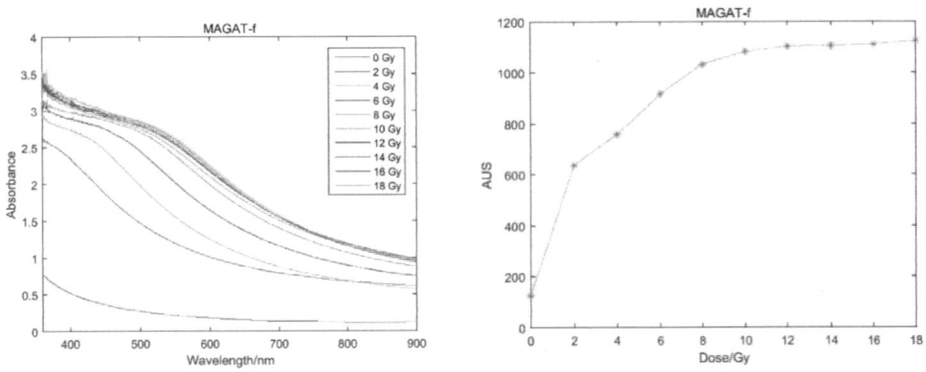

(a) (b)

Figure 2. (a) Absorbance of different scanning wavelengths and,
(b) integration of absorbance with wavelength (380 nm- 900nm) of different irradiation dose

3.2. Bragg peak

The depth-R_2 distribution obtained at HIRFL is shown in Fig.3 (c). The Bragg peak of carbon energy 80.55 MeV/u is measured with the gel. The carbon beam direction is perpendicular to the side wall of the gel container.

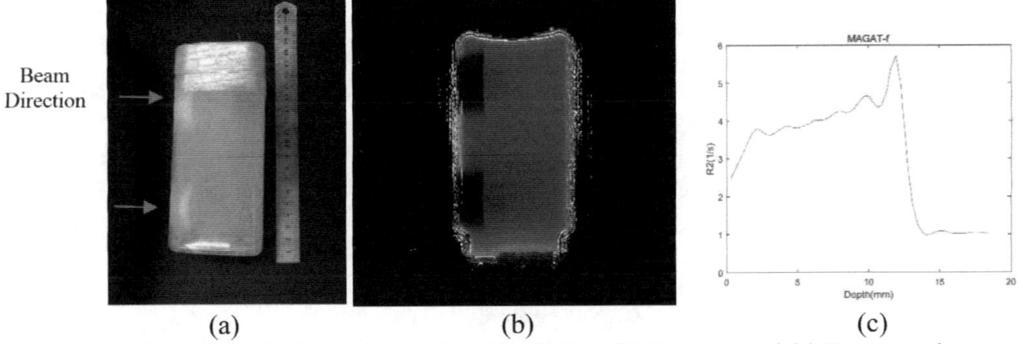

(a) (b) (c)

Figure 3. (a) MAGAT-f gel after carbon irradiation (b) T_2 map, and (c) Bragg peak curve

3.3. *Spot shape*

We measured the spot shape at different depths ranging from 0.3125 mm to 12.5 mm. The lateral dose distribution can be converted with R2-dose curve. Approximate Gaussian distribution is found for the lateral dose distribution at different depths.

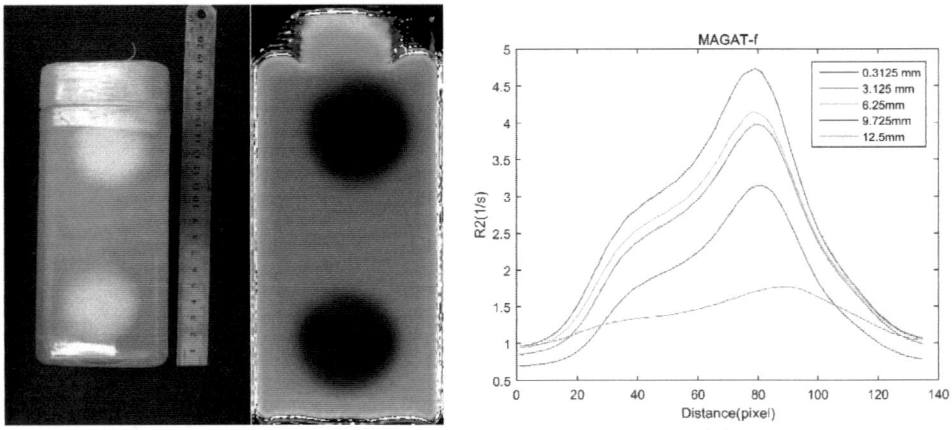

(a) (b) (c)

Figure 4. (a) Single spot measured with MAGAT-f gel, (b) T2-weight map (c), and lateral dose distribution at different depths

4. Discussion

We study the preliminary dose response of MAGAT-f gel after carbon irradiation. The irradiation energy is 80.55 MeV/u. The dependence of energy on the response curve is not included in this paper. A comprehensive study on more energies should be conducted. The lateral dose distribution at different depths is analysed. For further research, the polymer gel will be used for patient dose measurement. Furthermore, the effect of dose rate and fraction on the response curve should be investigated for carbon ion beam. All the gel preparation was conducted carefully. However, the reproducibility of the dose response is not investigated. This paper illustrates the preliminary feasibility of MAGAT-f gel dosimeter with carbon ion beam irradiation. The Bragg peak region demonstrates less response (saturation phenomenon). Proper correction method need be investigated for the saturation of Bragg peak.

5. Conclusion

Dose response curve of polymer gel dosimeter, MAGAT-f was investigated. R_2 and integration of absorbance as a function of dose are employed to characterize the dose response. An approximate linearity is found for the R_2-dose relationship, while an obvious saturation is found for the absorbance-dose curve after 10 Gy. The results illustrate the feasibility of utilizing polymer gel dosimeter for Bragg peak and spot shape measurement of carbon ion beam.

6. Acknowledgement

This work was jointly supported by the National Key Research and Development Program (No.2016YFC0105406) and National Natural Science Foundation of China (No.61571262, No.11575095, No.61171115). The authors appreciate Prof. Qiang Li of Institute of Modern Physics, Chinese Academy of Sciences for the experiments support and HIRFL for the irradiation experiments.

7. References

[1] Kantemiris I *et al* 2009 *J. Phys: Conf. Ser.* **164** 012055
[2] Maeyama T *et al* 2015 *Rad. Phys. Chem.* **107** 7-11
[3] Mizukami S *et al* 2021 *Gels* **7** 233

Journal of Physics: Conference Series **2630** (2023) 012017 doi:10.1088/1742-6596/2630/1/012017

[4] Furuta T *et al* 2015 *Phys. Med. Biol.* **60** 6531-46
[5] Hurley C *et al* 2005 *Appl. Radiat. Isot.* **63** 443-56

Journal of Physics: Conference Series **2630** (2023) 012018 doi:10.1088/1742-6596/2630/1/012018

End-to-end quality assurance for Volumetric Modulated Arc Therapy with Fricke-Xylenol orange-Gelatin gel dosimeters and dual-wavelength cone-beam optical CT scanner

A Rousseau[1], C Stien[1], S Dufreneix[1,2], G Boissonnat[1], JM Bordy[1], V Blideanu[1]
[1]Université Paris-Saclay, CEA, List, Laboratoire National Henri Becquerel (LNE-LNHB), Palaiseau, France
[2]Institut de Cancérologie de l'Ouest, Angers, France

Email: alice.rousseau@cea.fr

Abstract. The whole treatment process undergone by patients in clinics with Volumetric Modulated Arc Therapy (VMAT) can be tested by implementing 3D end-to-end (E2E) quality assurance with gel dosimetry. In this work, a 3D E2E test was performed in a head phantom for the verification of a VMAT treatment, using FXG (Fricke-Xylenol orange-Gelatin) gel dosimetry and a newly developed dual-wavelength reading method on a cone-beam optical CT scanner. This dosimetric method intends to enable accurate measurements in the out-of-field zone and in the tumor volume, with an effective dose range up to 10 Gy. CT images of the phantom with a gel flask were used to create a treatment plan with a brain tumor of complex shape. A very good agreement between 90 %, 80 %, 60 % and 40 % isodose curves and high 3D γ passing rates (2%/2mm) of 98.6 % and 96.7 % between measured and computed dose maps showed that E2E tests can be successfully implemented with this novel dosimetric method.

1. Introduction

Volumetric Modulated Arc Therapy (VMAT) is a technique developed to fit tightly the target volumes and spare as much as possible the surrounding healthy tissues during treatment. Used for stereotactic irradiations, VMAT can deliver complex dose distributions characterized by non-uniform doses and steep dose gradients. 3D end-to-end (E2E) quality assurance is therefore a necessary tool to test the whole treatment process that patients undergo [1]. Gel dosimeters are promising candidates to perform this test, as they are able to measure 3D dose distributions with a high spatial resolution. Among them, Fricke-Xylenol orange-Gelatin (FXG) gel with optical CT readout presents the advantage of being easy to implement. However, it is only accurate on a limited dose range up to 3 Gy [2,3]. Therefore, a novel dual-wavelength reading method of FXG gel dosimeters has been developed to increase the applicability to a dose range up to 10 Gy, and become clinically representative of dynamic radiotherapy applications [4].

In this study, FXG gel dosimetry and the associated dual-wavelength reading method applied to the cone-beam optical CT scanner Vista16™ (ModusQA) have been used to implement a 3D E2E test in a head phantom on a Novalis TrueBeam STx accelerator (Varian).

Journal of Physics: Conference Series **2630** (2023) 012018 doi:10.1088/1742-6596/2630/1/012018

2. Materials and Methods

2.1. FXG gel preparation

The FXG gel composition selected for this study provides a high-sensitive linear dose response up to 10 Gy [4]. The FXG gel dosimeters are prepared with ultrapure deionized water, 5 %wt gelatin from porcine skin (gel strength 300, Type A), sulfuric acid to reach a pH of 1.6, 0.3 mM ferrous ammonium sulfate and 0.09 mM xylenol orange. They are poured into Teflon-FEP cylinders of 3.9 cm diameter and 6.3 cm height. One batch of gel is prepared for a whole experiment. The gel dosimeters are prepared one day before their irradiation and are placed in the irradiation room at least eight hours prior to irradiation, for thermal stabilization and threshold dose lowering [4].

2.2. FXG gel dose calibration

Ten FXG gel flasks were irradiated at known doses on the 0.25 – 10 Gy dose range for calibration curves establishment. TRS 398 recommendations for irradiations in terms of absorbed dose to water under reference conditions were followed. A 67.5 x 64.5 x 56.0 cm^3 water tank and a 6 MV FFF (Flattening Filter Free) photon beam with a dose rate of 1400 MU/min were used with the Novalis accelerator. It was previously verified that the gel calibration coefficient was not dependent on the irradiation conditions (whether in a water tank or in an anthropomorphic head phantom).

2.3. Treatment planning and irradiations

The STEEV phantom (CIRS) was used for 3D E2E quality assurance tests (Figure 1 (a)). A water-equivalent plastic insert (outer dimensions 9.4 x 6.4 x 6.4 cm^3) was designed to fit a FXG gel flask into the phantom (Figure 1 (b)). CT images of the head phantom containing a gel flask, and covered with a thermoplastic mask (used for immobilization) were acquired using a CT scanner (Somatom, Siemens Healthineers). A 360° two-arc VMAT treatment plan with a calculation grid spacing of 1.25 x 1.25 x 1 mm^3 was created using Eclipse software (Varian) with the AcurosXB algorithm. A heterogeneous dose distribution was prescribed to a brain tumor of complex shape with three dose levels (4.9 Gy, 5.5 Gy and 6.5 Gy) to apply a dose gradient in the Planning Target Volume (PTV). The maximum dose delivered was 6.9 Gy (Figure 2). The phantom was precisely repositioned prior to irradiation using the ExacTrac system (Brainlab). Two gel flasks were irradiated with this treatment plan. Irradiations for FXG gel dose calibration and E2E tests were performed within one day to have the same thermal and spontaneous oxidation histories in the gel flasks.

(a) (b)

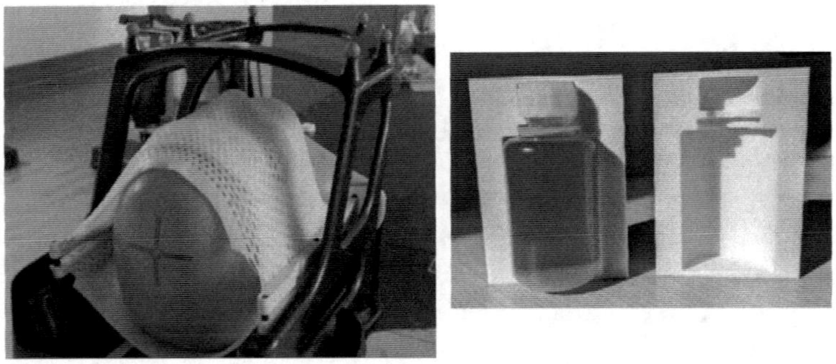

Figure 1. (a) STEEV phantom with thermoplastic mask for 3D E2E quality assurance and (b) water-equivalent plastic insert designed to fit the gel flask into the head phantom.

2.4. Dual-wavelength cone-beam optical CT readout

The new dual-wavelength reading method has been applied to the cone-beam optical CT scanner Vista16™ for the optical readout of the FXG gel flasks [4]. For the E2E QA measurements, the low dose values measured at 590 nm have been combined to the high dose values measured at 633 nm with

Journal of Physics: Conference Series **2630** (2023) 012018 doi:10.1088/1742-6596/2630/1/012018

a dose threshold between the two dose maps set at 2.5 Gy. A pre-irradiation reference scan and a post-irradiation data scan of the dosimeter were acquired at both wavelengths, each one with 2000 projections. The Feldkamp-Davis-Kress (FDK) algorithm and a Hamming filter were applied for the reconstruction with voxels of dimensions 0.5 x 0.5 x 0.5 mm^3. The matching liquid in the tank was a mixture of 10 %wt propylene glycol – deionized water.

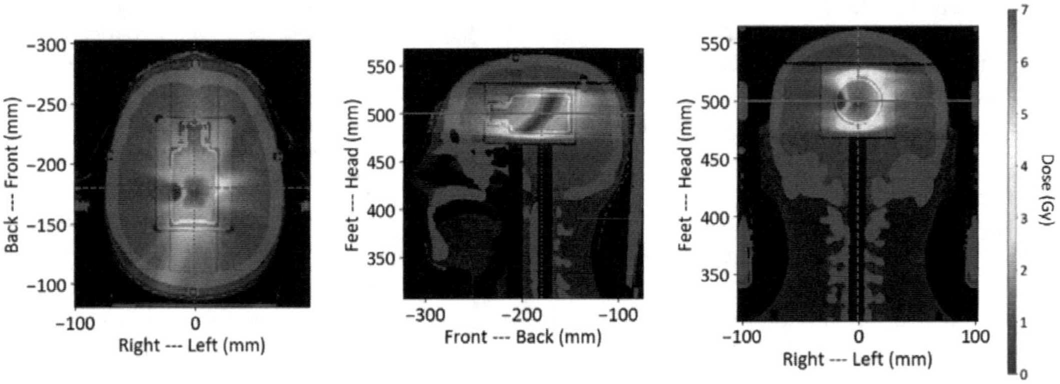

Figure 2. Axial, sagittal and coronal views of the computed dose distribution overlaid with the CT images of the head phantom and a gel flask.

2.5. Dose distribution analysis

Linear calibration curves $D = a_\lambda \times \Delta\mu + b_\lambda$ ($\lambda = 590$ nm or 633 nm) were established on the 0.25 – 4 Gy and 2 – 10 Gy dose ranges at 590 nm and 633 nm respectively. The optical attenuation coefficients $\Delta\mu$ in the previous equation correspond to the average values of the voxels contained in cubic regions of interest (ROIs) of 1 cm^3 at the center of the gel calibration flasks.

For the gel flasks irradiated with the VMAT treatment plan, a marking was made at their top to facilitate registration of the measured dose maps to the calculated one. Measured and computed dose distributions were analyzed with in-house programs developed using Python software. Cylindrical ROIs of 34 mm diameter and 53.5 mm height, corresponding to the volumes of gel inside the flasks, were selected in the dose maps. The comparison between measured and calculated dose distributions was performed by displaying isodose curves, dose profiles and Dose-Volume Histograms (DVHs) in the gel flasks, and by conducting 3D and 2D local γ-index analysis with 2%/2mm passing criteria. The PyMedPhys library was used in Python to conduct the γ-index analysis [5].

3. Results and Discussion

90 %, 80 %, 60 % and 40 % isodose curves are displayed in Figure 3 in the axial, sagittal and coronal slices passing through the tumor volume. A very good agreement between the isodose curves is observed, with only a slight discrepancy between 40 % isodose curves of maximum 2 mm. In addition, excellent 3D γ passing rates of 98.6 % and 97.6 % between computed and measured dose maps when using both gel flasks have been obtained. High 2D γ passing rates in the axial, sagittal and coronal slices passing through the tumor volume are also found between 92.8 % and 99.3 %. The failing points are mainly located in the regions where doses are below 3.5 Gy.

Figure 4 (a) represents an example of the measured and computed dose profiles along the front-back axis passing through the tumor volume. Low dose differences, below 2 %, are reported in the tumor volume between computation and measurement. However, higher dose differences (up to 7 %) are found for doses below 3.5 Gy. Nevertheless, the measured and computed DVHs of the gel flasks display a close agreement in the whole 0 – 7 Gy dose range (Figure 4 (b)).

The discrepancies found for doses below 3.5 Gy are not due to the dual-wavelength reading method as the change of reading wavelength was applied at 2.5 Gy, meaning that these discrepancies were observed at both wavelengths. They are not due to diffusion effects in the gel matrix either, as dose values below 0.5 Gy have been accurately measured in other experiments involving different treatment plans. The most probable reason may be that the treatment cannot be exactly delivered as planned in these regions, as significant dose constraints were applied to display a complex dose distribution. However, these discrepancies would not have been observed if less strict γ passing criteria had been selected (i.e. 3%/2mm with a global γ-index analysis), which are commonly used for QA tests [6]. Therefore, these discrepancies remain lower than the requirements for validating this treatment plan in clinical routine. Nevertheless, their cause will be investigated further with 2D film-based measurements.

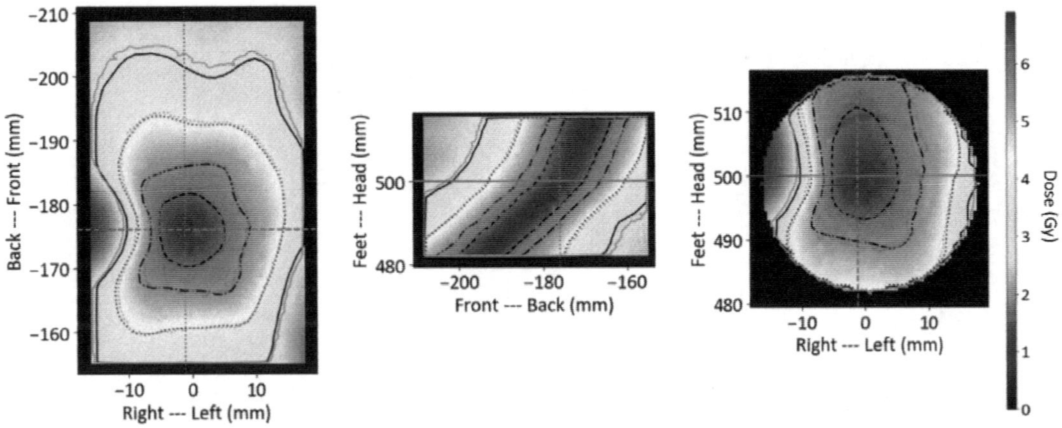

Figure 3. 90 %, 80 %, 60 % and 40 % isodose curves of the measured and computed dose maps in the axial, sagittal and coronal slices passing through the tumor volume in a gel flask (in background: measured dose distribution). In black: calculated isodose curves; in gray: measured isodose curves.

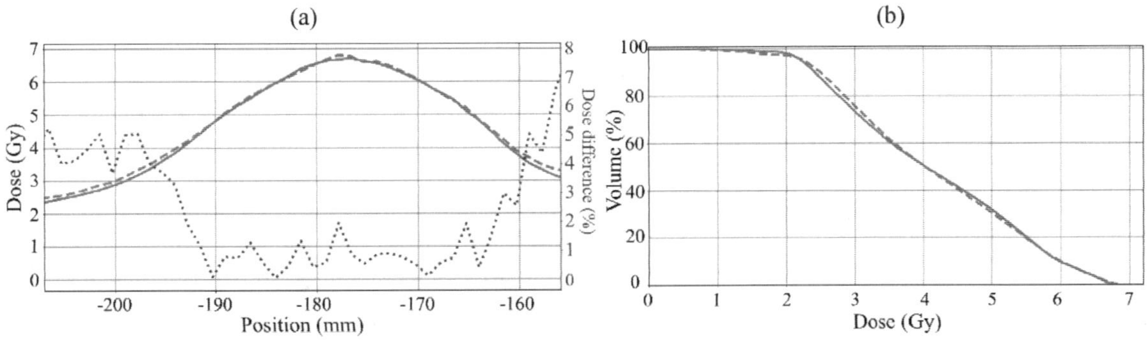

Figure 4. (a) Measured and computed dose profiles along the front-back axis passing through the tumor volume and corresponding dose differences between computation and measurement (in blue). (b) Measured and computed DVHs of a gel flask. Black lines: computation; red dashed lines: gel measurements.

4. Conclusions

The high γ passing rates and very good agreement between isodose curves, dose profiles and DVHs between computed and measured dose maps show that 3D E2E quality assurance for VMAT based on FXG gel dosimetry and dual-wavelength optical CT readout can be effectively implemented.

5. References

[1] Schreiner L J 2015 *J. Phys.: Conf. Ser.* **573** 012003
[2] Babic S *et al* 2008 *Int. J. Radiat. Oncol. Biol. Phys.* **70** 1281-91
[3] Alexander K M *et al* 2022 *J. Phys.: Conf. Ser.* **2167** 012022

[4] Rousseau A *et al* 2022 *Phys. Med.* **97** 1-12
[5] Wendling M *et al* 2007 *Med. Phys.* **34** 1647-54
[6] Miften M *et al* 2018 *Med. Phys.* **45** e53-83

Journal of Physics: Conference Series **2630** (2023) 012019 doi:10.1088/1742-6596/2630/1/012019

3D optically-stimulated-luminescence-based dosimetry using LYSO:Ce scintillators

Mads L. Jensen[1], **Rosana M. Turtos**[1], **Ludvig P. Muren**[2], **Brian Julsgaard**[1] **and Peter Balling**[1]

[1]Department of Physics and Astronomy, Aarhus University, Ny Munkegade 120, DK-8000 Aarhus, Denmark
[2]Department of Medical Physics, Aarhus University, Palle Juul-Jensens Boulevard 99, DK-8200 Aarhus, Denmark

E-mail: mlj@phys.au.dk

Abstract. The search for a reusable 3D dosimeter is ongoing and motivated by the impact it would have on development and verification of complex modalities in radiotherapy. We present a proof-of-concept 3D measurement of a proton-irradiated LYSO:Ce scintillator, using the resettable photon-emission mechanism known as optically stimulated luminescence and a novel optical readout system. Through this demonstration, we show that LYSO:Ce, in addition to being capable of real-time beam imaging, can be employed as a reusable post-irradiation 3D dosimeter with high spatial resolution.

1. Introduction

The complexity of radiation therapy is steadily increasing, with new strategies for tumour imaging and target definitions, treatment planning, and dose delivery being active topics of research and product development [1,2]. Therefore, the need to validate treatment plans for specific treatment modalities in three dimensions (3D) is likewise increasing. Such quality assurance can be provided by 3D dosimeters, which in most cases consist of gel-based materials with an irreversible chemical response to ionizing radiation that can be mapped using optical computed tomography [3-6]. A specific example is silicone-based radiochromic dosimeters, which have recently been used to investigate the effects of deformation during irradiation [7] and to demonstrate dose-response in the presence of the magnetic field in novel MR-guided radiotherapy [8]. Although much effort has been invested in solving challenges like quenching and dose-rate dependence [9], these dosimeters are by design one-time use dosimeters that require batch-specific calibration curves [10], limiting their feasibility in a clinical setting.

The search for a reusable 3D dosimeter is ongoing, and as previously suggested [11-14], optically stimulated luminescence (OSL) based dosimetry is a prime candidate. OSL is a photon-emission mechanism available in certain large-band-gap materials, which are able to trap and store information about energy deposition from ionizing radiation. Upon irradiation, free electrons and holes generated by the ionizing radiation immediately thermalize to the lowest available energy in their respective bands. From here, the electrons and holes may be captured to metastable trap states, energetically residing within the band gap of the material, effectively storing information about the deposited dose. This information can then later be retrieved by optically exciting the trapped electrons back to the conduction band, from where they may recombine radiatively through a recombination center. Embedding OSL-active nanoparticles in a transparent polymer matrix can potentially provide a reusable tissue-equivalent dosimeter. A strong candidate for such OSL-active nanoparticles is copper-doped LiF [15], and recent studies have demonstrated the potential of commercial LiF:Mg,Cu,P (MCP) powder for OSL-based 2D film dosimetry [16-18].

Scintillators represent the forefront of room-temperature radiation detection and spectroscopy, and have also found applications for real-time beam imaging in the form of liquid scintillators [19,20]. Moreover, OSL properties have been reported in cerium-doped compounds of the lutetium orthosilicate scintillator family $Lu_{(2-x)}Y_xSiO_5$ (LYSO:Ce) [21-23], indicating that these scintillators could also be interesting for dosimetry. A recent study [24] found the OSL yield of commercial LYSO:Ce crystals to be high enough to enable high-spatial-resolution dose imaging in 3D for photon and electron excitation. The aim of the presented study was to perform proof-of-concept 3D measurements of a proton-irradiated LYSO:Ce crystal using a novel optical readout system, to demonstrate the potential of OSL-based 3D dosimetry.

2. Methods and Materials

An optical readout system for OSL-based 3D dosimetry has been developed. In short, this system reads out the OSL-signal from inside the bulk of OSL-based dosimeters in a layer-wise manner, using a CCD camera and a moving light sheet. The light sheet and dosimeter are positioned on translation stages and move in a relative way, enabling the sheet to stimulate the dosimeter in a selected layer, emptying the traps and emitting OSL-photons. These photons are then imaged in 2D onto the SOPHIA-2048BR CCD camera from Princeton Instruments using an APO Macro objective from JenOptik and appropriate optical filters. Moving the light sheet provides the third dimension, and the 2D measurements can be stacked post measurement to yield a 3D distribution. The measurements presented in this contribution were acquired using a light-sheet width of 1 mm and a 4x4 binning of the CCD, yielding a voxel size of $0.18 \times 0.18 \times 1.00$ mm^3. Readouts were performed using an integration time of 2 s at each layer, scanning layers by moving the sheet towards the camera, and repeating this 40 times. As described in detail in a previous paper from our group [24], this readout method enables detection of estimated doses in LYSO:Ce crystals as low as 0.05 Gy with 2 % precision in voxel volumes of 1 mm^3.

The measurements presented in this contribution were executed using a $20 \times 20 \times 20$ mm^3 LYSO:Ce crystal from Crystal Photonics Inc. The crystal was irradiated with 4233 monitor units of 100 MeV protons (equivalent to $\sim 10^{10}$ 100 MeV protons according to simulations) at the Danish Center for Particle Therapy, Aarhus University Hospital, and read out using the described system an hour after irradiation. An inexpensive CMOS-camera was placed next to the crystal during this irradiation to qualitatively image the scintillation response. A 2 cm SolidWater® slab was placed in front of the crystal during irradiation to allow for higher beam energy with a close-to-minimum spatial divergence at the isocenter, and still ensure full enclosure of the Bragg peak inside the bulk of the crystal. The energy deposition of this irradiation was estimated using Monte Carlo-based Geant4 simulations [25,26] (FTPT_BERT physics list), tracking the step-wise energy deposition of individual protons from a collimated beam and scoring the post-step coordinates for each energy deposition. The results of this simulation were post-processed to yield a 3D dose-to-medium map, using custom-made software written in Python.

3. Results and Discussion

Figure 1a displays a true-color image of the scintillation from the proton-irradiated LYSO:Ce crystal during irradiation and clearly shows the Bragg peak of the proton beam within the crystal. This image confirms the abilities of these crystals as real-time beam imaging tools, using e.g. multiple high-quality cameras to image the scintillation response from different angles in a tomographic approach as was demonstrated in [19,20]

After the irradiation, the OSL emission was measured using the optical readout system, yielding the 3D distribution displayed in Fig. 1b. Here, the colormap has been adjusted to have an increasing transparency in regions of very low counts to make the Bragg peak more visible. Note that the light sheet scans in the z-direction and spans the xy-plane, entering the crystal from x = 20 mm. The sheet

Journal of Physics: Conference Series **2630** (2023) 012019 doi:10.1088/1742-6596/2630/1/012019

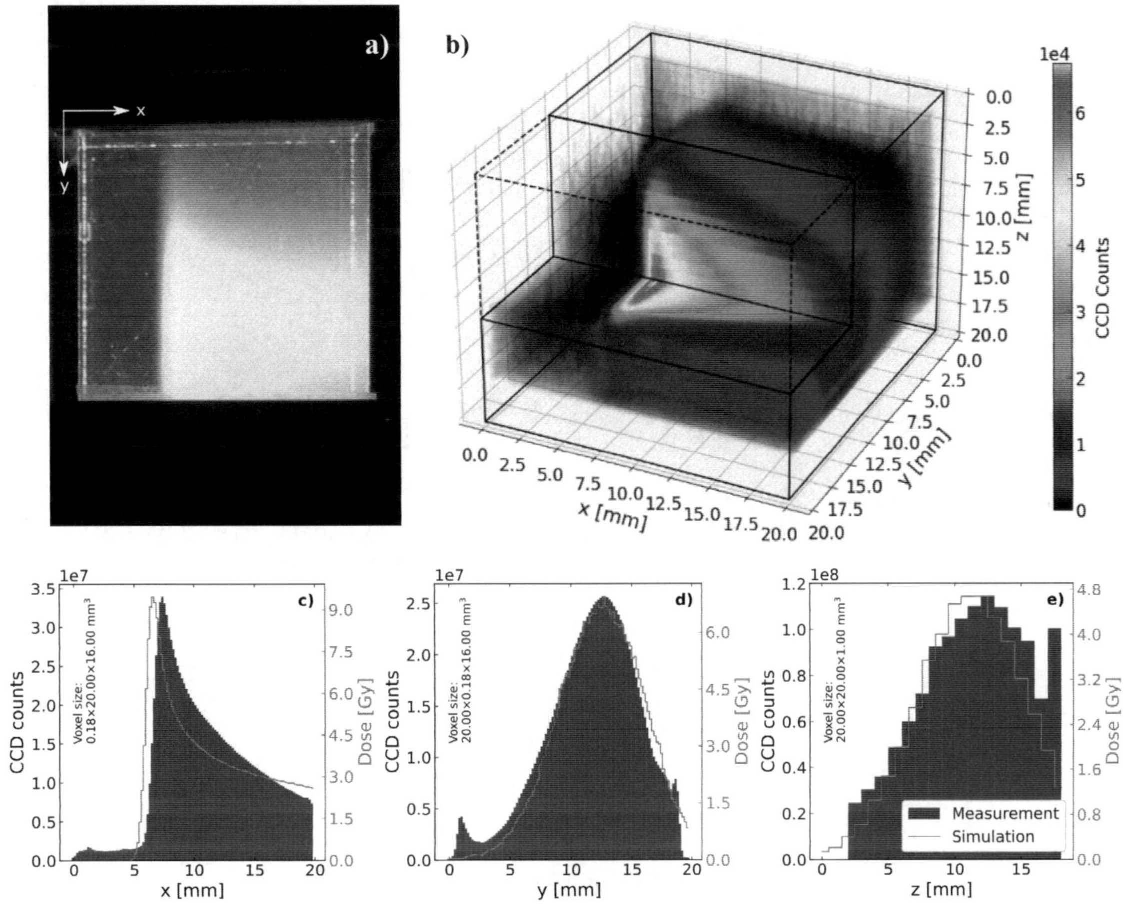

Figure 1. Readouts from a proton irradiated LYSO:Ce crystal. a) Image of the scintillation from the LYSO:Ce crystal during irradiation. Note that the camera angle gives rise to an optical artefact, exaggerating the slight displacement of the beam from the center of the crystal. **b)** 3D illustration of background-subtracted post-irradiation OSL readout of the crystal. The color-map has been adjusted to let voxels of very low intensity have increasingly high transparency to accentuate the Bragg Peak. Note that only the centremost 16 and 18 mm in the z-direction and y-direction, respectively, were stimulated. The full extent of the crystal (20x20x20 mm^3) is indicated with the black lines. The voxel size is 0.18x0.18x1.0 mm^3. **c-e)** Projections of the OSL readout displayed in b) onto each spatial axis along with a simulation indicating the estimated doses in the projected voxels with sizes indicated on the respective figures.

has been truncated to only stimulate the center-most 18 mm in the y-direction, to minimize scattering from the edges of the crystal which may cause premature readout of adjacent layers. A similar choice was made for the scanned z-range, where only sheet positions from 17.5 to 2.5 mm were read out, likewise to minimize scattering from the edges.

Projections onto each spatial axis of the signal can be seen in Fig. 1c-e along with projections of the simulated dose distribution. A good qualitative agreement is seen between data and simulation, with an especially good agreement in the falling edge of the Bragg peak in Fig. 1c, where the data and simulation have identical shapes. Moreover, the peak-to-entrance-ratios of the measurement and the simulation were estimated to be 4.2 and 3.7, respectively, varying by only 15 %. We attribute the sub-mm deviation in location (x-direction) of the Bragg peak to potential misalignments of the crystal during irradiation and/or readout and the simplified proton beam profile and material specifications used for the simulation. These measurements demonstrate the ability to read out information on the time-integrated dose distribution from OSL-based dosimeters exposed to proton-irradiation with sub-mm^3 spatial resolution in 3D. Moreover, the number of detected OSL photons per dose per voxel volume is in agreement with the value reported in [24], indicating the OSL light yield of LYSO:Ce to be within the same order of magnitude for proton and electron excitation.

The high density of LYSO:Ce (7.1 g/cm^3) inevitably makes these crystals less suitable for quality assurance in clinical radiotherapy. However, the presented results demonstrate the potential for 3D post-irradiation dose-imaging and serve as a proof-of-concept for the readout method and OSL-based 3D dosimetry.

4. Conclusion
We have demonstrated the application of LYSO:Ce scintillators as OSL-based reusable post-irradiation dosimeters for proton irradiation with high 3D spatial resolution. In accessing the information on the dose-distribution stored in these crystals, we have presented a proof-of-concept readout method for OSL-based dosimeters, and thus demonstrated the potential of OSL-based 3D dosimetry.

5. Acknowledgements
This work was funded by the Novo Nordisk Foundation. Lia Valdetaro and Peter S. Skyt are gratefully acknowledged for their help in planning, preparing, and executing the irradiation of the sample.

6. References
[1] Baumann M *et al* 2016 *Nat. Rev. Cancer* **16** 234-249
[2] Verellen D *et al* 2007 *Nat. Rev. Cancer* **7** 949-960
[3] Doran SJ 2009 *J. Phys.: Conf. Ser.* **164** 12020
[4] Schreiner LJ 2015 *J. Phys.: Conf. Ser.* **573** 012003
[5] Mijnheer B 2018 *Clinical 3D Dosimetry in Modern Radiation Therapy.* CRC Press
[6] Baldock *et al* 2010 *Phys. Med. Biol.* **55** 1-63
[7] Jensen SV *et al* 2020 *Phys. Imaging Radiat. Oncol.* **16** 134-137
[8] Jensen MB *et al* 2020 *Phys. Imaging Radiat. Oncol.* **16** 81-84
[9] Valdetaro LB *et al* 2021 *Phys. Imaging Radiat. Oncol.* **18** 11-18
[10] Jensen MB *et al* 2022 *Acta Oncol. (Madr)* **61** 264-268
[11] Sadel M *et al* 2017 *J. Phys.: Conf. Ser.* **847** 12044
[12] Balling *et al* 2018 *Eur. Pat. Appl. EP3264137A1*
[13] Nyemann JS *et al* 2020 *Radiat. Meas.* **138** 106390
[14] Nyemann JS *et al* 2022 *J. Phys.: Conf. Ser.* **2167** 012026
[15] Nielsen CL *et al* 2022 *Nano Lett.* **22** 1566-1572
[16] Nyemann JS *et al* 2023 *J. Phys.: Conf. Ser.* (This Edition)
[17] Sądel M *et al* 2021 *Sensors* **21** 6015
[18] Sądel M *et al* 2020 *Radiat. Meas.* **133** 106293
[19] Darne CD *et al* 2017 *Phys. Med. Biol.* **62** 5652
[20] Alsanea F *et al* 2019 *Nucl. Instruments Methods Phys. Res. Sect. A Accel. Spectrometers, Detect. Assoc. Equip.* **939** 16-21
[21] Twardak A *et al* 2014 *Radiat. Meas.* **71** 139-142
[22] Gartia RK *et al* 2011 *Nucl. Instruments Methods Phys. Res. Sect. B Beam Interact. with Mater. Atoms* **269** 30-33
[23] Ricci PC *et al* 2011 *J. Mater. Chem.* **21** 18425
[24] Jensen ML *et al* 2022 *Sci. Rep.* **12** 8301
[25] Turtos RM *et al* 2019 *Phys. Med. Biol.* **64** 185018
[26] Agostinelli S *et al* 2003 *Nucl. Instruments Methods Phys. Res. Sect. A Accel. Spectrometers, Detect. Assoc. Equip.* **506** 250-303

Journal of Physics: Conference Series **2630** (2023) 012020 doi:10.1088/1742-6596/2630/1/012020

Modified optical computed tomography scanner for low temperature scanning

Nathaniel Woolvett[1] and Kevin Jordan[1,2]
[1]London Regional Cancer Program, London Health Sciences Centre, London, Ontario, Canada
[2]Dept. of Medical Biophysics, Western University, London, Ontario, Canada

Email: kjjordan@uwo.ca

Abstract. Large radiochromic hydrogels, such as ferrous xylenol formulations require several hours to warm from the storage temperature of 278 K to 293 K for optical CT scanning and irradiation. This warm-up time can be avoided if gels are scanned at 278 K, resulting in a more practical 3D dose measurement. In this study a Vista16 optical CT scanner was modified to allow scanning at 278 K, below the temperature where condensation forms on the aquarium windows. The refractive index matching liquid was first cooled to 276K in order to cool the aquarium as it was filled. The modification involved pumping cool, dry air into the aquarium scanner compartment to provide a positive pressure, preventing condensation on the aquarium windows. A warming rate of 2 K per hour was achieved with passive cooling which relied on the liquid-filled aquarium's thermal mass. The radiochromic reaction rate decreased with temperature, requiring an increased post-irradiation wait time from 1 hour at 293 K to 1.5 hours at 278 K. In addition, initial sample transmission was greater by avoiding the auto-darkening that occurred as the samples were warmed from 278 K to 293 K over several hours. This increase in transmission resulted in a larger dynamic range for the dosimeter system.

1. Introduction

Radiochromic hydrogel dosimeter formulations such as ferrous xylenol orange (FX) and leuco crystal violet surfactant (LCV) are stored around 278 K to minimize darkening due to auto-oxidation. Before irradiation they are warmed to a uniform temperature, typically 293 K. For larger dosimeters, such as 15 cm diameter cylinders, this requires more than 5 hours in a water bath. A previous investigation reported reducing this time to 3 hours by active heating with a near infrared incandescent lamp [1]. If the gel dosimeter could be optically scanned and irradiated at 278 K this would reduce the time of the overall experiment. It would also result in higher transmission samples prior to irradiation since considerable darkening occurs during the 5-hour warm-up period. Higher pre-irradiation transmission results in a larger dynamic range for dosimeter system. In the case of the FX dosimeter, lowering the temperature also reduces the diffusion rate, which may be an additional advantage of low temperature scanning [2, 3]. Early investigations, in our laboratory, revealed that condensation on the aquarium windows, at the dew point temperature, was a problem for optical computed tomography (CT) scanning. Practically, this limited scans to temperatures of 283 K or greater. Also, some of the refractive index matching dye solutions that were optimized at 293 K would separate phases at lower temperatures. The earliest laser scanners were slow requiring several minutes per slice. Scan times in excess of one hour were typical. Maintaining a fixed low temperature for longer times was also another level of complexity. For these reasons, low temperature scanning was abandoned in favour of 293 K as a standard.

The Vista16, optical cone beam scanner, from ModusQA, can acquire a scan of 1000 projections in less than 5 minutes. If this scanner can be modified for 278 K scanning, then several hours per experiment could be saved, making 3D dosimetry of large gel dosimeters more practical. This investigation successfully scanned gels at 278 K with minor modifications to a Vista16 scanner which involved precooling the refractive index matching liquid and providing a positive pressure to the interior of the scanner with cool dry air to avoid condensation on the aquarium windows.

2. Methods

The Vista16 scanner has an aquarium which provides a 16 cm field of view and has a volume of 7.3 litres. The aquarium is constructed of aluminum face and top plates and plastic sides and bottom. It is aligned with the central compartment by sitting on 3 steel pins. Leaving a 0.5 mm gap between the aquarium top plate and scanner housing. There are 3 mm air spaces between the aquarium window face plates and aluminum bulkheads of the central compartment. Insulating foam sheets, 1 cm thick, were taped to the plastic sides and bottom to test if there was an increase in thermal stability. Drain ports in the bottom of the compartment were sealed with tape. Two ports were drilled to attach plastic tubing for pressurizing the compartment. The tubing was positioned to direct cool dry air into the 3 mm air spaces along the entrance and exit windows of the aquarium. Compressed lab air flowed through a heat exchanger constructed with 5 mm OD copper tubing (length = 6.3 m) coiled to fit into an insulated box and packed with ice. After the heat exchanger the tubing was connected to a water trap and routed through 6.5 mm OD nylon tubing to the Vista16 compartment. Pressure was verified by sensing cool air flow exiting the compartment above the aquarium windows.

The refractive index matching liquid was cooled to ~276 K prior to filling the aquarium. After cooling the aquarium, the liquid temperature stabilized at 278 K. Gel samples were transferred from the storage refrigerator to the aquarium for the pre-irradiation reference scan. The liquid temperature was monitored for stability. Following the reference scan, a jar filled with ice-water was substituted for the gel in the aquarium to provide passive cooling while the gel was being irradiated. Samples were irradiated in a 278 K water tank, with a 30 cm circular ^{60}Co beam at a source to surface distance of 60 cm. Maximum dose was 1 Gy, to ensure sufficient transmission for post irradiation scan. After a wait period for the radiochromic reaction to complete the gel was reinserted in the aquarium for the post scan. The laboratory temperature was maintained at 293 K.

The gel formulation (FX) consisted of 5% gelatin, 60 mM sulfuric acid, 0.5 mM tannic acid, 0.3 mM ferrous ammonium sulfate and 0.05 mM xylenol orange. Custom, cylindrical, polyethylene terephthalate vessels of 10 and 15 cm diameter were used for this study.

3. Results

Without cool, dry air to pressurize the scanner's aquarium compartment, condensation developed on the 278 K aquarium windows during summer months when the ambient building humidity was greater. Pressurizing the compartment had no impact on the aquarium warming rate. The insulating foam which covered only the plastic sides and base of aquarium also had no impact on the warming rate. The substitution of a jar of ice-water for the gel sample, when not scanning, was sufficient to maintain aquarium temperature at 278 K for greater than one hour. Sample transmission near the optic axis was 5% initially and 2.5% 85 minutes after 1 Gy irradiation relative to the liquid outside the vessel. A sequence of data scans was recorded to determine when the FX radiochromic reaction was completed. It was found that 1.5 hours was required to reach 95% completion at 278 compared to 1 hour at 293 K. While not the focus of this study, the auto-darkening of the samples was markedly reduced. Once the aquarium was filled with the 276 K solution the temperature stabilized at 298K after quickly cooling the aquarium. Without additional cooling the aquarium temperature increased 2 K per hour. The substitution of a jar of ice-water when not scanning was sufficient to maintain aquarium temperature at 278 K.

The transmission along the 15 cm diameter for a similar sample decreased to 60% compared to initial value when placed in the scanner at 278 K and allowed to warm to 293 K over 6 hours and 1.5 hours before final scan. This decrease in transmission due to auto-oxidation during warm-up and experiment could have been avoided if the 278 K scanning was employed. Figure 1 is a transverse slice from the 3D reconstruction at 1.5 cm depth in gel with dose of ~1 Gy in a 15 cm diameter vessel.

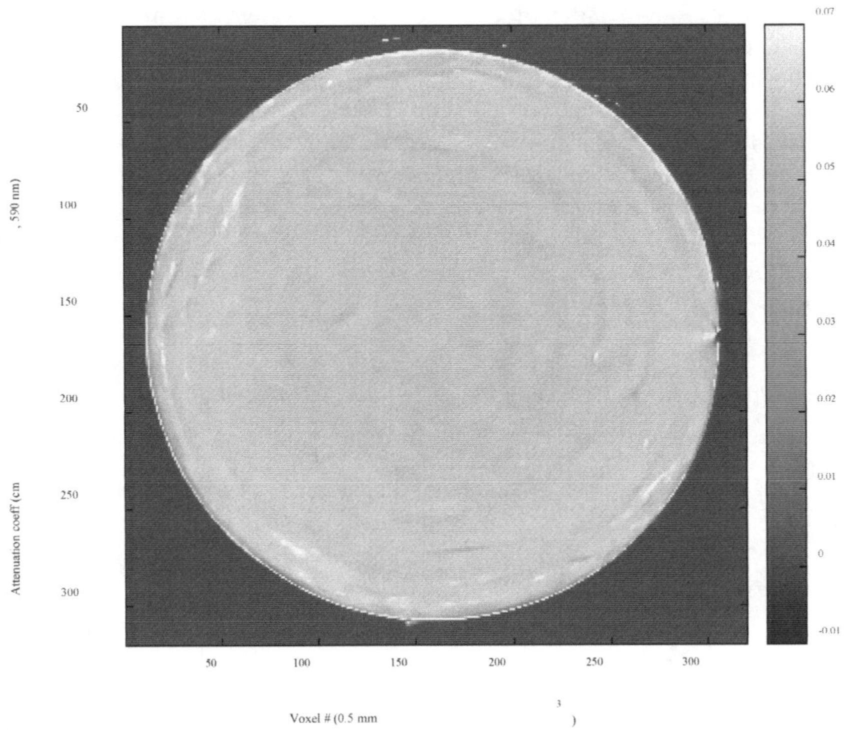

Figure 1. Optical CT reconstruction of scan at 278 K; transverse slice at 1.5 cm depth, 1 Gy dose, custom vessel diameter 14.7 cm.

4. Discussion

When the scanner aquarium was maintained at 278 K, condensation formed on the windows once the humidity level exceeded a threshold value. Specifically, in the warmer summer months the laboratory humidity level was above this threshold. Insulating the plastic sides and base had no effect on the heating rate. In principle, a low temperature aquarium could be designed without metal components and thermally insulated windows. However, increasing the number of glass-to-air interfaces would increase reflections present in the projection data. Degrading the accuracy of the reconstructions. Hydrophobic coatings could be applied to the windows as an alternate method to eliminate condensation. If the scanner was made air-tight and the aquarium re-designed then passively dehumidifying the interior of the scanner would also be an option.

Because the Vista16 scanner requires less than 5 minutes per pre or post irradiation scans, temperature drift during a scan is not an issue, primarily due to the large thermal mass of the 278 K, liquid-filled, aquarium. For this reason, active cooling was not required. Eliminating condensation enabled optical CT scanning at 278 K which avoided the 5-hour gel warm-up period for scanning at 293 K. However, it does introduce additional work and not all irradiation experiments can be easily performed at 278 K. Depending on the dosimeter properties and overall length of an experiment, there could be situations which warrant developing an active cooling approach. For example, flowing liquid from a larger insulated reservoir through the aquarium between scans could be implemented.

For this particular gel formulation, the radiochromic reaction rate decreased with lower temperature. With the practical consequence of increasing the post irradiation time required for the reaction to reach

95% or greater completion. An additional 0.5 hours was required to exceed this 95% completion period. This slower reaction rate negates the potential advantage of a lower diffusion rate [2]. Note, the tannic acid slowly reduces the ferric ions to ferrous, manuscript in preparation. The net effects of the tannic acid are increased time to maximum radiochromic response and lower auto-darkening. The main advantage of 278 K scanning of FX gels is the dramatic reduction in auto-darkening. Previously, the low initial optical transmission of FX gels has limited their use to samples smaller than 10 cm diameter. Coloured, index matching solutions have been used to compensate for the low initial transmission of the 10 cm samples. But often the coloured solution introduces drifts in transmission that become artefacts to the data sets. In this work, 15 cm diameter samples were scanned at 590 nm with a dose of 1 Gy. Typically, a 15 cm diameter, 278 K, FX gel will decrease in transmission more than 60% from initial placement in the 293 K scanner until 6 hours later when it has reached thermal equilibrium. By increasing the initial transmission of the gel samples at scan time, more dynamic range is available for writing dose distributions.

5. Disclosure

K Jordan has a licensing agreement with ModusQA related to dosimetry applications of optical computed tomography.

6. References

[1] Jordan K J *et al* 2017 *J. Phys.: Conf. Ser* **847** 012050
[2] Rae W I D *et al* 1995 *Med. Phys.* **23** 15-23
[3] Healy BJ *et al* 2003 *Med. Phys.* **30** 2282-91

Investigation of the dose rate and linear-energy-transfer of the signal quenching of radiochromic silicone-based dosimeters for different curing times and compositions

Lia Barbosa Valdetaro[1,2], Ludvig Paul Muren[1,2], Peter Sandegaard Skyt[1], Jørgen Breede Baltzer Petersen[4] and Peter Balling[3]
[1]Danish Centre for Particle Therapy, Aarhus University Hospital, Aarhus, Denmark.
[2]Department of Clinical Medicine, Aarhus University, Aarhus, Denmark.
[3]Department of Physics and Astronomy, Aarhus University, Aarhus, Denmark.
[4]Department of Medical Physics, Aarhus University Hospital, Aarhus, Denmark.

Email: liabar@rm.dk

Abstract. The response of radiochromic dosimeters based on silicone is influenced by the dose-rate quenching and also the linear energy transfer (LET). However, the impact of curing time and composition on quenching has not been fully characterized. We investigated two dosimeter compositions, with 5% and 9% curing agent, which cured for 1, 3 or 5 days and that subsequently were irradiated with an 80 MeV proton beam with three different beam currents. Monte Carlo simulations were utilised for the estimation of dose, dose rate and LET profiles. We found a significant decrease in LET-dependent quenching with curing time, but no significant decrease in dose-rate-dependent quenching.

1. Introduction

State-of-the-art proton therapy (PT) utilises the technique of spot-scanning, where the target is covered layer-by-layer by moving a thin monoenergetic proton beam [1]. Each energy layer is delivered with a dose rate determined by the cyclotron i.e., the machine-set dose rate, to ensure that the beam moves slowly enough to be steered by the magnets [1]. Consequently, the local dose rate can vary significantly in a single field. In addition to that, linear energy transfer (LET) increases towards the end of the proton range.

Three-dimensional (3D) dosimetry has the capability of providing high spatial resolution measurements [2–4], which can be valuable in the experimental validation of radiotherapy (RT). However, the demanding beam characteristics of spot-scanning PT challenge the use of 3D dosimeters, since they quench, i.e. under-respond [5–7].

In this study, radiochromic silicone-based dosimeters composed of leucomalachite green (LMG) were investigated, a radiosensitive dye and chloroform, which acts as an initiator for the chain process that converts the (colourless) LMG into malachite green (MG), both dispersed in a silicone matrix [8,9]. Given the matrix flexibility, this dosimeter garnered interest for the verification of motion and deformation studies[10]. However, complex and time-consuming correction models must be used in order to correct proton beam measurements of these dosimeters, since they present significant LET- and dose-rate dependent quenching, which cannot be realistically implemented in a clinical setting [7]. Previously, we found that curing time can have an impact on the dosimetric properties of radiochromic dosimeters, but only photons and low-LET protons (<2 keV/μm) were investigated [10]. Therefore, the aim of this study was to obtain a more comprehensive LET and dose rate parametrization of the dosimeter response.

2. Materials and methods

2.1. Fabrication and curing conditions
Dosimeters were fabricated from an silicone elastomer kit (SYLGARD 184 (DOW, Corning) which is commercially available, LMG and chloroform. Two different formulations were utilised in this study, containing different ratios of curing agent (CA) to silicone elastomer. The first formulation contained (in weight percentage) 5% CA, and the second, 9% CA. All formulations contained 0.26% LMG and 1.5% chloroform and the remaining weight percentage in silicone. The fabrication procedure was described in detail by Høye *et al.* [11]. The mixture was transferred with a large syringe into optical polystyrene cuvettes of dimensions 1×1×4.5 cm^3 and of a wall thickness of 1 mm. Whilst left curing at ~20°C, the cuvettes were protected from the light. Batches of dosimeters were cured for 1, 3 and 5 days prior to the irradiation day.

2.2. Dosimeter readout
Dosimeters were read out 1 hour before and 1 hour after irradiation with an optical 1D-scanner built in-house. The scanner uses a 635 nm laser diode with a 4.5mW beam, with width of 0.054±0.01 mm and height of 6 mm. Before the pre-irradiation read out, the dosimeters were removed from the cuvettes. The dosimeter was placed on a holder, which was moved in discrete 0.25 mm steps by a linear stage. The attenuation coefficient response of the samples, $\Delta\alpha = \alpha_{post} - \alpha_{pre}$, was defined as the difference between the dosimeter's attenuation coefficient measured before and after irradiation.

2.3. Proton irradiation
Dosimeters were irradiated at Aarhus University Hospital (Aarhus, Denmark) in the Danish Centre for Particle Therapy with protons using a Varian ProBeam isochronous cyclotron. A monoenergetic beam of 80 MeV that spot-scanned, delivering a 5x7 cm^2 uniform field was used. Varying dose-rate levels were achieved by requesting different cyclotron currents in Varian's beam delivery diagnostic tool. For this experiment, a spot separation of 5 mm was chosen, which was sufficient to achieve a homogeneous field according to the treatment planning systems (Eclipse version 15.6). The dosimeters were placed parallel to the beam direction, with 2 cm of solid water placed directly in front of them. Three dosimeters were irradiated for each curing time and composition for each cyclotron current.

2.4. Monte Carlo simulations
The distributions of proton dose and dose-averaged-LET (LET_d) were estimated using TOPAS (version 3.2) Monte Carlo (MC) in [12]. The silicone base of the dosimeter was defined using information from the manufacturer and using an expected chain length of polydimethylsiloxane (n = 362) [6,11]. A phase-space model used to simulate the beam was validated with an ionization chamber array (Giraffe, IBA dosimetry), a scintillator screen (XRV-4000 Beam Profiler, Logos Systems) and integral depth-dose measurements in a water phantom (MP3-PL, PTW) with a parallel plate ionization chamber (ROOS, PTW dosimetry). Local dose rates ($\dot{D}_{MC}$) were estimated from the MC dose distributions and the beam-on time, retrieved from irradiation log files.

2.5. Dose-rate and LET parametrization
We parametrized the dose-rate and LET_d dependent-quenching by the quenching correction factor (QCF) defined by an empirical formula [7]:

$$\frac{\Delta\alpha}{D_{MC}} = \frac{QCF^{-1}}{Gy{\cdot}cm} = exp\left(-a\frac{LET_d}{keV/\mu m} + b\right) + exp\left(-c\frac{\dot{D}_{MC}}{Gy/s} + d\right) \quad [1]$$

The dimensionless fit parameters were estimated using the Scikit-learn package in Python [13]. The change in the parameter-dependent coefficients *a* and *c*, which correspond, respectively, to the LET and dose-rate dependency, where used to evaluate the dosimeter quenching.

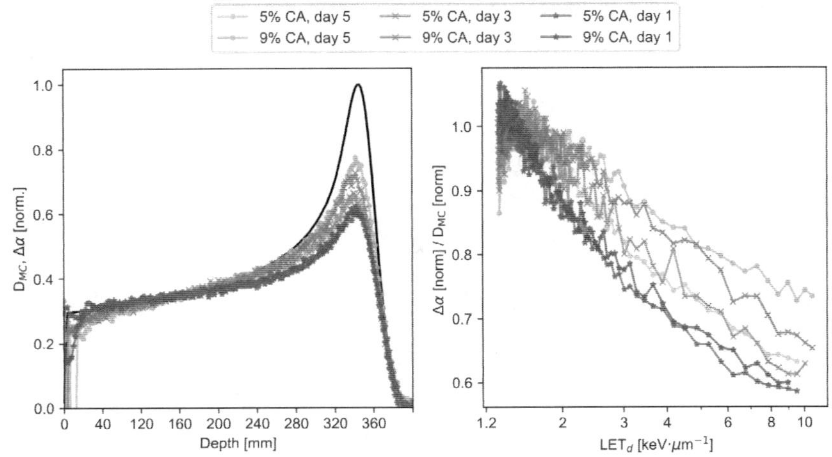

Figure 1. Optical densities (Δα) and calculated MC dose (shown in black for comparison) normalized 1 cm into the beam entrance, for the 5% and 9% CA dosimeter compositions, curing for 1, 3 or 5 days (Left panel). Ratio between (normalized) Δα and MC dose as a function of LET_d (Right panel).

3. Results

The surface fit resulted in R-squared values >0.96 for all datasets. The parameter-dependent coefficients (Table 1) indicated a significant decrease in LET quenching with longer curing times. The decrease in LET quenching is also clearly seen in Fig 1 and was even larger for 9% CA. Dose rate quenching did not decrease significantly between 3 and 5 days of curing for either of the two CA compositions (Fig 2, Table 1). There was a decrease by 13% and 34% in the dosimeter signal from 1 to 5 days for 5% CA and 9% CA dosimeters respectively.

Table 1. Parameter-dependent coefficients with 95% CI.

	5% CA, 1 day	5% CA, 3 days	5% CA, 5 days	9% CA, 1 day	9% CA, 3 days	9% CA, 5 days
$a \times 10^{-2}$	8.9±0.3	5.5±0.2	4.6±0.2	6.3±0.2	2.5±0.2	0.4±0.3
c	1.5±0.2	1.0±0.3	1.0±0.2	1.2±0.4	0.9±0.1	0.9±0.4

4. Discussion

The LET and dose-rate-dependent quenching for different curing conditions and compositions and for silicone-based radiochromic dosimeters were investigated. A significant decrease in dose-rate quenching was not found, which is consistent with the findings of Jensen *et al* [14], but there seemed to be a trend for decreasing dose-rate dependency with curing time for both curing compositions. The LET_d dependency decreased significantly with curing time for both curing compositions, although it was more pronounced for 9% CA dosimeters (Fig 1). Although extrapolating the current results could indicate a benefit in increasing the curing time even further, the decrease in dose response and continuous hardening of the silicone base [15] should be taken into consideration.

Although the main source or initiator of radical production of the chain process leading to the conversion of LMG into malachite green (MG) is believed to be chloroform [11,16], other not chemically cross-linked molecules could serve as an additional source of free radicals. This theory is supported by the higher dose response associated with dosimeters with less CA [14]. Since the 5% CA dosimeter compositions investigated in this study have a lower ratio of CA as specified by the manufacturer, not-cross-linked molecules are likely present. Additionally, dosimeters that have not been fully cured, as the batch curing for 1 day, were expected to have an even higher number of not-cross-linked molecules present.

Track structure theory could help explain the decrease of LET_d-dependent quenching with curing time [17]. Consider the dosimeter to be composed of discrete radiosensitive elements that are dispersed in a matrix, and that these elements can be "activated", i.e., have an interaction that leads to a measurable signal, by "hits" from ionizing radiation. Close to the end of the proton range (high LET), interactions alongside the ion track have a very short range, making it harder to reach a radiosensitive element, which lowers the probability of a "hit". In a dosimeter with a high dose response, more of these sensitive elements are quickly activated at lower doses, saturating the elements close to the particle track. Additionally, the large availability of radicals close to the particle track could increase the chance of radical recombination, and consequently, more chain processes would not generate a "hit", leading to an increase in LET-quenching [18]. A similar argument could be made to explain the dose-rate-dependent quenching; since proton beams with higher dose rate should also generate radicals at a higher rate, increasing the change of recombination.

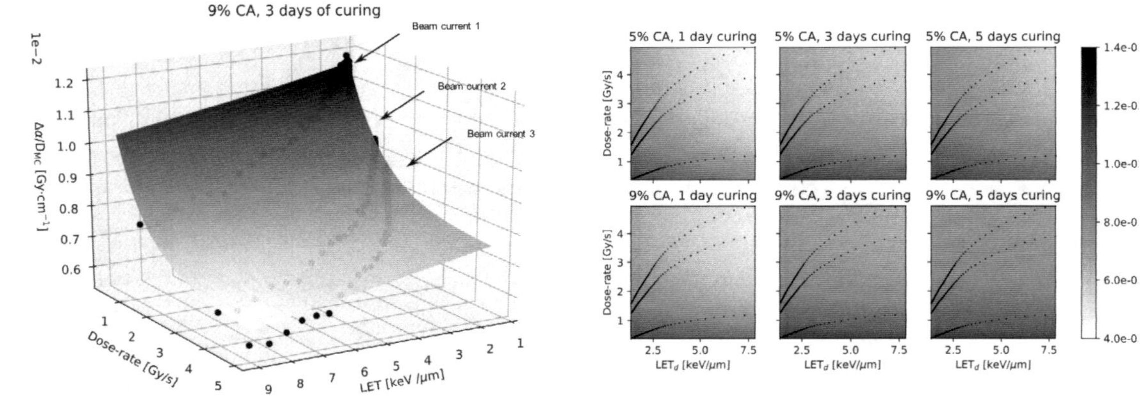

Figure 2. Surface fit to the 9% CA dosimeters cured for 3 days. The black dots represent the mean of the three cuvettes irradiated with each beam current, with the surface fit overlayed on top (left panel). The dosimeter quenching as a function of dose-rate and LET_d for different dosimeter compositions (upper panels: 5% CA, lower panels: 9% CA), and curing times (1, 3 or 5 days). Black markers indicate the measurement points, while the colormap was calculated from equation (1) (right panel).

By lowering the concentration of free radicals by either having a higher percentage of CA or longer curing, LET and dose-rate quenching were expected to decrease, which was indeed observed for the LET quenching. Regarding the dose-rate quenching, while this decrease has been observed for photons [14], it was not significant in this study (Figure 2). A potential explanation could be that, since the dose rate levels for protons are so much higher than photons, any potential improvement gained by lowering the availability of radical sources was not large enough to be significant.

In conclusion, it was found that longer curing times had the effect of reducing the LET_d-dependent quenching for both curing compositions, with a significantly larger effect on the 9% CA dosimeters. However, curing time had a very small impact on the dose-rate-dependent quenching for proton beams.

5. Acknowledgments

Mateusz Sitarz is kindly acknowledged for his assistance during irradiation. The Novo Nordisk Foundation funded the research.

6. References

[1] Kandula S *et al* 2013 *Med. Dosim.* **38** 390
[2] Baldock C *et al* 2010 *Phys. Med. Biol.* **55** R1

[3] Costa F *et al* 2019 *J. Phys: Conf. Ser.* **1305** 012009
[4] Schreiner L J 2015 *J. Phys: Conf. Ser.* **573** 012003
[5] Gustavsson H *et al* 2004 *Phys. Med. Biol.* **49** 3847
[6] Wheatley M J *et al* 2018 *Phys. Med. Biol.* **63** 215028
[7] Valdetaro L *et al* 2021 *Phys. Imag. Rad. Oncol.* **18** 11
[8] De Deene Y *et al* 2015 *J. Phys: Conf. Ser.* **573** 012025
[9] Høye E M *et al* 2015 *J. Phys: Conf. Ser.* **573** 012067
[10] Kaplan L P *et al* 2017 *Phys. Med. Biol.* **62** 5612
[11] Hoye E M *et al* 2017 *Phys. Med. Biol.* **62** N73
[12] Testa M *et al* 2013 *Med. Phys.* **40** 121719
[13] Pedregosa F *et al* 2011 *J. Mach. Learn. Res.* **12** 2825
[14] Jensen M B *et al* 2022 *Acta Oncol.* **61** 264
[15] Kaplan L P *et al* 2017 *Phys. Med. Biol.* **62** 5612
[16] Oldham M 2015 *J. Phys: Conf. Ser.* **573** 012006
[17] Butts J J *et al* 1967 *Radiat. Res.* **30** 855
[18] Horowitz Y S *et al* 2012 *Radiat. Prot. Dosim.* **150** 359

Proof-of-concept of proton and photon measurements for large-area 2D dosimeters utilising optically stimulated luminescence

Jacob S. Nyemann[1], Mads L. Jensen[1], Lia B. Valdetaro[2,3], Morten B. Jensen[2,3,5], Rosana M. Turtos[1], Brian Julsgaard[1,4], Jørgen B. B. Petersen[5], Ludvig P. Muren[2,3,5] and Peter Balling[1,4]
[1]Department of Physics and Astronomy, Aarhus University, Denmark
[2]Department of Clinical Medicine, Aarhus University, Denmark
[3]Danish Centre for Particle Therapy, Aarhus University Hospital, Denmark
[4]Interdisciplinary Nanoscience Center (iNANO), Aarhus University, Denmark
[5]Department of Medical Physics, Aarhus University and Aarhus University Hospital, Denmark

E-mail: jacob1soegaard@phys.au.dk

Abstract. Reusable high-resolution 2D dosimeters have a great potential for radiotherapy verification and quality assurance, not least due to the reduced cost and simpler calibration procedures implied by reusability. We have performed proof-of-concept measurements in both photon and proton beams with a reusable 2D dosimeter ($5 \times 7\ cm^2$; 1 mm thickness) based on optically stimulated luminescence, using a readout setup scalable to 3D. Our data indicated a strong linear-energy-transfer quenching for this type of dosimeter.

1. Introduction

Development of dosimetry systems capable of measuring dose in 3D has been pursued in the past decades to improve radiotherapy dose verification and quality assurance procedures [1, 2, 3]. Such systems would increase the dimensionality of the measured data when assessing the quality of both machines and patient treatment plans. Our research group is contributing to this development by introducing a reusable system for 3D dosimetry that is based on optically stimulated luminescence (OSL) [4, 5, 6]. Transparent OSL-active 3D dosimeters can be stimulated by laser sheets in contrast to the head-on stimulation used for 2D dosimeters [7]. Creation of such dosimeters demands a precise match between the index of refraction of the OSL-active powder and the used matrix, particularly if the size of the OSL-active powder is not reduced to the nanometer scale [8, 9]. Further efforts include studying the properties of LiF:Mg,Cu,P (MCP) for its usage as the OSL-active component in such a dosimeter [10].

At present, radiotherapy dose measurements are mainly performed using point and 2D dosimeters, where high-resolution 2D dosimeters, e.g., radiochromic films, generally are one-time use. Thus, a reusable high-resolution 2D dosimetry system with limited need for re-calibration may be a great asset for the daily routines at radiotherapy clinics. Finally, an OSL-based 2D dosimeter could facilitate a future scaling to OSL-based 3D dosimeters.

In the following, we present proof-of-concept 2D readouts of a large 2D dosimeter, based on an OSL-active composite material in the form of OSL-active MCP powder embedded in a silicone matrix. The use of OSL allows for multiple repetitions of irradiation and readout of deposited dose, i.e., the 2D dosimeter is reusable. We present dose maps from both photon and proton irradiations acquired using a single 2D dosimeter multiple times. The optical system used for readout is further scalable to 3D.

2. Materials and methods

2.1. Samples

The 2D dosimeter used was prepared as a composite material with 30 wt% of MCP powder (grain size of 80 – 200 μm) acquired from RadPro embedded into a silicone matrix (SYLGARD™ 184 Silicone Elastomer with silicone/curing agent ratio of 10:1). The composite material was cured and cut into an approximately rectangular shape of 5×7 cm^2 with a thickness of ~1 mm. We note that the mixing method used here was unable to yield a homogeneous particle distribution for the dosimeter.

2.2. Irradiations

Photon irradiations were undertaken at Aarhus University Hospital (Aarhus, Denmark) with a Varian TrueBeam linear accelerator with 6 MV beam quality, 10×10 cm^2 jaw-defined field, dose rate of 6.0 Gy/min at the isocenter, and a source-to-axis distance of 100 cm. The photon beam was equipped with a flattening filter to provide homogeneous irradiation within ± 3 %. The dosimeter was placed between two 5 cm Solid Water® slabs and centered in the field with a source-to-surface distance of 95 cm, and a dose of 5.0 Gy at the isocenter was given.

The proton irradiation was delivered by a Varian ProBeam Proton Therapy System at the Danish Centre for Particle Therapy (Aarhus, Denmark). A dose rate of 20000 MU/min (65 Gy/min at beam entrance) was used to deliver a single spot of 77 MeV protons, yielding a dose of 4.3 Gy at the Bragg peak. The rectangular 2D sample was placed between two silicone blocks to approximate charged-particle equilibrium at the dosimeter position. A Monte Carlo simulation representative of this set-up was performed with TOPAS (version 3.2) [11], with a validated beam model.

All doses and dose rates are presented as water-equivalent doses. Fairly high doses were chosen to ensure sufficient signal over the entire dosimeter range for these proof-of-concept measurements.

2.3. Readout and calibration

A 445 nm laser with a maximum output of ~4.5 W was used for optical stimulation of the sample during readout. The laser beam was manipulated to create a truncated Gaussian field with stimulation intensity ranging from 5 mW/cm^2 to 46 mW/cm^2 in the sample region (lowest near corners). With sufficient readout time, the decaying OSL signal will reach background level, even in low-intensity regions, and the time-integrated signal will be largely unaffected by the intensity profile of the stimulation.

Measurements were acquired by imaging the emitted OSL onto the CCD sensor of the camera (SOPHIA 2048B, Princeton Instruments). Imaging was achieved with a Jenoptik APO Macro lens system, with a field-of-view resulting in an imaged area of $(9.67 \pm 0.05)^2$ cm^2, with each of the 256×256 (8×8 hardware-binned) pixels representing an object size of $(\sim 0.38 \text{ mm})^2$. Elimination of unwanted light (e.g., stimulation light) was achieved with a combination of four Ø 2" filters placed in front of the imaging system. Each measurement consisted of 26 consecutive frames using an exposure time of 60 s. The time between irradiation and readout ranged from 1 hour to 3.25 hours. All measurements are presented after background subtraction. Background measurements were performed after bleaching for ~30 minutes under 385 nm LEDs (3 – 40 mW/cm^2) followed by 5 minutes under the stimulating laser (the reasoning behind this bleaching procedure is described elsewhere [12]).

To demonstrate the potential of this type of dosimeter, we performed a calibration measurement to correct for the inhomogeneous particle distribution. The calibration matrix was constructed by dividing the planned photon dose (5 Gy) with the observed total signal (time-integrated) for each pixel of the dosimeter area. The main objective of the calibration was to correct for inhomogeneities in the particle distribution, but by construction, it also converts from measured counts to dose levels. Thus, by using this calibration, the dose response is inherently assumed to be perfectly linear, unaffected by fading time, independent of dose rate, irradiation type (photon vs. proton), and linear energy transfer (LET). A clinically relevant calibration protocol must be more comprehensive to eliminate these assumptions.

3. Results and discussion

Figure 1 presents the detected OSL emitted from the 2D dosimeter after a homogeneous photon irradiation. The OSL signal curve of spatially integrated signal as a function of stimulation time, seen in Figure 1(a), shows a characteristic decaying behavior generally seen for OSL-based dosimeters as the OSL traps are emptied. The total OSL signal of the measurement is shown in Figure 1(b), found as the sum of all the frames taken during the measurement, yielding a signal map expected to be proportional to the given dose. However, with the dosimeter's inhomogeneous particle distribution, the signal response was inhomogeneous as well. Additionally, the (apparently) too short readout time used here caused an under response for corner regions (lowest stimulation intensity) unrelated to the particle distribution, an effect seemingly visible in especially the lower corners of Figure 1(b).

Figure 1(c) presents the same measurement after correction with the calibration matrix. After calibration, the dose map shows an almost uniform signal with a mean pixel value of (5.8 ± 0.4) Gy. Since the dose map in Figure 1(c) shows an elevated dose level (compared to the planned dose) and non-random variations, some (unidentified) systematic errors were not captured by our calibration with the current readout setup/procedure. This was most likely due to the many assumptions introduced with the simple calibration procedure used here.

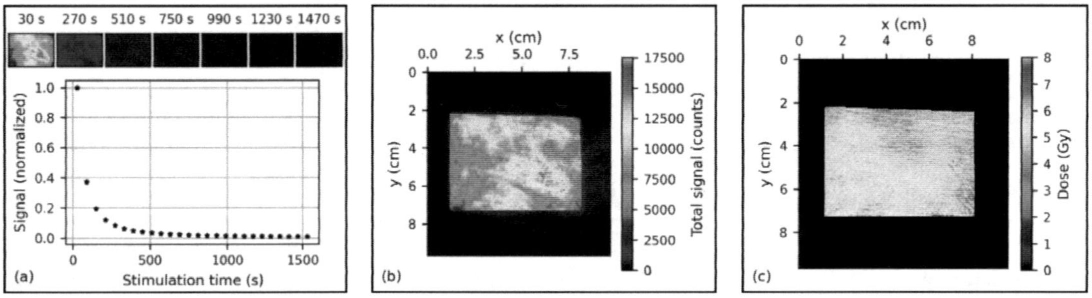

Figure 1. Spatially resolved 2D OSL readout of the rectangular 2D dosimeter uniformly irradiated by the photon irradiation. In (a), individual frames are shown on top, while the main graph shows a normalized OSL decay curve, where individual points are based on the sum of all pixels within the dosimeter area of a frame, e.g., the representative frames above. Time stamps are given as the centre of each captured frame. Panel (b) shows the measured 2D data given as a sum of all frames. Panel (c) presents the 2D measurement after applying the calibration matrix to the measured data.

In Figure 2, the calibrated 2D OSL distribution of the dosimeter following exposure to a single beam of protons is shown. The dose map shows the expected overall shape of a proton beam entering from the right with its fixed width, increased intensity in the central part of the beam, and increasing signal near the end of the track (the Bragg peak). With the line profiles shown in Figure 2(b), we see that these general trends are similar to that of a Monte Carlo simulation. However, the line profiles also highlight a clear under response near the Bragg peak. This is likely explained by quenching due to a high LET near the Bragg peak, which is a common issue encountered for proton dosimetry. This is consistent with the relatively strong quenching observed for the TL signal of MCP dosimeters [13]. Despite challenges with MCP particle distribution (possibly contributing to the distorted Bragg peak) and LET quenching, Figure 2 clearly shows the potential for a composite dosimeter material with readout based on OSL. In particular, the present measurements highlight the scalability of 2D OSL dosimeters with our 5×7 cm^2 dosimeter and 9.67×9.67 cm^2 readout area, significantly larger than previously demonstrated [14, 15].

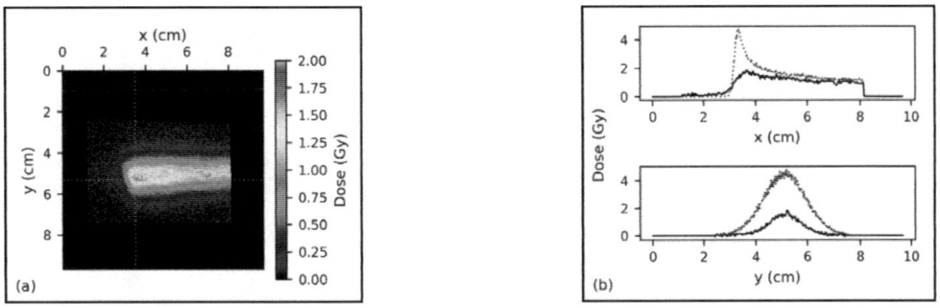

Figure 2: Panel (a) shows a 2D dose map of the rectangular 2D dosimeter irradiated by the described proton beam after adjustment with the calibration matrix. Horizontal and vertical line profiles of the 2D dose map (black lines) are shown in (b) alongside representative line profiles from a Monte Carlo based simulation, which are shown for comparison in blue.

4. Conclusion

We have shown proof-of-concept 2D measurements of photon and proton irradiation profiles using a composite MCP-silicone dosimeter. The proof-of-concept measurements were performed on a 5×7 cm^2 dosimeter, and read out with a setup capable of 9.67×9.67 cm^2 dosimeters. Finally, we showed that the OSL signal from micrometer-sized MCP powder exhibits significant quenching in high-LET regions of a proton beam, i.e., near the Bragg peak. Future research in this direction will include the study of LET quenching, improved calibration procedures, and consistency of readout.

5. Acknowledgements

This work was funded by the Independent Research Fund Denmark, Technology and Production Sciences and the Novo Nordisk Foundation and the. P S Skyt is acknowledged for his assistance regarding dose delivery from the proton beam facility at the Danish Centre for Particle Therapy.

6. References

[1] Doran SJ 2009 *Appl. Radiat. Isotopes* **67** 393-398
[2] Schreiner L J 2015 *J. Phys.: Conf. Ser.* **573** 012003
[3] Mijnheer B (ed) 2018 Clinical 3D dosimetry in modern radiation therapy
[4] Balling P *et al* 2018 European Patent Application EP3264137A1
[5] Nyemann J S *et al* 2022 *J. Phys.: Conf. Ser.* **2167** 012026
[6] Jensen M L *et al* 2022 *Sci. Rep.* **12** 8301
[7] Jensen M L *et al* 2022 *J. Phys.: Conf. Ser.* (This volume)
[8] Nielsen C L *et al* 2022 *J. Phys.: Conf. Ser.* (This volume)
[9] Nielsen C L *et al* 2022 *Nano Lett.* **22** 1566-1572
[10] Nyemann J S *et al* 2020 *Radiat. Meas.* **138** 106390
[11] Perl J *et al* 2012 *Med. Phys.* **39** 6818-6837
[12] Nyemann J S *et al* 2022. (In prep.)
[13] Olko P *et al* 2020 *Radiat. Prot. Dosim.* **192** 165-177
[14] Sadel M *et al* 2020 *Radiat. Meas.* **133** 106255
[15] Sadel M *et al* 2020 *Radiat. Meas.* **133** 106293

Journal of Physics: Conference Series **2630** (2023) 012023 doi:10.1088/1742-6596/2630/1/012023

Optimizing the transparency of nano-LiF:Cu/silicone nanocomposites for 3D optically stimulated luminescence dosimetry

Camilla L Nielsen[1], Rosana M Turtos[1], Brian Julsgaard[1,2], Ludvig P Muren[3,4] and Peter Balling[1,2]
[1]Department of Physics and Astronomy, Aarhus University
[2]Interdisciplinary Nanoscience Center (iNANO), Aarhus University
[3]Danish Centre for Proton Therapy, Aarhus University Hospital
[4]Department of Medical Physics, Aarhus University & Aarhus University Hospital

E-mail: cln@phys.au.dk

Abstract. Nanoparticles displaying optically stimulated luminescence embedded in a transparent polymer matrix have been proposed as a reusable high-spatial resolution 3D dosimeter. We measured the refractive indices of LiF:Cu nanoparticles and silicone and found a mismatch of 0.03-0.05 for the relevant wavelengths, explaining the transmission loss through 1 cm-sized nanocomposite dosimeters. We propose to bridge the refractive index gap by coating the LiF nanoparticles with a shell of SiO_2. Initial studies show successful SiO_2 growth, although more work is required to produce core-shell nanoparticles with an optimized refractive index.

1. Introduction

Recently, there has been significant progress in the development of a reusable, high-fidelity 3D dosimeter based on optically stimulated luminescence (OSL) [1–3]. OSL is a well-known technique for integrating dosimetry based on charge carriers excited by ionizing radiation becoming trapped in metastable dosimetric trap states within the band gap of an insulator. To read out the trapped charge – which is ideally proportional to the deposited dose – a light source is used to promote the carriers to the conduction (or valence) band. Once mobile, the charges can recombine radiatively giving the designated OSL signal. To reduce cost and at the same time enable large dosimetric volumes, a composite dosimeter consisting of OSL active nanoparticles embedded in an elastomer matrix has been proposed. Copper-doped LiF nanocubes (nLiF:Cu) of ~50 nm side length were recently identified as a viable candidate material for the nanoparticles [3]. The particles show a high OSL signal emitted in the UV region, which can be stimulated with visible light such as green and blue lasers. Furthermore, they seem close to ideal for embedment in silicone (polydimethylsiloxane) thanks to the near match of their refractive indices. This is also illustrated by the high transparency of mm-thick 2D sheets of nLiF in silicone [3].

An important next step in scaling up the nanocomposite dosimeters from 2D to 3D is optimizing the transparency to secure the necessary transmission of light through large volumes. Light scattering can be minimized for example by lowering the particle size below the wavelength of the light to reduce Mie scattering and by matching the refractive indices perfectly. This can be done by adjusting the refractive index of either the matrix or the nanoparticles. The latter can for instance be done by employing the observation that the effective refractive index of a core-shell particle to a good approximation (for small particles and relatively thin coatings) will be the volume-average of the pure substances [4].

In this contribution, we will report the refractive index mismatch between nLiF:Cu and a silicone matrix and present preliminary investigations of using a SiO_2 shell on the LiF nanoparticles to adjust their refractive index.

2. Materials and methods

Copper-doped LiF nanoparticles were prepared by mixing precursors carrying the different constituents in a home-built flow setup, as described in detail in [3]. The refractive index of nLiF:Cu was measured by dispersing a known amount of nanoparticles in solutions of known refractive index and measuring the transparency of these suspensions at the relevant wavelengths. This mimics the embedded state, and the value found for the refractive index of nanoparticles in this way is assumed similar to that exhibited by nLiF:Cu in a silicone nanocomposite. All transmission measurements were performed on a Perkin Elmer Lambda 1050 spectrophotometer in total transmission mode using an integrating sphere.

The nanocomposite samples were made by embedding a certain amount of nLiF:Cu particles in Dow Corning Sylgard 184 using a base to curing agent ratio of 10:1 and room temperature curing. The refractive index of pure silicone was measured using an LLG Labware uniREFRACTO 2 refractometer as it cured under the same conditions applied to the dosimeters.

SiO_2 shells were grown on the LiF:Cu nanoparticles using the modified Stöber process [5]. Using NH_3 in water to catalyze the hydrolyzation of tetraethyl orthosilicate (TEOS) to SiO_2, two samples were prepared with nominal shell thicknesses of 5 and 20 nm. The resulting nanoparticles were characterized by powder x-ray diffraction (PXRD, Rigaku Smartlab) and a combined scanning electron microscopy (SEM, FEI Nova NanoSEM) and energy dispersive x-ray spectroscopy (EDS, Ametek EDAX) study. Furthermore, the OSL and radioluminescence yield were found for uncoated and coated nanoparticles with a $^{90}Sr/^{90}Y$ β-ray source and a blue laser [3].

3. Results & Discussion

The wavelength dependent transmission of various concentrations of nLiF:Cu in silicone shown in Fig. 1a is consistent with the $1/\lambda^4$ relation expected for Rayleigh scattering [6]. There is a clear indication that the refractive indices are not matched perfectly, since the dropoff in transmission towards lower wavelengths is exacerbated by higher load levels of nanopowder in silicone. Since the OSL signal from the nanoparticles is at a wavelength of $\sim$325 nm, where the transparency is already poor even at low concentrations of LiF, Fig. 1a clearly illustrates the inadequacy of the as-synthesized nanoparticles together with the standard silicone as a high signal-to-noise-ratio macroscopic 3D dosimeter.

Fig. 1b shows total transmission curves for various solutions with dispersed nanoparticles denoted by their refractive index at the wavelength of OSL emission. From these curves, the refractive index of the nanoparticles can be extracted. The observation that different solutions peak in transparency at different wavelengths (Fig. 1b) reflects the difference in dispersion between the nanoparticles and the solutions. The large band gap of nLiF:Cu results in an almost flat dispersion across a large wavelength range, whereas the organic solvents (methanol and ethylene glycol) used to prepare the reference solutions have a much more marked dispersion towards short wavelengths, in fact close to that of silicone. By using the maximum of each of these curves as an estimate for the wavelength at which the refractive index matches between the organic solvent and the nanoparticles, the colored points seen in Fig. 1c can be plotted. A fit to a first-order Sellmeier equation yields the solid black line. Also drawn is the tabulated dispersion curve of bulk LiF [7] (dotted black), which slightly exceeds that of nLiF. Note the identical color map between 1b and 1c, identifying the dispersion curves in Fig. 1c with the refractive index of each solution in Fig. 1b.

The refractive index of Sylgard 184 depends on the ratio of curing agent to base polymer and the curing temperature [8]. The measurement of the silicone used in the current work yielded a refractive index rising slightly with curing time before stabilizing at 1.4126(1) after 4 days (measured at 589 nm). Assuming a dispersion similar to the one measured in [9], the refractive index of silicone can be plotted

(Fig. 1c). The measurement in [9] was done on a temperature-cured sample of Sylgard 184 (at 150°C) the dispersion of which is not necessarily identical to that of a room-temperature cured sample. It can, nevertheless, provide a guide to the mismatch as a function of wavelength. The refractive index mismatch of 0.029 at a stimulation wavelength of 445 nm and 0.049 at an OSL emission wavelength of 325 nm indicates a potentially more serious problem than first anticipated from the transmission curves in Fig. 1a. These curves were measured using total transmission, thus also taking into account forward diffuse scattering, which is not included in the Rayleigh theory. Direct transmission measurements are therefore necessary to accurately determine the extent of the problem.

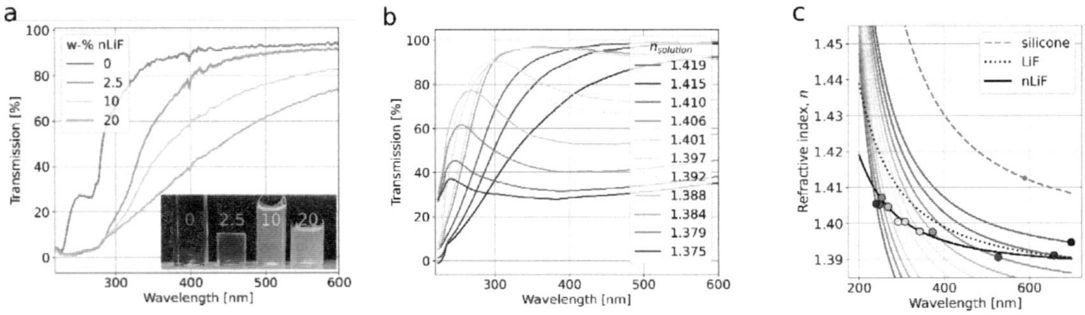

Figure 1. (a) Total transmission curves for four different cuvette-sized silicone molds with various weight fractions of nano-LiF powder. (b) Total transmission of 10 w-% nLiF:Cu powder dispersed in solutions of various refractive indices (n_{solution}) and (c) the refractive index of the nanoparticles as function of wavelength found by assuming a perfect match between the nanoparticles and the solution at the peak of the transmission curves in (b). The colored curves are dispersion curves of the solutions (see legend in (b)) obtained by interpolation between those of the pure solvents (methanol and ethylene glycol) measured in [10]. The refractive index of silicone shown in gray is based on the measurement at 589 nm extended to other wavelengths using tabulated values for the dispersion curve [9]. Also shown is the refractive index of nLiF (fit in solid black) and bulk LiF (from [7] in dotted black).

The mismatch between the refractive indices of the nanocrystals and silicone found above poses as a challenge for achieving the required transparency in a 3D OSL dosimeter. As a way of compromising for the too-low refractive index of the nanoparticles, a shell consisting of a material with a larger refractive index can be grown. SiO_2 has a moderately higher refractive index $n_{SiO_2} \sim 1.47$ at 400 nm, which motivates an attempt to grow LiF/SiO_2 core-shell nanoparticles.

The PXRD of the two different variations of core-shell nLiF:Cu/SiO_2 nanoparticles with nominal shell radii of 5 and 20 nm as well as the uncoated nLiF particles confirm the existence of LiF nanoparticles of ~50 nm both before and after the shell growth (see Fig. 2a). The coated samples in addition show a broad peak at $2\theta \approx 23°$ corresponding to amorphous nano-SiO_2 [11], which is rising with increasing SiO_2 content. The combined SEM and EDS study confirmed the presence of both F, Si, and O in the coated samples (note that Li cannot be detected by EDS), suggesting a successful synthesis. In fact, the Si/F ratios agree relatively well with the expected values assuming all TEOS to be hydrolyzed. The measured ratios give expected 3 and 27 nm shells as opposed to the nominal 5 and 20 nm. The SEM images of all samples reveal cubic nanoparticles with a very monodisperse size distribution around 51 nm with a standard deviation of 6 nm (see Fig. 2b for the 5 nm coated sample). This is seemingly at variance with the desired core-shell geometry. The morphology might be elucidated by EDS mapping using scanning transmission electron microscopy. This will be the focus of future study. The methodology for measuring the refractive index of nLiF presented in this contribution can be used on the coated samples as well, giving a straightforward way of identifying the optimal core-shell particles.

The effective dilution of the nLiF:Cu particles is reflected in scaled down intensities of both radioluminescence and OSL for the coated particles as compared to uncoated, while the emission wavelength and decay time remain the same. The observation that SiO_2 on and around the LiF:Cu nanoparticles does not seem to extinguish the OSL signal is encouraging for future applications of the Stöber process in modifying the refractive index.

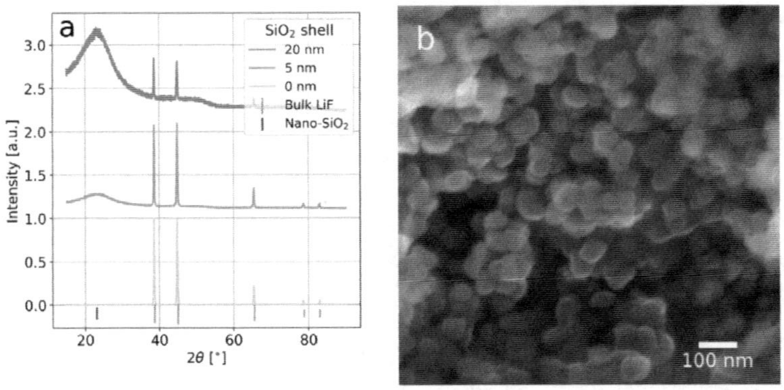

Figure 2. (a) PXRD data of uncoated and coated (with shells of nominally 5 and 20 nm SiO_2) nLiF:Cu. Expected peak positions for bulk LiF and nanosized amorphous SiO_2 are shown as bars on the bottom. (b) SEM image of the 5 nm SiO_2 coated nLiF:Cu sample. A large number of ~50 nm nanocubes can be seen.

A conceptually different strategy involves fully utilizing the nanosized nature of the particles. If the OSL properties can be conserved while lowering the particle size, the nanoparticles will practically disappear optically according to Rayleigh scattering theory. This would also mitigate the need to obtain transparency at two wavelengths (the stimulating and the emitted OSL light), which is challenged by the difference in dispersion between the two materials. Possibilities in controlling the particle size, for example by varying the precursors and/or solvents of the synthesis, will be the focus of future investigations.

4. Conclusion

The transparency loss caused by the mismatch in refractive index between the nanoparticles and the silicone matrix must be resolved before cm-sized dosimeters can be realized. Even though the as-synthesized LiF nanocubes have a refractive index relatively close to silicone, transmission measurements of nanocomposite dosimeters show a considerable scattering of light – especially at low wavelengths. An initial pilot synthesis showed formation of SiO_2, however not necessarily as a shell around the LiF nanocubes. The search for a successful synthesis yielding the optimal coating thickness is still ongoing.

5. References

[1] Nyemann J S *et al* 2022 *J. Phys. Conf. Ser.* (In press)
[2] Jensen M L *et al* 2022 *Sci. Rep.* **12** 8301
[3] Nielsen C L *et al* 2022 *Nano Lett.* **22** 1566-1572
[4] Kuzma A *et al* 2012 *J. Appl. Phys.* **112**
[5] Yang X *et al* 2013 *J. Mater. Chem. C.* **1** 3359-3366
[6] Jin Y et al 2016 *J. Mater. Chem. C.* **4** 3654-3660
[7] Li H H 1976 *J. Phys. Chem. Ref. Data.* **5** 329-529
[8] Santiago-Alvarado A *et al* 2020 *Mater. Res. Express.* **7** 45301
[9] Schneider F *et al* 2009 *Sensors Actuators, A Phys.* **151** 95-99
[10] Kozma I Z *et al* 2005 *J. Opt. Soc. Am. B.* **22** 1479-1485
[11] Sun J *et al* 2017 *Nanomaterials* **7** 1-15

Journal of Physics: Conference Series **2630** (2023) 012024 doi:10.1088/1742-6596/2630/1/012024

Bare spherical gel dosimeter with optical computed tomography scanning

Olivia Moluchi[1], Matt Mulligan[1] and Kevin Jordan[1,2]
[1]London Regional Cancer Program, London Health Sciences Centre
[2]Dept. Of Medical Biophysics, Western University, London, Ontario, Canada

Email: kevin.jordan@lhsc.on.ca

Abstract. Radiochromic polyvinyl alcohol-iodide (PVA-I) hydrogel dosimeter with crosslinking by glutaraldehyde (GTA) has been reported. Because of the transparent gel's mechanical strength, a study to cast samples in spherical shapes was initiated. The bare sample, 47 mm in diameter was mounted on a funnel and held by suction in a custom, 96 mm diameter, cylindrical vessel for 3D optical CT scanning. A solution of the same formulation without GTA provided refractive index matching inside the vessel for optic scanning and eliminated concentration gradients. The gel sphere remained in the vessel after draining the solution for the 'in-air' irradiation. A 20 Gy, maximum dose to the gel sphere was delivered by a single, 6 MV, x-ray beam. Comparison of central axis depth doses for the gel reconstruction and Monte Carlo calculation revealed similar results, indicating accurate 3D dosimetry within 1 mm of the gel surface will be possible. The gel recorded slightly greater dose in build-up region and slightly lower dose in the exit hemisphere relative to the calculation.

1. Introduction

Recently a formulation of a radiochromic polyvinyl alcohol-iodide (PVA-I) hydrogel dosimeter with crosslinking by glutaraldehyde (GTA) was introduced by Hayashi et al [1]. The dosimeter has several advantages for optical computed tomography (CT) readout. These include; high transparency, very low diffusion, no dose rate effect, high melting point and reusability following thermal annealing. Crosslinking results in a relatively strong and elastic hydrogel. While investigating formulations with lower concentrations of potassium iodide for 3D dosimetry at orthovoltage energies, we decided to revisit 3D gel dosimetry for gels outside of protective vessels. In principle, if the vessel can be eliminated, 3D optical CT scanning is improved since refractive index matching between gel and liquid can be achieved. The plastic vessels typically have refractive indices (n) near 1.5 while for gel formulations n ~ 1.35. This refractive index mismatch leads to missing data near the vessel walls at higher incident angles. An initial investigation with FX-5% gelatin dosimeters determined they were mechanically too weak to handle without the support of the vessel. It was also realized that a vessel wall was needed to provide a barrier to oxygen (O_2) diffusion since the gel was consuming O_2 through auto-oxidation. Without the vessel barrier, a 3D O_2 gradient would form causing spatial artifacts in dosimeter response. The O_2 dependence of the PVA-I gels has not been reported. Several years later, a genipin crosslinked gelatin dosimeter without a vessel was evaluated, but at the degree of crosslinking appropriate for optical transmission, this formulation was again too mechanically fragile [2]. The performance of transparent latex balloons as thin-walled vessels for optically scanning radiochromic gels was also examined [3,4]. The thin walls allowed accurate dosimetry to within 2 mm of the gel perimeter. Also, the balloon acted as a skin so that non-crosslinked gelatin gels could deform without tearing. Other groups have made bare dosimeters for optical CT scanning. For example, silicone gel dosimeters have been cast into a cylindrical shape for deformable dosimetry and plastic Presage dosimeters have been cast in the shape of a mouse [5,6].

In this study, we investigate casting PVA-I hydrogels in a spherical shape and irradiating the sphere in-air to provide a simple experiment for comparison with Monte Carlo calculations. The spherical shape was chosen based on symmetry and availability of hollow plastic spheres to serve as moulds.

2. Methods

The PVA-I gel formulation included: 8% PVA (Mowiol 18-88, Sigma-Aldrich #81365) [7], 20 mM potassium iodide, 200 mM gluconic acid lactone, 100 mM fructose and 5 mM GTA. This solution was poured into clear polymethyl methacrylate (PMMA) hollow spheres purchased from Amazon (item description, clear fillable plastic balls), OD=50 mm, ID=47 mm. The inner surfaces of the PMMA spheres were coated with alpine ski wax and polished to a smooth finish to act as a mould release agent. Samples were stored for 3 days at 20 °C in the dark for the crosslinking reaction to complete. Attempts at crosslinking at higher temperatures such as 45 °C sometimes resulted in bubble formation. Samples prepared with 4% PVA were more flexible and deformed in air due to gravity and were excluded for this project.

Following the removal of gel spheres from their respective moulds, they were stored in a solution of this same formulation minus the GTA providing no net diffusion of mobile solutes. This solution also provided refractive index matching for the optical CT scans. Gel spheres were held by suction to a funnel mount in the custom 9.6 cm diameter, cylindrical vessel for optical CT scanning, see Fig. 1.

Figure 1. PVA-I hydrogel sphere transmission image pre irradiation, pixel size = 0.11x0.11 mm.

This liquid was drained for 'in-air' irradiation, leaving the gel sphere secured to the funnel. The liquid was returned to the vessel for post irradiation scans. This approach avoided trying to remount the gel sphere in the same orientation on the funnel for post irradiation scans. The gel sphere had been irradiated with an 8 Gy uniform dose and this same single beam irradiation, optically scanned and then thermally annealed 3 days prior to this experiment. The scanner was a Vista16 (ModusQA) modified with a custom 1.5 cm, ~530 nm source [8]. Outside of the custom vessel the aquarium was filled with ~22% propylene glycol – water solution, to maximize field of view. The iterative CT algorithm, OSC-TV with 0.25x0.25x0.25 mm voxels was used to reconstruct the 1000 projection data set.

The Monte Carlo calculation was performed with DoseXYZnrc for a 47 mm diameter sphere of water with mass density of 1.027 g-cm^{-3}, irradiated in-air at 100 cm SAD with jaw-defined 8x8 cm, 6 MV, Varian TrueBeam x-ray beam. Voxel size was 1x1x1 mm with estimated calculation error of 1% for the maximum dose voxel. The centre of the gel sphere was placed at linac isocentre and irradiated to 2000 monitor units (MU, ~20 Gy) at 600 MU/minute with a horizontal beam. The sample was again scanned after a 60 minute radiochromic development time. Note this sample had been previously irradiated to a similar dose and then annealed at 47 C for 3 hours.

3. Results

Handling the gel sphere in the solution was challenging since it was nearly invisible in the solution and had similar density. For this data set, the equator of the gel sphere was tilted from the horizontal plane and the data in Fig. 2 represents a horizontal plane ~1 mm above the highest position of equator. The equator is the line where the two halves of the spherical mould connect.

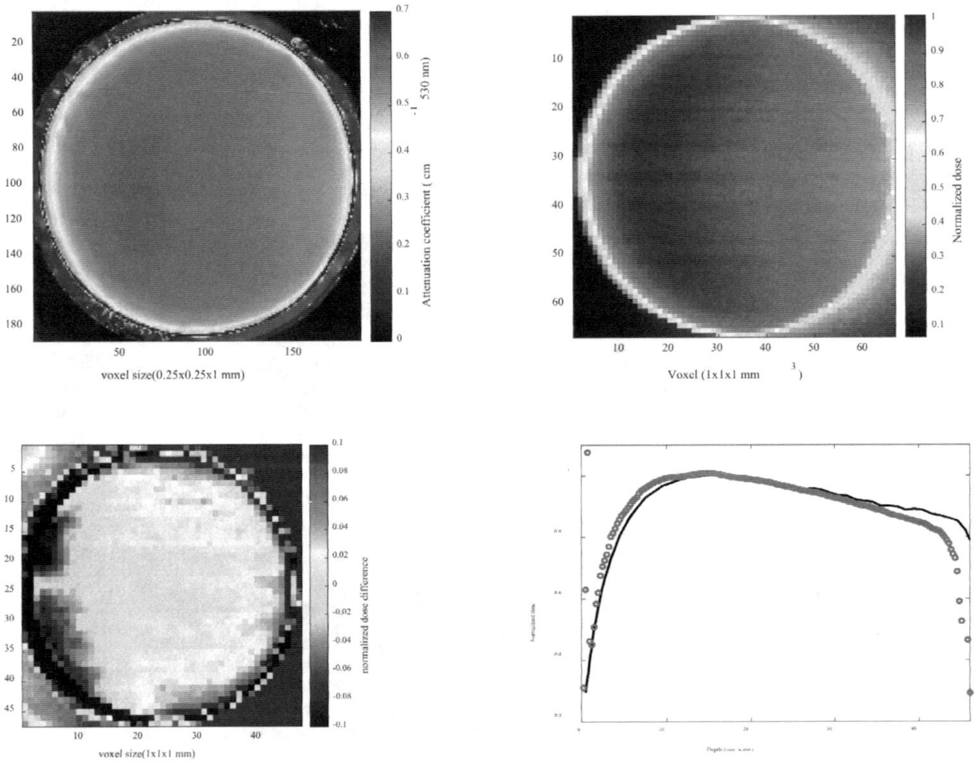

Figure 2. Upper left; Reconstructed attenuation coefficient, average of 4 central horizontal slices, 6 MV beam incident from left (1 mm^3 standard deviation <1% of mean). **Upper right**; Monte Carlo calculation, normalized to maximum dose. **Lower left**; Planar normalized dose difference, MC-gel, note gel reconstruction at voxel=1x1x1 mm not shown. **Lower right**; Plot of central axis, normalized depth doses, gel (red circles, voxel=1x1x0.25mm), Monte Carlo (black line, voxel=1x1x1mm, dose error 1% at maximum dose).

In Fig. 2, the upper panels are central slices from the 3D arrays of reconstructed gel attenuation coefficient and Monte Carlo calculated dose normalized to maximum dose. Note the gel image is the average of 4 adjacent 0.25 mm slices to match the slice thickness of calculation for comparison purposes.

The 6 MV x-ray beam was incident from the left and the gel centre was located at the TrueBeam machine isocentre. From the optical CT reconstruction, the gel diameter was 46.5 mm while the plastic mould had an inner diameter of 47 mm. Possibly the sphere relaxed in size when extracted from the mould. In the central axis depth dose plot the voxels of 1x1x0.25 mm were obtained by averaging voxels from the 0.25x0.25x0.25 mm reconstruction in order to compare with the calculation. In Fig. 2, the lower right panel is a dose difference plane containing the beam central axis, calculation-gel. Both MC calculation and gel measurements were normalized at maximum dose and attenuation coefficient, respectively. In this case the gel was reconstructed with 1 mm isotropic voxel size, (not shown) which enhanced the apparent differences in the high gradient surface voxels. The planar dose difference shows the same gel under-response in the exiting hemisphere as seen for the central axis depth dose graph.

4. Discussion

This study has demonstrated that PVA-I hydrogels can be cast in a 3D, clear, colourless shape such as a sphere. The mechanical strength of this 8% PVA formulation allowed the spherical shape to be maintained in air while being held by a suction, funnel mount. Accurate positioning of the sphere was a problem and we are currently investigating casting the sphere with an indexed mounting post embedded within the gel. The agreement between calculation and measurement suggests that accurate dosimetry to within 1 mm of the surface will be achievable. Since, the edge of the gel sphere is observed in Figure 1, transmission image refractive index matching could be improved. This would reduce reconstruction artefacts at gel surface. A better agreement between the incident surface and central axis maximum dose would be obtained by shifting the gel depth ~1 mm. However, additional experiments are required to determine if this difference is real or an artefact. One confounding factor is that the gel was irradiated while still in the empty vessel. The 0.5 mm thick PFA Teflon wall may have had a measurable effect that was not included in the calculation. Also note the lower response of the gel in the exit hemisphere. Possibly, the previous dose history and annealing contributed to the differences observed. A more thorough evaluation will involve, irradiating the sample with a uniform dose to verify isotropic response and ensuring linear dose response before the in-air irradiation. If the uniform dose response shows near-surface artefacts, oxygen gradient and evaporation effects will require assessment. When handling the bare gel sample, it is anticipated that temporarily covering the gel with a thin plastic film will provide protection. Other data in larger vessels indicate a threshold dose of ~5 Gy before linear response is achieved. This gel formulation is elastic due to the crosslinking and quantitative deformable 3D dosimetry should be possible.

In conclusion, 3D dosimetry of a PVA-I, 47 mm, gel sphere with optical CT readout has been demonstrated and the initial results are similar to a Monte Carlo calculation after normalizing the 3D data sets.

5. Disclosure

K. Jordan has a licensing agreement with ModusQA related to dosimetry applications of optical computed tomography.

6. References

[1] Hayashi S *et al* 2022 *J. Phys.: Conf. Ser.* **2167** 012014
[2] Jordan K 2009 *J. Phys.: Conf. Ser.* **164** 012027
[3] Jordan K 2013 *J. Phys.: Conf. Ser.* **444** 012039
[4] Jordan K *et al* 2013 *J. Phys.: Conf. Ser.* **444** 012081
[5] De Deene Y *et al* 2015 *Phys. Med. Biol.* **60** 543-1563
[6] Bache S T *et al* 2015 *Med. Phys.* **42** 846-855
[7] Lazzeri L *et al* 2019 *Phys. Med. Biol.* **64** 0850515
[8] Dekker K H *et al* 2018 *Phys. Med. Biol.* **63** 21NT02

Rheology modifiers in Fricke – gelatin dosimeters

Kalin I. Penev[1,2], **Kibret Mequanint**[1,3]
[1]Department of Chemical and Biochemical Engineering, The University of Western Ontario, London, Canada
[2]Modus Medical Devices Inc., London, Canada
[3]Biomedical Engineering Program, The University of Western Ontario, London, Canada

E-mail: kmequani@uwo.ca

Abstract. Fricke gel dosimeters, while very sensitive and versatile, continue to suffer from one major issue – the diffusion of the dose signal. To tackle that problem, here we present the use of inert, linear polymers with very high molecular weights as rheology modifiers that can slow the diffusion in weak gels, based on 4% gelatin. At 0.2% loading of polyacrylamide (with mol. wt. 5×10^6 Da), a moderate, 20% decrease of the diffusion coefficient was achieved. This effect is compared to that of adding a crosslinker, such as glyoxal or increasing the concentration of gelatin by 1%.

1. Introduction
The development of new ferrous hydrogels dosimeters (Fricke gels) continues to be an active area of research (e.g. [1]) due to their relative ease of preparation, high sensitivity, and versatile applications in the three-dimensional dosimetry of ionizing radiation. Upon irradiation in an acidified aqueous medium, ferrous ions (Fe^{2+}) are oxidized to ferric ions (Fe^{3+}); the resulting dose signal can be read by either magnetic resonance imaging or, in the presence of an appropriate metal indicator, as a color change by optical computed tomography. One unresolved issue with Fricke gels is the tendency of the dose signal to dissipate over time due to the diffusion of the ferric ions in the hydrogel. To counteract this effect a variety of approaches have been proposed, such as the introduction of chelating agents and other ligands [1-3], changing the gelling agent [4-6], chemical functionalization of the gelling agent [5], and introduction of heterogeneous inclusions [7]. Each approach comes with its own set of benefits and drawbacks such as high background signal, lower sensitivity, fast auto-oxidation, *etc.* What we propose here is to act on the liquid phase of the gel by addition of high molecular weight rheology modifiers – polymers that increase the viscosity of solution at very low concentrations. Similar work has been reported in the past [8], showing that introduction of low molecular weight rheology modifiers such as glycerol and saccharides limits the signal diffusion in Fricke gels. However, the glycerol and saccharides were too reactive and required too high concentrations to be of a practical use. By contrast, herein we demonstrate the use of two nearly inert, linear polymers, poly(ethylene oxide) (PEO) and polyacrylamide (PAM), with a very high molecular weight of *ca.* 5×10^6 Da, and concentrations not higher than 0.2%, tested within a ferrous – methylthymol blue – gelatin gel dosimeter, based on our earlier work [9].

2. Experimental
Gelatin (Gl) (Type A, bovine, 300 Bloom) was obtained from Rousellot (USA), methylthymol blue (MTB) from TCI America (USA); ferrous ammonium sulfate (FAS), ferric ammonium sulfate

($F^{III}AS$), 2.00N sulfuric acid (SA), 40% glyoxal (Gx), polyacrylamide (PAM) (product no. 178042500), and polyethylene oxide (PEO) (product no. A1553636) were obtained from Fisher Scientific (USA). PAM and PEO were dissolved to 1% solutions by dispersion in ice-cold deionized water and mixing at room temperature for a day, gelatin was dissolved with gentle heating, the rest of the chemicals were first dissolved in 100 mM SA immediately prior to their use. The order of addition of the individual components is important: to 2N sulfuric acid was added the PEO or PAM stock solution and mixed thoroughly, followed by addition of acidified solutions of MTB and FAS, and finally Gx, and deionized water. After homogenization of the other components, the gelatin solution, at around 30 °C, was added last. For all formulations, the concentrations of SA, FAS and MTB were kept constant at 35 mM, 0.1 mM and 0.1 mM, respectively. In the preliminary studies, gelatin was used at 4% and in the follow-up studies at 4 and 5%. The concentrations of the additives are given separately in each study.

For dosimetric evaluation, the gels were poured into 4-mL polymethylmethacrylate cuvettes, capped with polypropylene caps. Irradiations were performed on a Co-60 source (Eldorado 6, Atomic Energy of Canada Ltd.) to doses of 2.5, 5.0, and 10.0 Gy. The auto-oxidation was evaluated for 14 days, as unirradiated samples aged in the fridge. Optical readout was performed on a GENESYS 10S spectrophotometer (Thermo Scientific). For preliminary diffusion estimation, individual samples were semi-shielded with a led block and irradiated to 10.0 Gy, but for statistical purposes and to create sharp gradients, for each studied formulation a triplicate of cuvettes were prepared by first pouring into each cuvette 1.5 mL of gel without FAS or MTB, followed by setting at fridge temperature for 1-2 hours, addition of ~2 mL of gel prepared with $F^{III}AS$ instead of FAS, and setting in the fridge for another 2 hours. Once a diffusion gradient was formed by either approach, the gels were brought to room temperature and scanned on a flatbed scanner Expression 12000XL (Epson) every 20 to 30 minutes for 2-4 hours. Image analysis was performed in the red channel using ImageJ ver. 1.53n [10], while curve fitting and statistical analysis were done in SigmaPlot ver. 11 (SPSS Inc.). The diffusion profiles were fitted to the complementary error function to determine the 'curvature parameters' n (mm^2), which were fitted vs time to determine the diffusion coefficients D (mm^2/h), according to [11], as shown in Figure 1.

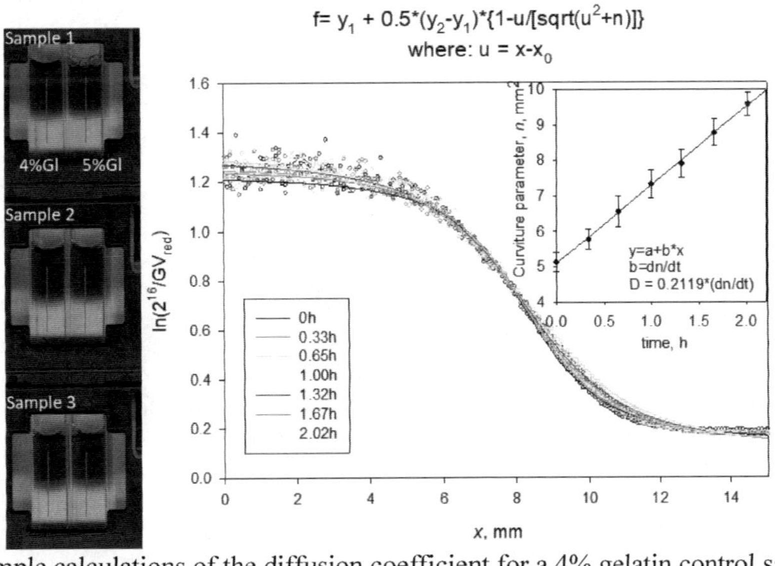

Figure 1. Example calculations of the diffusion coefficient for a 4% gelatin control sample. The image from the flatbed scanner is analyzed in ImageJ (left panel) and analyzed in SigmaPlot (right panel).

3. Results and discussion

A preliminary set of experiments (data not shown) using 4% gelatin gels revealed that paradoxically at concentrations below 0.1% the presence of either PEO or PAM caused increased diffusivity relative to the control. This effect was most pronounced at PEO, which at 0.05% caused a 40% increase of diffusivity. Above 0.1% the rheology modifiers decreased the diffusivity, more pronounced in the case of PAM which at 0.2% loading caused approximately 20% reduction of the diffusion coefficient. At low concentrations, the additives likely disrupt the native gelatin matrix, increasing the disorder in the system (entropy) and therefore increasing the diffusivity of the ferric complex. At higher polymer concentration, more "free" water is bound to its chain, overcoming the entropic factor, and decreasing the diffusivity. Further, PEO is a polymer with no side groups, while PAM has the relatively bulky C-formylamide side group (–C(=O)NH_2) which can form strong hydrogen bonding with the Fe^{III}-MTB chelate, decreasing diffusion on a molecular level. In the follow-up study at 4% gelatin, the following formulations were evaluated: 0 or 0.2% PAM, and 0 or 10 mM Gx, plus 35 mM SA, 0.1 mM FAs and 0.1 mM MTB, against a control with 5% gelatin.

The optical attenuation in the control formulation (4% Gl) is given in Figure 2A, showing a maximum of the dose sensitivity at 620 nm, as can be expected for ferrous-MTB gels. The behavior of the other formulations is similar and only their background attenuation is given here (Figure 2B): all additives increase the background attenuation, but a more careful evaluation of that effect (Figure 2B, insert) shows that the scatter at 620 nm in a gel with 4%Gl and 0.2% PAM is not higher than that a 5%Gl gel at 585 nm – which is typical for ferrous – xylenol orange gel dosimeters. Therefore, the PAM-containing gels likely have acceptable level of optical scatter.

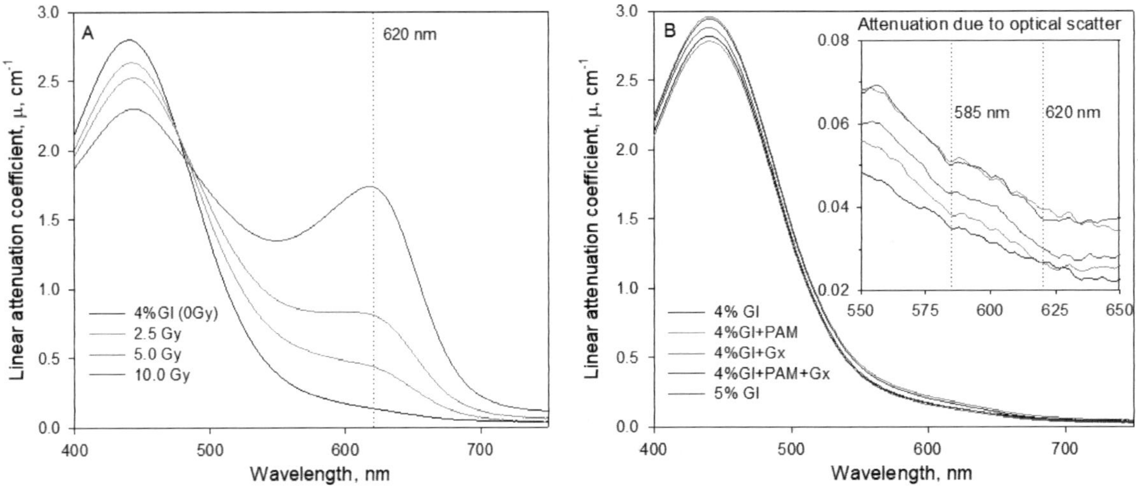

Figure 2. Optical attenuation, dose sensitivity and scatter in the studied formulations.

Figure 3 provides a visual representation of properties of the different formulations. Addition of 0.2% PAM provided an approximately 20% decrease in diffusivity, practically identical to the effect of the addition of 10 mM Gx or increasing the gelatin concentration from 4% to 5%; there was no synergistic effect on the diffusion between PAM and Gx. Among the formulations with decreased diffusivity, the background attenuation (optical scatter) was highest in those that contained PAM, as discussed above; however, the auto-oxidation rate is lowest, signifying potentially extended storage period for these gels. Another possible advantage to using PAM or Gx in a 4% gelatin gel is that the resulting gels have relatively low total suspended solids loading, which will likely accelerate the kinetics of the signal-forming reaction and therefore minimize any dose rate effects in the dosimeter, as we have shown in [9].

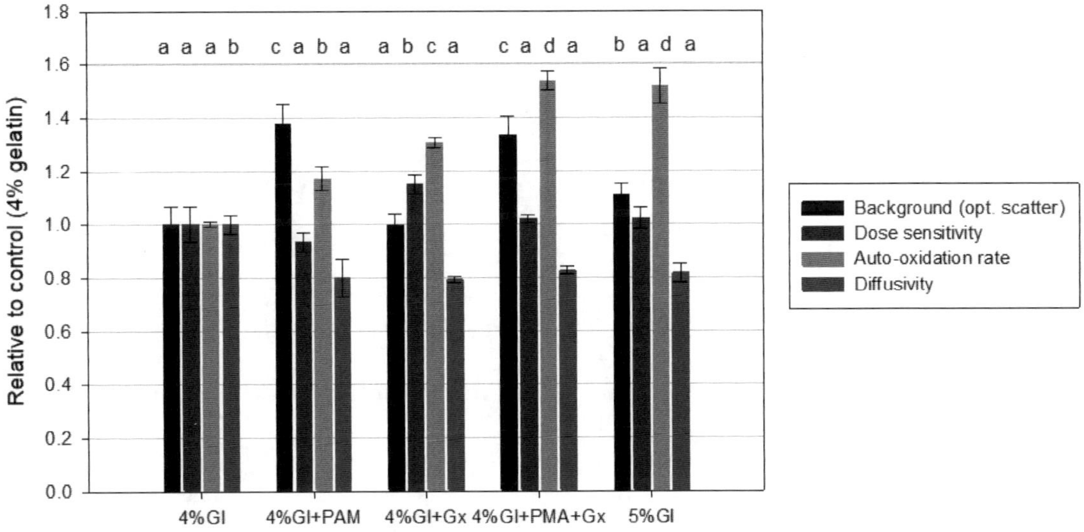

Figure 3. Dosimetric properties of the tested formulations, relative to the 4% gelatin control. The roman letters "a" to "d" above each column indicate statistically equal values for the different formulations at $p < 0.05$; lower letters correspond to lower values. Reference values for the control: background attenuation ($10^3 \times \mu_0$) = 123±8 cm^{-1}, dose sensitivity ($10^3 \times \Delta\mu$/dose) = 162±11 cm^{-1}Gy^{-1}, autooxidation rate ($10^3 \times \Delta\mu$/time) = 18.4±0.2 cm^{-1}day^{-1}, diffusivity (D) = 0.547±0.019 mm^{2}h^{-1}.

4. Conclusions

We have successfully demonstrated the use of very high molecular weight of (5×10^6 Da), generally inert, linear polymers such as poly(ethylene oxide) (PEO) and polyacrylamide (PAM) can be used as rheology modifiers in Fricke gels to slow the diffusion of the ferric ions. A moderate decrease of 20% of the diffusion coefficient was seen in weak, gelatin-based gels. The only drawback of using such polymers appears to be increased optical scatter, which is still manageable. Future work may evaluate the practical applications of these findings in large samples, particularly for small-beam irradiations and prescriptions with variable dose rates. The wider implications of the current study can be found in the rheological modification of other gel dosimeter chemistries, where decreased total suspended solids loading may be warranted.

5. References

[1] Lazzaronia S *et al* 2020 *Rad. Phys. Chem.* **171** 108733
[2] Rae W I D *et al* 1996 *Med. Phys.* **23** 15
[3] Penev K I *et al* 2013 *Phys. Med. Biol.* **58** 1823
[4] Chu K C *et al* 2000 *Phys. Med. Biol.* **45** 955
[5] Smith ST *et al* 2019 *Phys. Med. Biol.* **64** 205017
[6] Piotrowski M *et al* 2022 *Materials* **15** 984
[7] Maeyama T *et al* 2014 *Rad. Phys. Chem.* **96** 92
[8] Jordan K *et al* 2010 *J. Phys.: Conf. Ser.* **250** 012048
[9] Penev K I *et al* 2017 *Med. Phys.* **44** 1948
[10] Schneider C A *et al* 2012 *Nat. Methods* **9** 671
[11] Kron T *et al* 1997 *Magn. Reson. Imaging* **15** 211

Journal of Physics: Conference Series **2630** (2023) 012026 doi:10.1088/1742-6596/2630/1/012026

A 3D printed patient-specific vaginal mould for brachytherapy

R. Wilks[1,2], S.B. Crowe[1,2,3,4], P. Chan[1,5], R. Cheuk[1,5], T. Kairn[1,2,3]
[1]Cancer Care Services, Royal Brisbane and Women's Hospital, Herston, Qld, Australia
[2]School of Information Technology and Electrical Engineering, University of Queensland, St Lucia, Qld, Australia
[3]School of Chemistry and Physics, Queensland University of Technology, Brisbane, Qld, Australia
[4]Herston Biofabrication Institute, Metro North Hospital and Health Service, Herston, Qld, Australia
[5]Faculty of Medicine, Queensland University of Technology, Brisbane, Qld, Australia

Email: t.kairn@gmail.com

Abstract. Patient specific applicators are needed for vaginal brachytherapy treatments in cases where conventional cylindrical applicators are unsuitable or unusable. These applicators are often produced using a time-consuming and comparatively imprecise moulding method. This proof-of-concept study used Monte Carlo calculations to investigate potential dosimetric effects from creating applicators using several common 3D printing materials. A sample mould was then replicated using a fused deposition modelling (FDM) technique, which allowed catheter channels to be precisely placed with reference to treatment goals, before 3D printing from thermoplastic using a consumer-grade 3D printer. The Monte Carlo results indicated that several FDM filaments caused substantial dose depletions (up to 6%) within the model applicators while having a minor effect (less than 1%) on dose in surrounding tissue. Compared to the sample mould, the 3D printed applicator achieved superior dosimetry in terms of target coverage, while also passing manual tests of smoothness and usability. This study demonstrated an overall process by which 3D printing could replace an imprecise and time-consuming manual process and potentially achieve improved dosimetry in brachytherapy treatments of irregular vaginal anatomy.

1. Introduction

Brachytherapy is well-established as a treatment for gynaecological cancer [1,2], with intracavitary treatments for cervical cancer dating back to 1903 [1]. High dose rate (HDR) brachytherapy treatments for vaginal cancers have conventionally used cylindrical applicators that are commercially available in varying diameters, with a single catheter through the centre of the cylinder. There are two major dosimetric limitations of this design. The first is that the single central catheter can produce only cylindrically symmetric dose distributions that produce sub-optimal treatments in cases where the required dose coverage is far from cylindrical [3]. This limitation has been addressed by integrated shielding or the increasing use of cylindrical applicator designs that incorporate multiple catheters or the use of interstitial needles in combination with cylindrical applicators, so that comparatively complex dose distributions can be sculpted to cover specific regions [3,4]. The second limitation is the cylindrical design of the applicator itself. Generic cylinders and rods commonly introduce air gaps and produce limited dose coverage of the target [5,6], especially for irregular anatomy such as for larger tumours or surgical resection sites [7,8]. This substantial geometric limitation can require replacement of simple cylindrical applicators with entirely new and patient-specific designs [7,8].

As an alternative to commercial applicators with generic geometry, mouldable materials have been developed to produce a geometry specific to the patient anatomy [7,8]. For example, Fricotan (Reid Technology, Auckland, New Zealand), a mouldable material designed for audiology applications, has been used to produce patient-specific moulded applicators that reduce the impact of air gaps while providing more optimised catheter placements [7]. Optimisation of catheter placement is often achieved via an iterative fabrication-scanning-planning-refabrication process, where catheters are positioned in the mould by hand, with reference to the tumour shape, location and dimensions [7,8]. Although the dosimetry is improved, the imprecise method of catheter placement within the mould will not guarantee a dose distribution that is fully optimised to the patient anatomy. Additionally, multiple attempts at optimising the source line placement can lead to resource wastage such as time and materials.

A further consideration when using moulded applicators is the potential non-water-equivalence of the moulding materials [7,9]. This is a particular concern if the brachytherapy dose calculation formalism of the AAPM Task Group 43 is used, while uncertainties in material composition may also be a concern if model-based brachytherapy dose calculation algorithms are used [10,11].

These uncertainties in the radiological properties of the material and the imprecise placement of the source lines are trade-offs for reducing air gaps, increasing treatment accuracy and potentially increasing patient comfort [7] through the use of moulded patient specific applicators. This proof-of-concept study therefore investigated a potential 3D printed alternative where catheter channels are precisely embedded in a material of known composition and density, to generate a more optimal brachytherapy plan, improving the plan quality further by increased coverage of the target volume.

2. Method

For this proof-of-concept study, Monte Carlo simulations were used to investigate the dosimetric effects of surrounding an Ir-192 brachytherapy source with several commonly used fused deposition modelling (FDM) 3D printing materials. EGSnrc and DOSRZ codes were used to calculate brachytherapy dose with and adjacent to cylinders of polylactic acid (PLA), polycarbonate (PC), polyethylene terephthalate glycol (PETG), acrylonitrile styrene-acrylate (ASA), and two Nylon filaments, Nylon and Nylon-X. All simulations used the nominal densities of the 3D printing materials, although simulations were repeated with reduced densities for PC and Nylon when samples of these two materials printed at 100% in-fill were found to have densities noticeably lower than nominal. Despite this issue, PC and Nylon were retained in this study due to their high melting points and likely ability to survive autoclave sterilisation. A 3.6 mm long line source was modelled in these simulations using the spectrum of an encapsulated Nucletron Ir-192 source, distributed with EGSnrc. Results were compared against calculations in an identical geometry, where the 3D printing material was replaced with water.

To demonstrate the proposed 3D printing workflow for gynaecological brachytherapy, a sample mould was replicated using an FDM 3D printing technique. The sample mould had been created using a fabrication process described by Nilsson et al [7], where a positive cast was created from Fricotan by an oncologist with gynaecologic brachytherapy experience, a negative was created using dental alginate, and then brachytherapy catheters were shaped into the desired configuration and supported by a small thermoplastic jig before being positioned in the negative and surrounded by Fricotan to produce the desired applicator design. Under this process, catheter positioning is constrained by the challenge of fixing the catheters in the desired shape without being able to visualise their position in the completed mould.

To create a 3D printed applicator with the same external geometry as the sample mould, the positive cast was CT scanned at high resolution (using a process previously described for 3D printed bolus quality assurance testing [12]) and imported into the MIM Maestro contouring software suite (MIM Software Inc, Cleveland, USA), where the Fricotan cast was contoured exported as a stereolithography (STL) structure. This structure as then smoothed using Meshmixer (Autodesk Inc, San

Rafael, USA) and imported into Blender software (Blender Foundation, Amsterdam, Netherlands) where catheter positions were placed, based on the intended dose to the treatment target as specified by the clinician. Three catheters were placed within the structure, as this was the number used in the original mould. The resulting design was printed from PLA using an Ultimaker 2 Extended+ 3D printer (Ultimaker BV, Utrecht, Netherlands). Finally, the 3D printed applicator was CT scanned at high-resolution so that a treatment plan could be created in the Varian Eclipse treatment planning system (Varian Medical Systems, Palo Alto, USA) and compared with the result from the sample mould.

3. Results

Figure 1 shows the results of the Monte Carlo calculations. The result for PLA is highlighted with filled black data points, since this material was selected for use in the workflow demonstration due to its low cost, broad availability and ease of use. The Monte Carlo result for PLA was similar to the results for ASA and PC (shown in Figure 1), with dose calculated within the applicator differing substantially from water although these differences had minimal effects on dose to surrounding tissue.

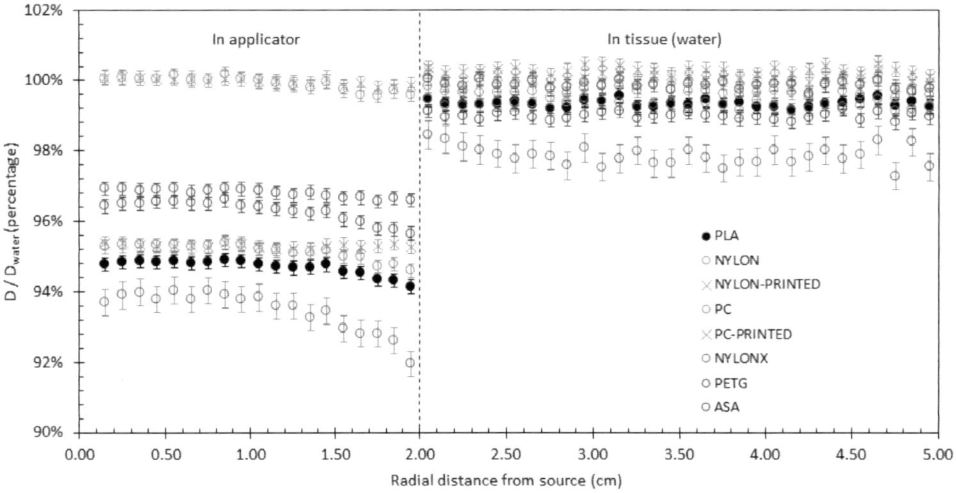

Figure 1. Dose calculated within and around a 3D printed applicator, relative to the dose calculated in water, using a worst-case thickness of 2 cm of 3D printed material between the Ir-192 source and the surrounding tissue, modelled as water.

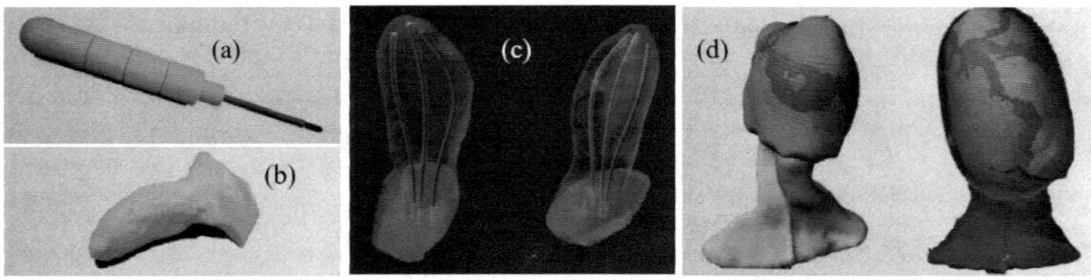

Figure 2. (a) Photograph of cylindrical applicator. (b) Photograph of moulded patient-specific applicator. (c) 3D print design with internal catheter geometry. (d) Dosimetric comparison between Fricotan mould (left) and 3D printed applicator (right), where the cyan structure represents the treatment volume to be irradiated by the prescribed dose (blue) and the 90% coverage dose (magenta).

Figure 2 shows key stages in the proposed process of fabricating and using a 3D printed applicator including: identification that a standard cylindrical applicator (Figure 2(a)) would be unsuitable for the patient, creation of a patient-specific mould (Figure 2(b)), use of the mould to design a 3D print with optimised catheter channel positions (Figure 2(c)), printing and use of the 3D printed applicator to achieve an improved treatment dose distribution compared to the manual method of physically placing catheters inside the mould (Figure 2(d)). In this case, the 3D printed applicator also passed a visual inspection and standard manual tests [7] for clinical suitability in terms of surface smoothness and easy insertion of catheters and clear transit of a dummy source.

The major limitation of this proof-of-concept was the use of PLA as the 3D printing material. PLA has a comparatively low melting point (often quoted as between 170 and 200 degrees), which makes it unsuitable for autoclave sterilization. Results in Figure 1 suggest that use of Nylon FDM filament might be advisable for producing improved water-equivalence while also being potentially sterilizable at high temperatures. However, the FDM printing process also has the potential to produce a very slight surface roughness which, despite feeling smooth to the touch, might represent an opportunity for increased bacterial growth. For these reasons, it is recommended that further work in this promising area focus on the use of high-resolution printing technologies such as stereolithography, PolyJet or selective layer sintering. Nonetheless, this study demonstrated an overall process by which 3D printing could replace an imprecise and time-consuming manual process and potentially achieve improved dosimetry in brachytherapy treatments of irregular vaginal anatomy.

4. Conclusion

The Monte Carlo results indicated that several FDM filaments caused substantial dose depletions (up to 6%) within the model applicators while having a minor effect (less than 1%) on dose in surrounding tissue. A dosimetrically superior plan was achieved using the proposed 3D printing process, highlighting the importance of catheter placement in brachytherapy. In addition to providing the patient with a treatment plan optimised to their anatomy, a 3D printing process combined with the educated, precise placement of catheters can conserve time and materials, potentially streamlining the patient journey.

5. References

[1] Kemikler G 2019 *Turk. J. Oncol.* **34** (Supp 1) 1–10
[2] Banerjee R *et al* 2014 *Int. J. Women's Health* **6** 555-564
[3] Glaser S M *et al* 2015 *J. Contemp. Brachytherapy* **7** 312-318
[4] Westerveld H *et al* 2020 *Lancet Oncol.* **21** e157-e167
[5] Maxwell Á *et al* 2016 *Brachytherapy* **15** 832-838
[6] Sapienza L G *et al* 2019 *Brachytherapy* **18** 620-626
[7] Nilsson S *et al* 2015 *Brachytherapy* **14** 267-272
[8] Magné N *et al* 2010 *Brachytherapy* **9**(3) 274-277
[9] Crowe S B *et al* 2018 *Australas. Phys. Eng. Sci. Med.* **41** 731-737
[10] Rivard M J *et al* 2004 *Med. Phys.* **31** 633-674
[11] Beaulieu L *et al* 2012 *Med. Phys.* **39** 6208-6236
[12] Kairn T *et al* 2021 *Phys. Eng. Sci. Med.* **44** 1187-1199

Journal of Physics: Conference Series **2630** (2023) 012027 doi:10.1088/1742-6596/2630/1/012027

Loss and reintroduction of the radical initiator into the FlexyDos3D silicone dosimeter for 3D printing

M J Wheatley[1] and Y De Deene[2]
[1]School of Engineering, Macquarie University, Sydney, Australia
[2]School of Science, University of Western Sydney, Sydney, Australia

E-mail: Morgan.Wheatley@hdr.mq.edu.au

Abstract. 3D printers allow for the rapid construction of complex 3D objects that would be very time-consuming with traditional casting techniques. Patient-specific objects have been created for years in the fields of dentistry, prosthetics, and surgical guides. However, 3D printed objects using materials that also serve a functional purpose, biologically or chemically, are now finding bio-medical applications. A custom 3D printer has been made that is able to print the FlexyDos3D silicone dosimeter, but the dosimeter's sensitivity is severely decreased after printing. Testing was performed to determine if chloroform is lost during printing and techniques were tested to reintroduce chloroform into the dosimeter after printing to improve the dose response of the dosimeter. Results showed that the chloroform was completely evaporated from the dosimeter when exposed to conditions similar to that during printing and that chloroform could be reintroduced by soaking or exposure to chloroform vapours which led to an increase in the dose sensitivity. Further testing of the reintroduction of chloroform into the dosimeter is ongoing.

1. Introduction

Three-dimensional dosimeters have many benefits over one- and two-dimensional dosimeters, but importantly they can record radiation dose contiguously in the three spatial dimensions instead of, at best, a sparsely sampled set of dose values in a 3D space. 3D dosimeters can also be made into the shape of an object for treatment validations, such as a human head or breast [1, 2]. Most 3D dosimeters are cast in moulds to make the shape of the object. However, casting techniques can be very time consuming in initially making the mould, and they are not able to make more complex objects with hollow cavities in a single casting, such as a heart or lungs. As such, making patient-specific dosimeters with casting techniques is not feasible for patient-specific treatment validations.

Additive manufacturing techniques (commonly referred to as 3D printing) allow for the construction of an object by depositing material in successive layers to create a 3D object. As such, more complex shapes can be created that can include cavities. The silicone dosimeter, FlexyDos3D, has been shown to be capable of being 3D printed with the use of a custom-built 3D printer [3]. It can print the dosimeter material in its initial uncured state and form a solid shape by extruding the material through a narrow nozzle into a high temperature chamber within the printer that causes the printed material to cure rapidly.

Simple objects, including both solid and hollow cubes and cylinders have been created with the printer but the dosimetric properties of the dosimeter have been found to be far less sensitive than dosimeters created with casting techniques. The difference in the manufacturing techniques involves much higher temperatures during 3D printing of the material to rapidly cure the material, and the much higher surface area to volume ratio involved when extruding the dosimeter through a narrow nozzle. As such, it is believed that the chloroform within the dosimeter, which acts as a radical initiator for the radiation dose response, is likely evaporating out of the dosimeter to a much larger degree than previous tests performed on smaller surface area to volume tests. This would lead to a reduction in the dose-response [4]

Investigations into the evaporation of chloroform from the dosimeter material during printing were performed. However, these tests could not be performed by printing the material as measurements of the mass of the dosimeter material before printing could not be measured. The mass could not be accurately measured because not all of the dosimeter material could be extruded out of the tubing completely. Instead, tests were to be performed on thin castings (roughly 1mm thick) at the temperature used during printing as these could be measured accurately before and after curing at the temperatures used in the 3D printer heated chamber.

Tests were also performed to determine if chloroform could be introduced into the dosimeter after manufacture instead of during and if this would improve the dose response sensitivity towards that of the FlexyDos3D dosimeter with chloroform added during manufacture.

2. Methods and Materials

2.1 Dosimeter recipe

The standard FlexyDos3D dosimeter uses a silicone elastomer (Sylgard® 184, Merck) as its major constituent with 93% w/w base silicone and 4% w/w curing agent, 3% w/w chloroform, and 0.03% w/w leucomalachite green (LMG). However, to investigate the response of the dosimeter without chloroform and with reintroduction of chloroform into the dosimeter after manufacture, another formulation of the FlexyDos3D dosimeter without chloroform was made. It consisted of the same ratio of silicone base, curing agent, and LMG but no chloroform. The FlexyDos3D formulation without chloroform is referred to as 'un-activated FlexyDos3D' and the FlexyDos3D formulation with chloroform is referred to as 'standard FlexyDos3D'.

For the standard FlexyDos3D, the chloroform is first added to the LMG and mixed until the LMG is dissolved. Then for both samples, the chloroform and LMG or just the chloroform is mixed into the silicone base mixture and then the curing agent is mixed in. Bubbles are introduced into the mixture by the stirring and so a vacuum chamber is used to pull the bubbles from the mixture until there are no remaining bubbles.

The samples are then cured in an oven or the 3D printer oven chamber to rapidly cure the dosimeter.

2.2 Chloroform evaporation

The silicone dosimeter cures rapidly from a liquid state to solid when exposed to high temperatures. To be able to print the material in layers without flowing into a puddle, 160°C is used within the printer's oven chamber. However, chloroform has a relatively low boiling temperature (~60°C), and this would likely result in a loss of the chloroform within the dosimeter due to the higher temperatures used.

Two 5g samples of FlexyDos3D dosimeter were made, one of standard FlexyDos3D and the other un-activated FlexyDos3D. Each of the samples were poured into glass petri dishes, which gave a height of roughly 1mm of dosimeter, and then cured in the 3D printer at 160°C for two hours. Two hours was chosen as it currently takes the printer about two hours to print 60mL of dosimeter and so this is how long the first printed layer of dosimeter would likely be exposed to the high temperatures. The samples were weighed before curing and again after curing to determine any mass lost during curing. To determine if the mass loss was the chloroform and not part of the silicone material, a comparison was made between the mass lost in the standard and the un-activated FlexyDos3D castings.

2.3 Chloroform penetration

Previous research into the FlexyDos3D dosimeter had shown that chloroform was able to evaporate out of the dosimeter over time, even at room temperature and that the chloroform concentration affected the dose response of the dosimeter. The opposite case was considered where a FlexyDos3D dosimeter without any chloroform in it could be exposed to chloroform, either by soaking in chloroform or by sealing the dosimeter in a chamber with chloroform vapour, and if the FlexyDos3D dosimeter would then absorb this chloroform and cause it to exhibit a dose-response.

One sample of standard FlexyDos3D and 3 samples of un-activated FlexyDos3D were made each consisting of 20g of their respective dosimeter in a plastic cylindrical vial measuring roughly 25 mm on its inner diameter. Each sample was cured in a 60°C oven for two hours. The lower temperature was used for curing the samples as the plastic containers would melt at higher temperatures and the container had to be broken away from the dosimeter after casting for the later testing. After curing, two of the un-activated samples, were removed from the cylinders. Each sample was weighed again before each sample was placed in a separate sealed container. One sample was suspended in the container above chloroform amounting to 3% w/w of the samples mass (0.6g). The other sample was immersed completely in chloroform for two minutes, before being removed from the chloroform, patted dry, and then weighed to determine the amount of chloroform absorbed in the dosimeter. Both treated dosimeters were then sealed into separate air-tight containers. The un-activated FlexyDos3D sample and the standard FlexyDos3D sample were used as controls and were not exposed to chloroform after manufacture but were placed in sealed containers. All the samples were also stored in an opaque container to eliminate exposure to light sources before irradiation.

The samples were irradiated with 10Gy each with a 1cm x 1cm square beam. After irradiation, all the samples were again stored within the opaque container.

Readout of the samples was performed with a modified dual-wavelength cone-beam optical scanner based off a Modus Vista™ scanner (Modus Medical Devices Inc, London, Canada). The samples were rotated within the optical scanner through a full rotation under two different wavelengths of light and 256 images were taken in each revolution with a resolution of 0.12 mm / pixel at the centre of the scanner's rotation.

A filtered back-projection reconstruction was performed in the Matlab software (The Math Works, Inc).

3. Results and Discussion

The masses of the evaporation samples cured at 160°C for 2 hours are shown in Table 1 for both before and after irradiation.

Table 1. Change in mass of cured samples

	Before Irradiation Mass [g]	After Irradiation Mass [g]	Change in Mass [g]
Un-activated FlexyDos3D	5.069	5.024	-0.045
Standard FlexyDos3D	5.105	4.939	-0.166

The un-activated FlexyDos3D did not have any chloroform and showed a minimal change in mass as compared to the standard FlexyDos3D sample which did have 0.175g of chloroform within the sample before curing. The mass of the standard FlexyDos3D sample changed by roughly the mass of chloroform added to it after curing at high temperatures within the printer oven.

For the irradiation of the un-activated FlexyDos3D samples, a significant dose response was seen after the delivery of 10Gy in both the soaked and vapour exposed samples as compared to the sample with no chloroform treatment. The standard FlexyDos3D sample showed the greatest optical density change due to radiation. The reconstructed optical CT images of the optical density difference can be seen in Figure 1 and the optical density at the maximum optical density change for each sample is shown in Table 2.

Table 2. Optical density at Dmax in the treated samples

	Control	Vapour	Soak	Standard FlexyDos3D
Optical density [cm^{-1}]	0.020	0.034	0.049	0.108

The greater response seen in the FlexyDos3D sample as compared to the treated samples is likely due to a few factors. The LMG in the un-activated samples was unable to dissolve fully within the silicone mixture and could be observed visually as opaque clumps within the silicon matrix. As there was less LMG dispersed throughout the cured sample, it would seem reasonable that the dose response of the colour change would also be lessened.

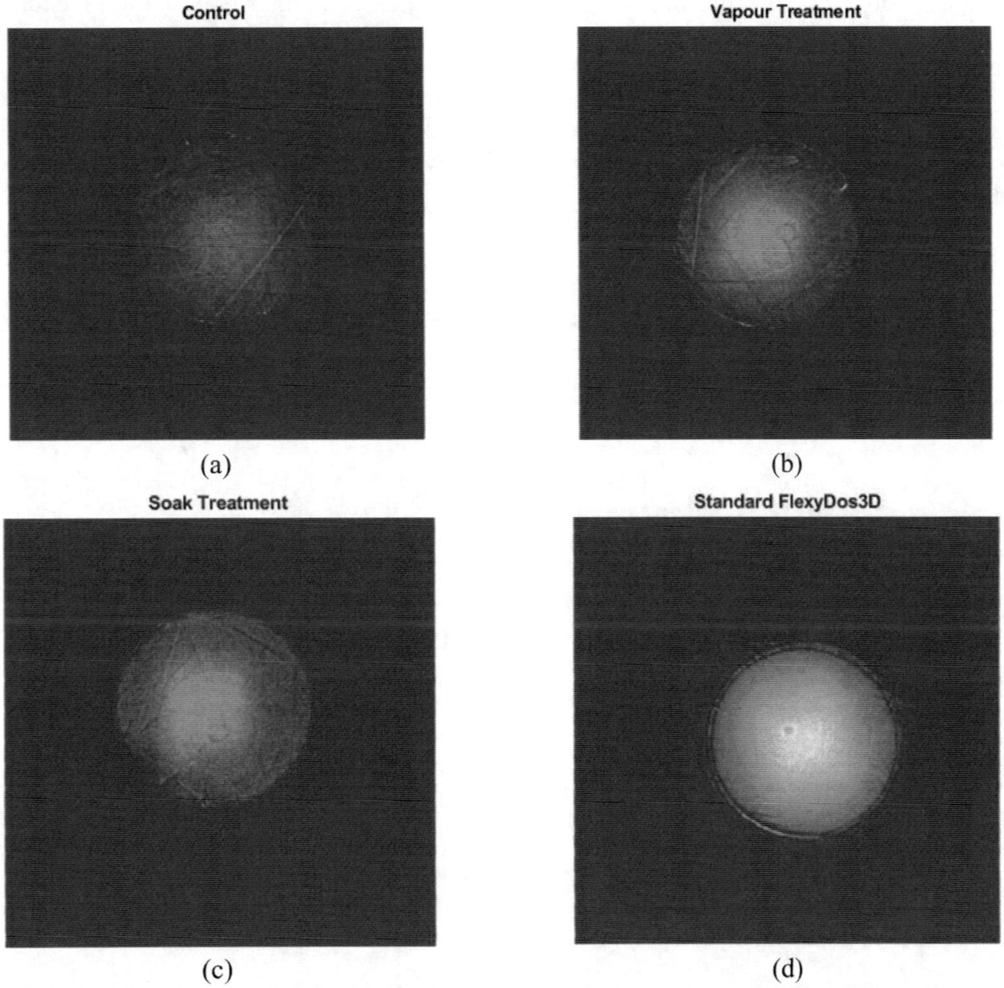

Figure 1. Optical CT reconstructions of the irradiated samples. Each sample was irradiated with a 1cm x 1cm square beam to 10Gy (Dmax). (a) A FlexyDos3D sample with no chloroform added, (b) chloroform was added by exposing the dosimeter to chloroform vapours for 12 hours, (c) chloroform was added by soaking the dosimeter in chloroform for 2 minutes, (d) a standard FlexyDos3D sample with the chloroform added during manufacture. The streaking artifacts seen in (a – c) are due to the LMG clumps that were present throughout the samples because the LMG was unable to dissolve completely within the silicone mixture during manufacture without chloroform.

4. Conclusion

The FlexyDos3D dosimeter loses its chloroform component when heated to high temperatures and having a high surface area to volume ratio. The chloroform is also likely lost during the 3D printing of the dosimeter where the extrusion of the material through a 0.6mm nozzle into a high temperature chamber meet the conditions of high temperature and high surface area to volume ratio.

However, whilst the testing had shown that chloroform was able to evaporate out of the dosimeter, further tests showed that chloroform could also be absorbed by the dosimeter when the dosimeter was briefly immersed in chloroform or if in a sealed container with chloroform vapours. This method could be used for dosimeters produced by the 3D printer that would lose their chloroform during printing to restore their dosimetric properties. The LMG will likely need to be initially dissolved into some chloroform as the LMG did not dissolve within the dosimeter even after soaking or exposure to chloroform vapours. The inhomogeneity of LMG led to imaging artifacts in the current study and may also lead to dose response errors if it is not homogeneously dispersed within the dosimeter.

Further testing will investigate the homogeneity of the chloroform penetration in larger phantoms and quantitative measurements of the dose response for the chloroform penetrated samples. The time required to expose the samples to chloroform will also be investigated to ensure homogeneity of chloroform distributed in the dosimeter.

5. References

[1] De Deene Y *et al* 2020 *Phys. Med. Biol.* **65** 225031
[2] Iqbal K *et al* 2014 *J. Appl. Clin. Med. Phys.* **15** 4531-4531
[3] Wheatley M J *et al* 2022 *J. Phys.: Conf. Ser.* **2167** 012010
[4] De Deene Y *et al 2015 Phys. Med. Biol.* **60** 1543-1563

Journal of Physics: Conference Series **2630** (2023) 012028 doi:10.1088/1742-6596/2630/1/012028

Lung radiotherapy verification with a 3D printed thorax phantom and an ionisation chamber array

T. Kairn[1,2,3], L. Jessen[1,4], J. Bodnar[1], P. H. Charles[2,3,5], S.B. Crowe[1,2,3,5]
[1]Cancer Care Services, Royal Brisbane and Women's Hospital, Herston, Qld, Australia
[2]School of Information Technology and Electrical Engineering, University of Queensland, St Lucia, Qld, Australia
[3]School of Chemistry and Physics, Queensland University of Technology, Brisbane, Qld, Australia
[4]Medical Radiation Physics, Lund University, Lund, Sweden
[5]Herston Biofabrication Institute, Metro North Hospital and Health Service, Herston, Qld, Australia

Email: t.kairn@gmail.com

Abstract. In this study, a 3D printing error inspired the development of a novel method for using a sagittally-sliced 3D printed thorax phantom to perform dosimetric verifications of lung radiotherapy treatment methods using a 2D ionization chamber array. A full-size thorax model was designed for 3D printing with multiple tissue densities including lung and bone and printed as a series of 2.4 cm sagittal slices using a Raise 3D Pro dual nozzle printer (Raise 3D Technologies Inc, Irvine, USA). An error introduced midway through printing resulted in one half of the phantom being printed at unrealistically high densities. A method was therefore devised whereby the entire phantom was used to plan two lung treatments, one conventionally fractionated and one hypo-fractionated, which were then verified via measurements using an Octavius 729 ionisation chamber array (PTW-Freiburg GmbH, Freiburg, Germany) in combination with several correctly-printed slices of the phantom. The measurements allowed dose distributions in planes through the target, adjacent to the target and at the location key of organs at risk to be verified, for both treatment plans. This method has the potential to be adapted for use with other phantoms and other dosimetry arrays to allow efficient evaluation of future treatment techniques.

1. Introduction

Dosimetric verification of radiotherapy treatment dose using a humanoid phantom containing lung-like heterogeneities is widely recommended when implementing new lung treatment modalities, and is a component of credentialling for several clinical trials of lung stereotactic ablative radiotherapy (SBRT/SABR) [1,2]. These recommendations and requirements have arisen from the known fallibilities of radiotherapy treatment planning systems when calculating dose in low-density media [3,4].

While several generic commercial phantoms are available to facilitate these verification measurements, demands for adaptability and patient-specificity have recently led to a growth in the development of bespoke thorax phantoms [1,2] including phantoms produced using 3D printing techniques [5,6,7,8]. Since several groups had produced 3D printed thorax phantoms that had empty space in the lung region, to allow moving components to be placed within [7,8], the initial aim of this study was to produce a 3D printed thorax phantom with a moving component where lung-equivalent material was included in both the moving component and the rest of the pulmonary cavity. However, when this work was interrupted by the COVID-19 pandemic and affected by a 3D printing fault [9], we took the opportunity to instead develop a novel method for using a sagittally-sliced 3D printed thorax phantom to perform dosimetric verifications of lung treatment methods using a 2D ionization chamber array.

2. Method

A full-size female thorax model was created from CT images of the XCAT computational human phantom [10] by using thresholding to delineate tissues of sufficiently different density to warrant 3D printing with different in-fill percentages. The phantom was divided into 2.4 cm sagittal slices, as suggested by Craft and Howell to minimise thermal warping during fabrication [8]. The phantom was also designed with an elliptical cavity and matching insert, which was originally intended to become the moving component.

All 3D printing was completed using a Raise 3D Pro dual nozzle printer (Raise 3D Technologies Inc, Irvine, USA), using different densities of poly-lactic acid (PLA) filament (Shenzhen Esun Industrial Co Ltd, Shenzhen, China) to model all soft tissues including lung and using StoneFil Terracotta filament (Formfutura BV, Nijmegen, Netherlands) to model all bone. A gyroid in-fill pattern was used for lung as recommended by Tino et al [11], while a grid pattern was used to maximise print speed for all denser tissues. In-fill percentages required to replicate the densities of the identified tissue types were derived by 3D printing test cubes at different in-fill percentages and performing transmission measurements with the nominal 6 MV treatment beam, via the method described by Dancewicz et al [12]. Complete printing of the 20 cm long thorax section was expected to require 11 kg of PLA and 1 kg of StoneFil (costing approximately € 250) and take 600 h total printing time.

The printing process was interrupted for several months at approximately the half-way point, due to the local impacts of the COVID-19 pandemic, and then recommenced after a firmware update had reset the extruder calibration (e-step) to a default value, resulting in an increase filament extrusion per mm of print [9].

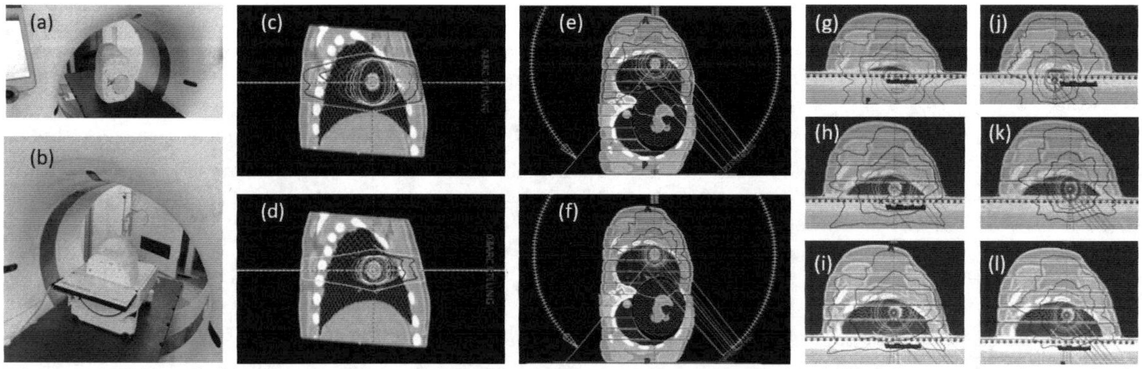

Figure 1. (a) and (b) photographs of CT setup for planning scan and verification scans. (c) and (d) sagittal views of conventionally fractionated and SBRT/SABR treatment plans. (e) and (f) transverse views of conventionally fractionated and SBRT/SABR treatment plans. (g), (h) and (i) transverse views of verification plans for conventionally fractionated treatment. (j), (k) and (l) transverse views of verification plans for SBRT/SABR treatment.

In order to salvage this faulty print and produce useful results, one new slice was printed with a faux tumour close to the right chest wall and the phantom was CT scanned in a left lateral decubitus position to allow a novel dosimetry study to be undertaken. Treatments were planned for the decubitus phantom and an experimental method was devised where measurements were performed with an Octavius 729 ionisation chamber array and lower part of an Octavius 2 phantom (PTW-Freiburg GmbH, Freiburg, Germany) replacing the left hand (incorrectly printed) parts of the phantom.

Treatment planning followed standard local procedures, with the unusual orientation of the phantom handled by assuming the "patient" was lying supine and simply rotating the arc start and stop positions by 90 degrees, so that the contralateral lung was avoided as usual. The treatment plans and verification plans are shown in Figure 1, alongside photographs showing the setup of the phantom during CT scanning in its treatment planning configuration and an example verification configuration.

Two treatment plans were created. One was a conventionally fractionated treatment (60 Gy in 30 fractions) to a PTV generated by applying a 10 mm margin to the visible tumour in all directions except superior-inferior where a 20 mm margin was used, as though planning with a 3D CT scan of a breathing patient. The other plan was a SBRT/SABR treatment (48 Gy in 4 fractions) to the visible tumour plus a 5 mm margin on all sides, as though the visible tumour represented a maximum intensity projection image from a 4D CT series. Both treatments were planned using a volumetric modulated arc therapy (VMAT) method, using the Varian Eclipse treatment planning system (Varian Medical Systems, Palo Alto, USA). Inverse planning objectives differed between the two plans, taking into account the two different prescriptions as well as the dose tolerances of the heart, lung, spinal canal, rib and liver OARs.

The verification measurement method consisted of using the Octavius 729 ionisation chamber array to measure dose planes in various sagittal slices. In order to compare the array measurements to the planned dose, the Octavius 729 array was CT scanned three times, with different numbers of phantom slices placed on top. To achieve measurements indicative of the dose to the centre of the target, the left periphery of the target, and the heart and spinal canal OARs, the rightmost four, five and six slices of the phantom were carefully placed on top of the array and aligned using lasers to duplicate their arrangement during the phantom treatment planning scan (see Figure 1). The two lung treatment plans were each copied onto these three array scans and all doses were recalculated and exported for measurement on a Varian iX linac.

3. Results and discussion

Figure 2 shows the results of the verification measurements for both arcs from the two treatments that were planned for the faux tumour in the 3D printed phantom. The results from measuring with four phantom slices on top of the ionization chamber array confirm the dose calculation accuracy within the target and also illustrate the difference in relative dose falloff between the conventionally fractionated treatment and the SBRT/SABR treatment. Other than the larger size of the conventionally fractionated target, the main contributor to this is the comparative priority of sparing the surrounding lung from the very high (12 Gy) prescribed dose per fraction for the SBRT/SABR target (note the different vertical scales in Figures 2(a)-(b) and (c)-(d)).

The effect of target size is also apparent in the results for five phantom slices, also shown in Figure 2, where dose from the SBRT/SABR treatment plan has started to decrease, while the central region of the array is still encompassed in the prescription dose for the conventionally fractionated treatment.

Beneath six slices of 3D printed phantom, the results in Figure 2 show that the dose from the two treatment plans have substantially decreased relative to their prescriptions, although in this region the Octavius 2 phantom beneath the ionization chamber array is evidently producing increased out-of-field scatter that would not be present in the patient or in a fully realised 3D printed phantom.

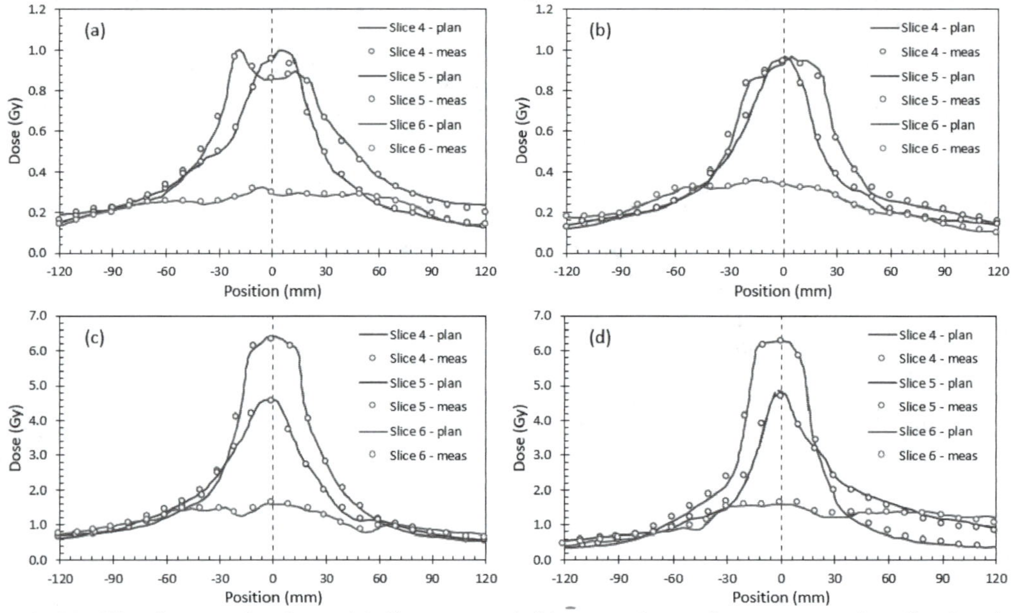

Figure 2: Verification results from (a) first arc and (b) second arc, from conventionally fractionated plan, and (c) first arc and (d) second arc, from SBRT/SABR plan, showing results for verification plans with four, five and six slices of phantom on top of the dosimeter array.

4. Conclusion

By using a 3D printed phantom, a novel method for verifying lung radiotherapy treatment plans using an ionization chamber array was developed. This method has the advantage of allowing instantaneous dose measurements of beams largely delivered though the right-hand slices of heterogeneous humanoid phantom, after being planned with reference to a more complete phantom scan that included relevant OARs. This method has the potential to be adapted for use with other phantoms and other dosimetry arrays to allow efficient evaluation of future radiotherapy treatment techniques.

5. References

[1] Clark C H *et al* 2017 *Phys. Medica* **44** 171-176
[2] Kron T *et al* 2018 *Br. J. Radiol.* **91** 20170737
[3] Dunn L *et al* 2015 *Phys. Medica* **31** 435-441
[4] Crowe S B *et al* 2012 *IFMBE Proc.* **39** 1803-1806
[5] Tino R B *et al* 2022 *Med. Phys.* **49** 52-69
[6] Kairn T *et al* 2020 *Phys. Eng. Sci. Med.* **43** 701-710
[7] Mayer R *et al* 2015 *Rev. Sci. Instr.* **86** 074301
[8] Craft D F *et al* 2017 *J. Appl. Clin. Med. Phys.* **18** 285-292
[9] Charles P H *et al* 2022 *Phys. Eng. Sci. Med.* **45** 377
[10] Segars W P *et al* 2009 *Proc. IEEE* **97** 1954-1968
[11] Tino R *et al* 2019 *Phys. Med. Biol.* **64** 21NT05
[12] Dancewicz O L *et al* 2017 *Phys. Medica* **38** 111-118

Journal of Physics: Conference Series **2630** (2023) 012029 doi:10.1088/1742-6596/2630/1/012029

Fading correction for calibration of a novel 2D OSL-film dosimeter

M. Caprioli[1], L. Delombaerde[1,2], M. De Saint-Hubert[3] ,L. de Freitas Nascimento[3], S. S. Galvez Febles[3], K. Himschoot[4], B. van der Heyden[3], D. Vandenbroucke[4], R. De Roover[2], Leblans[4], W. Crijns[1,2]
[1]Department of Oncology, KU Leuven, Herestraat 49, Leuven, 3000, Belgium
[2]Department of Radiation Oncology, University Hosptals Leuven, Herestraat 49, Leuven, 3000, Belgium
[3]Research in Dosimetric Application group, Belgium Nuclear Research Centre (SCK CEN), Boeretang 200, Mol, 2400, Belgium
[4]Corporate Innovation Office, Agfa N.V., Septestraat 27, Mortsel, 2640, Belgium

Email: marco.caprioli@uzleuven.be

Abstract. A new film dosimeter based on optically stimulated luminescence (OSL) was calibrated for high-resolution 2D dosimetry in the context of radiotherapy patient-specific quality assurance (PSQA). The OSL-film signal is linear with the dose and decays with time, in a process called fading. Two models were proposed to characterize the fading: (1) a linear model independent from the delivery and of practical use thanks to the straightforward signal-to-dose conversion, and (2) a delivery-dependent model, to investigate whether a more simplified model may be sufficient to accurately determine the fading independently from the specific-irradiation conditions. The models were used in a nonlinear regression over an experimental dataset acquired with a 6MV photon beam in standard calibration settings. Both models were accurate and showed an adjusted-R^2 of 0.98 (1) and 0.99 (2). The residual analysis on both models allowed to define the minimum scanning time to have a discrepancies between the models <1%. Under this condition the OSL fading can be assumed independent on the irradiation parameters. As a consequence, within these boundary conditions, the linear model can be used for the OSL-film calibration.

1. Introduction

Advanced radiation therapy (RT) treatments provide highly modulated dose distributions with rapid dose falloff outside the target volumes. The increased complexity of treatment plan ballistics and delivery, stresses the importance of PSQA. The role of dosimetry in PSQA is to verify with multidimensional and high-resolution dose measurements the degree of agreement with respect to the calculated dose distribution. New OSL dosimeters are investigated by the QUARTEL-project (Quality Assurance in RadioTherapy by Stimulated and Excited Luminescence Dosimetry) as an alternative to standard dosimeters (e.g., radiochromic films). These novel devices are capable of performing 2D dose measurements down to low doses with sub-mm spatial resolution defined by the readout system. Moreover, the same OSL-film is reusable after a complete signal erasure and can be applied for multiple dose measurements. The characterization of the new film will allow investigation of the OSL signal for 2.5D dose measurements. This application will require the use of a stack of the same OSL-film.

The OSL-film has a linear dose-to-signal[1], [2] measured as luminescence produced after an external light stimulation that induced electron-hole pairs recombination. Luminescent materials suffer from a time-dependent response caused by spontaneous electron-hole pairs recombination and resulting in a reduction of the measured signal intensity. This process is also called **fading**. A power law physical model was developed to describe the fading according to the probability of spontaneous recombination of excited shallow traps via quantum-mechanical tunneling effect[3]. This is the case of some luminescent material such as $BaFBr:Eu^{2+}$, or $Zn_2SiO_4:Mn$, while Jursinic et al.[4] justified the exponential decay for an $Al_2O_3:C$ phosphor material on the basis of room-temperature instability of the shallow traps[4].

When used for radiotherapy applications, fading affects the signal-to-dose conversion resulting in an error in the evaluation of the delivered dose. Moreover, the degree of signal fading depends on treatment-specific irradiation conditions such as dose rates, and irradiation times. The current work characterized the dose response specific for a $BaFBr:Eu^{2+}$-based film suitable for PSQA. The signal-to-dose conversion used a calibration function, modeling both the dose dependence and the time-dependent signal loss caused by fading. This function can be represented as a 2D calibration surface as a function of two variables: delivered dose (D) and the waiting time from the start of the exposure and the film scanning (t_{scan}). Two models were proposed. The first is characterized by a linear signal build-up followed by a power-law fading independent of the specific irradiation conditions, irradiation time (t_{irr}), and dose rate($\dot{D}$). This linear model allows a simple signal-to-dose conversion independently on the point-to-point difference in the effective delivery time. The second, instead, has a signal loss that depends on t_{irr}. A residuals error analysis was executed to evaluate the possible error arising when the fading is assumed to be independent of the irradiation parameters. The residual error analysis was used to define the best t_{scan} for three radiotherapy schemes, i.e., VMAT, IMRT, and Helicoidal Tomotherapy (HT), with different beam-on times, 3min 10.5min, 20min[5], [6].

2. Methods and Material

Two different power-law fading models were used to characterize the signal of a novel $BaFBr:Eu^{2+}$ -based OSL-film. First, a linear model was considered. It consists of a linear signal-to-dose relation lowered by a power law factor dependent on the moment at which the film was scanned (t_{scan})

$$(1)\ S(\dot{D}, t_{irr}, t_{scan}) = A \cdot \dot{D}_{mean} \cdot t_{irr} \cdot (t_{scan} + \tau)^{-n}$$

In Eq.1, the delivered dose is given as the product of constant dose rate, $\dot{D}_{mean}$, and the duration of the irradiation t_{irr}. The proportionality depends on the film's sensitivity A, while the fading is characterized by material-specific parameters; τ and n. The second model is delivery-dependent. Here each infinitesimal signal contribution has a different degree of fading depending on when it was formed during irradiation. The signal contribution at the beginning of the exposure is higher than the one formed at the end of irradiation

$$(2)\ S(\dot{D}, t_{irr}, t_{scan}) = \frac{A \cdot \dot{D}_{mean}}{1-n} \cdot [(t_{scan} + \tau)^{1-n} - (t_{scan} + t_{irr} + \tau)^{1-n})]$$

The linear model can be derived as a Taylor first-order approximation of the delivery-dependent model in Eq.2.

An experimental dataset of 86 measurements was acquired by irradiating the OSL-film on a TrueBeam™ STx using a 6MV 10x10cm² open field in reference conditions; i.e., 0° gantry angle, 90cm SSD and at 10cm depth. The measurements consisted of three constant dose-rates ($\dot{D}$) combined for six irradiation times (t_{irr}) of 0.2,1, 2, 4.5min at 600MU/min, 12min at 300 MU/min and 23min at 200MU/min, and various readout times (t_{scan}) between 4 min and 1440 min. The dose-rates were varied to keep the resulting dose levels in the range from classic fractionation to single-fraction SRS doses. The irradiations were executed in dimmed light condition to prevent external stimulation and signal loss due to excessive light exposure. Next, the film was scanned with a computed radiography (CR15-X,

Agfa Healthcare) reading system. The reader produced the typical OSL luminescence by exciting the trapped electrons with a solid-state laser of 80mW with a light spectrum centred at 665nm. The net signals, obtained as the mean raw value of a 1x1 cm² ROI minus the correspondent background, were fitted via a robust nonlinear regression method according to the models in Eq.1 and2 and with respect to the variables $\dot{D}, t_{irr}, t_{scan}$. The residual error analysis was performed by considering the relative difference map between the regression models.

3. Results

The linear and delivery-dependent models fitted the experimental data with an adjusted-R^2 respectively of 0.98 and 0.99. Coherently, the logarithm of the root-mean-square error expressed as $\ln\left(\sqrt{\sum_{i=1}^{n}\frac{(M(\dot{D},t_{irr},t_{scan})-S(\dot{D},t_{irr},t_{scan}))}{n}}\right)$ and relative to the difference between the measured and the estimated signal, decreased from 0.016 for the linear model to 0.013 for the delivery-dependent one. The estimated models` coefficients obtained via robust nonlinear regression are reported in Tab.1 together with the upper and lower 95% confidence intervals. With the exception of an outlier that has a 5% discrepancy with respect to the fitted values, a regression analysis on the residuals confirmed the accuracy of the models. The model's residual error map (%), depicted in Fig.1b, is always lower than 3.05 and for longer irradiation, the difference between the models increases at shorter t_{scan}.

Table 1. Regression coefficient for both models: (1) Linear model, (2) Delivery-dependent model. The 95% confidence intervals are also reported.

Regression Coefficients	Linear Model	Delivery-dependent Model
A	2769 [2672-2867]	2863 [2776-2951]
τ	18 [15-22]	20 [17-23]
n	0.157 [0.151-0.164]	0.162 [0.157-0.168]

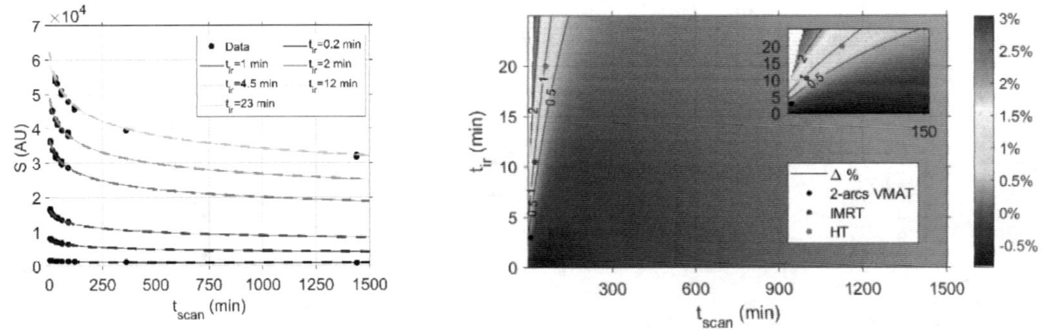

Figure 1. (Left) Different fading curves obtained for six irradiation times with a robust nonlinear regression. The regression was performed according to a linear (solid line) and the delivery-dependent model (dashed line). (Right) Relative difference map (%) between the linear and delivery-dependent fading models. The minimum t_{scan} with a difference < 1% for three IMRT, VMAT, and SBRT treatment modalities is reported.

4. Discussion

The results in Fig.2 (right) showed that the error when assuming the fading as independent from the specific irradiation conditions increases with the irradiation times (Fig.2) up to 3.04% for t_{ir}=23 min. This error arises by neglecting the fading contribution that occurs during the exposure. However, longer waiting times, t_{scan}, decrease the error <1% and allows the use of the approximated linear model for the signal-to-dose conversion in RT application. The characteristic t_{irr} for different classes of SRS and SBRT treatments were considered according to their different beam on time: 16 patients treated with (1)

2-arcs VMAT and (2)sliding window IMRT. In addition (3) one peripheral lung SBRT treatment with Helicoidal Tomotherapy was also considered[5], [6]. These treatments have different average beam-on times, respectively 3min to 10.5min and 20min. The minimum readout time to have a model difference <1% is 4min for the multi-arcs VMAT plan, 25min for the IMRT, and 63 min for the longest HT treatment.

The linear model is preferred to a more complex delivery-dependent model because the signal-to-dose conversion function does not require to know a priori the time of exposure for point within the treated region, $t_{irr}(x,y)$. This information is not easily accessible in clinical practice. Additionally, the linear model can be inverted to obtain an analytical expression for the signal-to-dose conversion, when $D = \dot{D} \cdot t_{irr}$.

(3) $D = \frac{M(D,t_{scan})}{A(t_{scan}+\tau)^{-n}}$

Where M is the measured OSL signal. The dose calibration surface accounting for the signal fading (Fig.2) has residual dose uncertainties <0.005mGy for t_{scan}>1min.

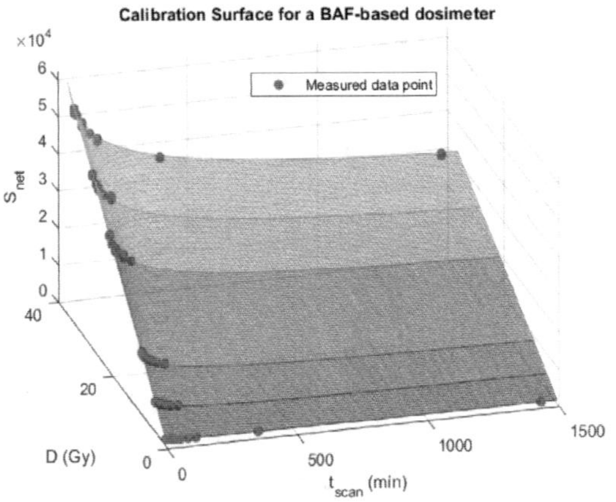

Figure 2. Dose calibration surface for a BAF-based OSL dosimeter affected by signal time decay.

5. Conclusion

The calibration of a novel OSL-film for RT application was corrected for fading with two different models. If t_{scan} > 63min than the fading can be considered approximately independent on the specific RT treatment irradiation conditions as the signal intensity decreases by a fading factor solely dependent on t_{scan}.

6. Acknowledgement

The research is financed in the QUARTEL-project (Quality Assurance in RadioTherapy by Stimulated and Excited Luminescence Dosimetry VLAIO [VLAIO HBC.2020.3003]), a joint financing of the Flemish government and AGFA N.V.

7. References

[1] Kry S F *et al* 2020 *Med Phys.* **47** e19-e51
[2] Yukihara E G *et al* 2008 *Phys. Med. Biol.* **53** R351-79
[3] Huntley D J 2006 *J. Phys. Condens. Matter* **18** 1359-1365
[4] Jursinic P A 2007 *Med. Phys.* **34** 4594-4604
[5] Weyh A *et al* 2013 J. Appl. Clin. Med. Phys. **14** 4065
[6] Roa D E *et al* 2012 **37** 257-264

Journal of Physics: Conference Series 2630 (2023) 012030 doi:10.1088/1742-6596/2630/1/012030

Characterisation and application of three-dimensional silicone-based radiochromic dosimetry in 1.5 T magnetic resonance imaging-guided radiotherapy

Morten B Jensen[1,2,3], Peter Balling[4,5], Hans L Riis[6,7], Anders S Bertelsen[6], Simon V Jensen[1,2], Simon J Doran[8], Jørgen B B Petersen[3] and Ludvig P Muren[1,2]

[1]Danish Centre for Particle Therapy, Aarhus University Hospital, Aarhus, Denmark
[2]Department of Clinical Medicine, Aarhus University, Aarhus, Denmark
[3]Department of Medical Physics, Aarhus University Hospital, Aarhus, Denmark
[4]Department of Physics and Astronomy, Aarhus University, Aarhus, Denmark
[5]Interdisciplinary Nanoscience Center, Aarhus University, Aarhus, Denmark
[6]Laboratory of Radiation Physics, Odense University Hospital, Odense, Denmark
[7]Department of Clinical Research, University of Southern Denmark, Odense, Denmark
[8]The Institute of Cancer Research, London, UK

E-mail: mobjje@rm.dk

Abstract. The magnetic field in magnetic resonance imaging-guided radiotherapy (MRgRT) systems influences the three-dimensional (3D) dose deposition and hence the delivered dose distributions. The aim of this study was to investigate the dose-rate dependency and dose response of silicone-based radiochromic dosimeters for photon irradiation in the precense of a magnetic field using a 1.5 T MRgRT system. Additionally, the study aimed to provide a proof of the concept of radiotherapy treatment based on a treatment plan calculated on magnetic resonance imaging rather than a computed tomography (CT) scan. The delivered dose was read out in 3D with an optical CT scanner and the dose distribution was verified using gamma analysis. We found an insignificant dose-rate dependency for dose rates ranging from 3.2 to 5.1 Gy/min and a linear dose response up to 20 Gy. A 3D 3%/3mm gamma analysis showed a pass rate of 95.1%. The dosimeter showed clinical potential for 3D dose verification of MRgRT delivery.

1. Introduction

Trajectories of secondary electrons are affected by the magnetic field inherently present in magnetic resonance imaging-guided radiotherapy (MRgRT) systems, resulting in changes of the three dimensional (3D) dose distribution [1,2]. Therefore, current dosimetry systems and quality assurance procedures used for conventional radiotherapy (RT) must be revisited and appropriately adjusted to be applicable for these hybrid imaging/treatment machines.

During the recent years, there has been increasing interest in 3D dosimetry due to the high conformity of new treatment modalities such as intensity-modulated radiation therapy and volumetric modulated arc therapy [3,4]. A radiochromic silicone-based 3D dosimeter was first introduced in 2015 [5–8] and has been characterized for conventional megavoltage X-ray RT [8] and megavoltage X-ray RT in the presence of a magnetic field using a 0.35 T MRgRT system [9]. Furthermore, it has been characterized for proton therapy [10] and has shown potential for high resolution dose verification in three dimensions [8]. The dosimeter is deformable and can be used to investigate the influence of deformation during irradiation [11]. Furthermore, it can be molded into complex phantom geometries with embedded air cavities to investigate dose deposition at air-tissue interfaces which is of special interest for MRgRT systems due to the electron return effect [12].

However, the magnetic field strength will affect the severity of any magnetic field dose effect [13] and hence, characterization of the dosimeter in the presence of the magnetic field originating from a 1.5 T MRgRT system is a prerequisite for using the dosimeter in such systems. Therefore, the aim of this study was to characterise the dose-rate dependency and dose response of silicone-based radiochromic dosimeters in the presence of a magnetic field. And additionally, the study aimed to provide a proof of the concept of RT treatment based on a treatment plan calculated on magnetic resonance imaging (MRI) rather than a computed tomography (CT) scan using a 1.5 T MRgRT system. The delivered dose was read out in 3D using an optical CT scanner and the dose distribution was verified using gamma analysis.

2. Material and Methods

2.1. Fabrication

Dosimeters were fabricated from 93.2% silicone elastomer (SE), 5% curing agent (CA), 0.26% leucomalachite green (LMG) and 1.5% chloroform (all in weight percentages). SE and CA came from the commercially available SYLGARD 184 silicone elastomer (DOW) kit. LMG was dissolved in chloroform before SE and CA were added and thoroughly mixed using an electric mixer. Air bubbles were removed from the mixture using a vacuum desiccator before dosimeters were cast and left to cure protected from light at room temperature. Two batches were fabricated. Batch 1: 60 cuvettes (1 cm × 1 cm × 4.5 cm) curing for 62 hours (from casting to pre-irradiation read-out); Batch 2: a cylindrical dosimeter (diameter 5 cm and height 5 cm) curing for 69 hours.

2.2. Irradiation conditions

All dosimeters were irradiated at Odense University Hospital using an Elekta Unity (Stockholm, Sweden) 1.5 T MRgRT system with a 7 MV flattening-filter-free beam quality, source-to-axis distance of 143.5 cm and a pulse repetition frequency (PRF) of 275 Hz. Cuvettes were irradiated with a 10 cm x 10 cm field with a gantry angle of 90 degrees and centred in the radiation field by placing them standing on a water-equivalent RW3 (PTW, Freiburg, Germany) platform. The dosimeters were placed between RW3 build-up and backscatter slabs with a constant source-to-surface distance of 133.5 cm. While backscatter slabs of 5 cm were used for all irradiations, the build-up slabs varied among 4.5, 9.5 and 14.5 cm to obtain different dose-rate conditions.

The dosimeters were irradiated to doses of 2, 5, 10, 15 and 20 Gy and the number of monitor units needed to deliver a given dose was calculated on a virtual water phantom in the treatment planning system (TPS – Monaco, v. 5.51.10). The number of photon pulses (#pulses) was measured for each irradiation with a TF930 frequency counter (Aim-TTi, GB), and local dose rates were given as (dose × PRF) / #pulses which corresponded to 3.2, 4.1 and 5.1 Gy/min for build-up slabs of 14.5, 9.5 and 4.5 cm, respectively (electron gun duty cycle was 100%). Two dosimeters were irradiated concurrently with a total of four dosimeters irradiated per dose level for each dose rate (60 in total). The alignment of the dosimeters was verified with electronic portal imaging device (EPID) images acquired while the treatment beam was on (i.e., no additional dose given).

The cylindrical dosimeter was placed standing on a polystyrene platform to locate the irradiation isocentre within the cylinder. A T1 weighted MRI reference scan was acquired and imported to the TPS and the external boundary of the dosimeter was delineated and the volume was assigned a relative electron density of 1.0. An ellipsoid of 0.66 cm^3 was defined as target, and inverse treatment planning was performed using a half-arc 5-field equally spaced step-and-shoot beam configuration with a 10 Gy target mean prescribed dose. The treatment plan was exported to the Elekta Unity system and the reference image was matched to the online MRI scan based on a rigid translation registration. The body outline of the dosimeter was verified on the online MRI and the online plan was created using the same field configuration as the reference plan; see Bernchou et al. [14] for a more detailed description. The online plan was exported for data analysis with an 1 mm dose grid resolution and statistical uncertainty of 1% in the Monte Carlo dose results.

2.3. Pre- and post-irradiation optical read-out

The cuvette-sized dosimeters were read out optically at 627 nm using a spectrophotometer (Spectroquant Pharo 100) 9 hours pre- and 5 hours post-irradiation. The dose responses were given as the slope of the difference in linear attenuation coefficient ($\Delta\alpha = \alpha_{post} - \alpha_{pre}$) plotted against dose.

The cylindrical dosimeter was read out 7 hours pre- and 5 hours post-irradiation at 635 nm with the Vista™ 16 Optical CT scanner (Modus Medical Devices, Canada) using 1000 2D projections over 360-degrees rotation. Data reconstruction was performed with software provided by the manufacturer as an integral part of the optical CT scanner using an OSC-TV algorithm with a voxel volume of 1.0 mm^3; see Valdetaro et al. for a more detailed description [10].

2.4. Gamma Analysis

The optical CT read-out and the exported TPS calculation were normalized and compared using a global 3%/3 mm as well as 2%/2 mm 3D gamma analysis. The measured and simulated data were aligned by applying an image registration in MATLAB. Creep-up effects (i.e. non-irradiation induced gradual increase in absorption over time) [15] in the 3D dosimeter result in a background offset. A gamma analysis was performed with an offset subtracted, the same for all voxels, from the optical CT read-out prior to normalization. The offset was chosen so that voxel values within the low-dose bath of the normalized optical CT read-out were comparable to voxel values of the low-dose bath of the normalized TPS dataset. The data was normalized by selecting an identical high dose region in both datasets. Only voxels at 10% or more of the maximum value in the TPS dataset were included in the gamma analysis. Furthermore, slices of the measured dataset containing many artefacts were disregarded.

3. Results

We found dose responses in cuvettes (including 95% confidence interval) of $(20.4 \pm 0.6) \times 10^{-3}$ cm^{-1} Gy^{-1} for a dose rate of 3.2 Gy/min, $(20.5 \pm 0.5) \times 10^{-3}$ cm^{-1} Gy^{-1} for a dose rate of 4.1 Gy/min and $(20.1 \pm 0.5) \times 10^{-3}$ cm^{-1} Gy^{-1} for a dose rate of 5.1 Gy/min. For the cylindrical phantom, the 3%/3 mm gamma pass rate was 95.1% and the 2%/2 mm pass rate was 82.0%.

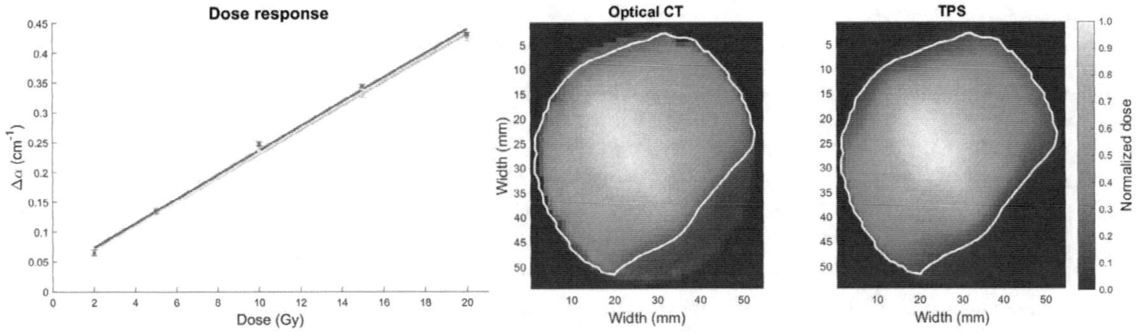

Figure 1: Left: For cuvette exposures, red, yellow and green correspond to dose rates of 3.2, 4.1 and 5.1 Gy/min, respectively. Data points represent the mean value and the error bars represent the 95% confidence intervals of the four dosimeters irradiated per dose level for each dose rate. $R^2 > 0.99$ for all three regressions. Right: Normalized dosimeter dose distributions in the frontal plane (perpendicular to the longitudinal axis of the cylinder) for optical CT read-out and TPS dose calculation. The magnetic field is perpendicular to the cylindrical axis. The white line represents the 10% cut off.

4. Discussion

In the presence of a magnetic field originating from a 1.5 T MRgRT delivery system, cuvette-sized silicone-based radiochromic dosimeters showed a linear dose response up to 20 Gy with no significant variation of response as a function of dose rate in the range from 3.2 to 5.1 Gy/min. Acceptable gamma pass rates were obtained when subtracting an offset from the 3D read-out.

The dose response for the applied radiochromic dosimeter vary across studies [16,17], attributed to factors such as a slightly different curing temperature (room temperature varies with season of the year), read-out wavelength, batch-to-batch dependence, ambient light exposure, time between pre- and post-irradiation due to fading (i.e. temporal absorption changes for irradiated dosimeters) and creep-up effects [15] as well as different curing times [16]. To assess if the magnetic field affects the dose response, dosimeters prepared from the same batch must be irradiated concurrently under the same conditions with and without the presence of a magnetic field.

The insignificant dose-rate dependency is in line with previous findings in the absence of a magnetic field based on a previous established protocol [7,16] as well as for a 0.35 T MRgRT system (9% CA, 89.2% SE, 0.26% LMG and 1.5% chloroform) [9].

The limitations of the study include the method of image registration based on optical CT read-out and the TPS dose matrix instead of embedded markers in the dosimeter that could be aligned with the MRI DICOM images. Furthermore, it is an approximation to apply the same offset for all voxels for the optical CT read-out prior to normalization as temporal changes of optical density are dose dependent [15]. This emphasizes the importance of minimizing the time between irradiation and read-out as this reduces creep-up and fading effects [18]. Additionally there is a lack of explicit connection between the cuvette-sized and cylindrical dosimeters. Further experiments must be conducted to assess volume and surface effects (i.e. if dosimetric properties depend on the volume and surface area) and hence, if dosimetric properties of the cuvette-sized dosimeters can be directly translated into larger samples.

In conclusion, this study of the silicone-based radiochromic dosimeter in the presence of a magnetic field using a 1.5 T MRgRT system showed an insignificant dose-rate dependency in the range from 3.2 to 5.1 Gy/min and a linear dose response up to 20 Gy. A treatment plan calculated on an MRI reference image was verified in 3D with a 3%/3 mm gamma pass rate of 95.1%. The dosimeter showed clinical potential and applicability for 3D dose verification for photon irradiation of MRgRT delivery.

5. Acknowledgements

Adriaan Fietje is gratefully acknowledged for his assistance during the conduction of experiments. This research was funded by the Novo Nordisk Foundation (Grant number NNF18OC0034718).

6. References

[1] Raaymakers B W *et al* 2004 *Phys. Med. Biol.* **49** 4109-18
[2] Lee H J *et al* 2017 *Radiother. Oncol.* **125** 426-32
[3] Baldock C *et al* 2010 *Phys. Med. Biol.* **55** R1-63
[4] Oldham M 2015 *J. Phys. Conf. Ser.* **573** 012006
[5] De Deene Y *et al* 2015 *J. Phys. Conf. Ser.* **573** 012025
[6] De Deene Y *et al* 2015 *Phys. Med. Biol.* **60** 1543-63
[7] Høye E M *et al* 2015 *Phys. Med. Biol.* **60** 5557-70
[8] Høye E M *et al* 2015 *J. Phys. Conf. Ser.* **573** 012067
[9] Jensen M B *et al* 2020 *Phys. Imaging Radiat. Oncol.* **16** 81-4
[10] Valdetaro L B *et al* 2021 *Phys. Imaging Radiat. Oncol.* **18** 11-8
[11] Kaplan L P *et al* 2017 *Phys. Med. Biol.* **62** 5612-22
[12] Raaijmakers A J E *et al* 2005 *Phys. Med. Biol.* **50** 1363-76
[13] Raaijmakers A J E *et al* 2008 *Phys. Med. Biol.* **53** 909-23
[14] Bernchou U *et al* 2021 *Phys. Med. Biol.* **66** 045034
[15] Du Y *et al* 2020 *Radiat. Phys. Chem.* **168** 108574
[16] Jensen M B *et al* 2022 *Acta Oncol.* **61** 264-8

[17] Hoye E M *et al* 2017 *Phys. Med. Biol.* **62** N73-89
[18] Valdetaro L B *et al* (in press)

Journal of Physics: Conference Series **2630** (2023) 012031 doi:10.1088/1742-6596/2630/1/012031

Different aspects of plan complexity in prostate VMAT plans

E. Terzidis[1,2], F. Nordström[1,2], J. Götstedt[1,2], A. Bäck[1,2]
[1]Department of Medical Radiation Sciences, Institute of Clinical Sciences, Sahlgrenska Academy, University of Gothenburg, Sweden
[2]Department of Therapeutic Radiation Physics, Medical Physics and Biomedical Engineering, Sahlgrenska University Hospital, Gothenburg, Sweden

Email: emmanouil.terzidis@gu.se

Abstract. In this work we evaluated VMAT plan complexity by using different methods and approaches: complexity related to the aperture shape, the nature of the dynamic delivery, the impact of delivery variations as well as complexity analyses based on measurements. Prostate cancer treatment plans with different levels of complexity were created for three different patient cases. The plans intended to be most complex were also scored to be the most complex according to the evaluation methods used. The results for the other plans were more diverse. The methods that included 3D spatial information on complexity gave additional information important for the analysis of clinical relevance. In order to use specific complexity estimation methods for taking clinical decisions it is essential to understand how different aspects of complexity are incorporated in each method.

1. Introduction

The absorbed dose distribution delivered to a patient using VMAT is determined by the treatment machine, the treatment plan characteristics, and the patient geometry. The delivered dose distribution is calculated using an algorithm in the treatment planning system (TPS). The calculation algorithm includes limitations and approximations and, furthermore, assumes a perfect treatment delivery and most often one single static patient geometry. This means that the calculated dose distribution is not a perfect description of the delivered dose distribution. Difference between doses calculated and the delivered doses to the patient can be referred to as the dosimetric uncertainty of the treatment.

The dosimetric uncertainty of a treatment plan, i.e., without taking the patient geometry into account, is related to the plan complexity [1]. Plan complexity includes three sources of dosimetric uncertainty:

1. *Dose calculation errors*
   Limitations and approximations in the dose calculation, for example the definition of the beam model in the TPS.
2. *Delivery variations*
   Differences between planned and actual position of treatment machine parameters, such as for example collimator jaw and multi leaf collimator (MLC) positions.

3. *Dynamic delivery uncertainties*
   Uncertainties related to the dynamic nature of the VMAT delivery, e.g., depending on acceleration and deceleration of MLC leaves.

Earlier studies have shown that VMAT plan uncertainties can be quantified by complexity metrics [2]. Suggested metrics focus on specific aspects of plan complexity and are therefore not generally interconvertible but could be complementary. In this work different aspects of complexity were investigated by using four different approaches: the edge area metric (EAM) which is related to the aperture shape and designed to include both dose calculation errors and delivery variations [3], the modulation index MI_T designed to include dynamic delivery uncertainties [4], theoretical simulations of delivery variations, and finally quasi-3D measurements that include all aspects of plan complexity.

2. Methods and Materials

2.1. *Treatment plans*

For the purpose of this study, three VMAT plans for prostate cancer have been selected to represent different patient geometries. The treatment plan used for the actual treatment of the patient ("clinical plan") was reoptimized using the Eclipse TPS (Varian Medical Systems) to create one plan with reduced complexity ("simple plan") and one of higher complexity ("complex plan"). These two additional plans were generated by altering the monitor unit (MU) objective and the aperture shape controller (ASC) level [5]. These are the only available tools designed to regulate plan complexity, within Eclipse. The use of the ASC tool has been shown to be able to reduce plan complexity without altering the plan quality metric, including a set of DVH-metrics [6]. The weight of the aperture shape penalty can be adjusted by the user and in five steps, ranging from "Very Low" to "Very High". The highest level of ASC was used when creating the "simple plans", while for the creation of the complex plans ASC was turned off. Besides altering MU objective and ASC, the number of arcs was also varied to regulate plan complexity. Based on the clinical plan, that was created with two arcs in all cases, the number of arcs were reduced by one when creating the simple plan and increased by one for the creation of the complex plan. All plans were created for 6 MV photons and a TrueBeam with a Millennium MLC (Varian Medical Systems). Effort was put into creating dose distributions similar to the clinical treatment plan for the case of each patient. Photon Optimizer (PO version 16.1.0) was used for the optimisation of all treatment plans and the dose was calculated with the Anisotropic Analytical Algorithm (AAA version 16.1.0) with a 2.5 x 2.5 mm calculation grid. In total 9 treatment plans were incorporated in this study, including the 3 clinical plans.

2.2. *Evaluations based on complexity metrics*

Complexity scores were calculated according to the complexity metrics EAM and MI_T. The EAM describes the aperture shape by defining a complex region close to the MLC edges in relation to the open part of the aperture[7]. Small and/or irregular aperture shapes results in an EAM score closer to 1, while larger less complex apertures are described by a lower score. The calculation of the EAM was performed with a complexity region of 2.5 mm inside and outside of the MLC edge.

The modulation index MI_T takes into account the MLC speed and the acceleration/deceleration for each MLC leaf and gantry rotation as well as dose rate variations. Should the variation in the MLC speed or acceleration for a single leaf between two consecutive control points exceed a certain threshold value the modulation increases. This threshold value was defined by the observed values standard deviation in the actual plan (i.e. speed or acceleration) multiplied with a spectrum function z(f) as defined by Webb [8]. In this work a maximum value of $f = 0.5$ with an interval of 0.01 was used.

2.3. *Delivery simulations*

The effect of delivery variations was estimated with the aid of an inhouse developed MATLAB (MathWorks, Natick, MA, USA) script which introduced systematic (i.e., same for the whole plan) and random (i.e., specific for each control point) offsets in certain treatment machine parameters. The variation of a machine parameter value was randomly generated consistent with the probability described by a normal distribution centred at the value specified in the treatment plan and truncated at 3σ corresponding to the maximum value of variation for this parameter. More specifically, a maximum systematic offset (3σ) of ±2 mm and ±1 mm were applied for the Y and X jaw position respectively. For the MLC position, a systematic offset with a maximum of ±1 mm was applied for all leaves in both banks, and an additional systematic shift of maximum ±0.3 mm was applied for one of the banks. A random offset of a maximum of ±0.35 mm was applied for each individual leaf. The collimator and gantry angle for each field were allowed to vary to a maximum of ±0.6 and ±0.3 degrees respectively. This MATLAB simulation script yielded a treatment plan, where the dose was recalculated with small variations in machine parameters. To evaluate the impact of these delivery variations, we performed twenty simulations for each plan. The output dose distribution of these simulations were inputted in an Eclipse Scripting Application Programming Interface (ESAPI script) to create a combined 3D distribution, where instead of voxel dose we displayed the standard deviation (SD) of dose over all simulations in that voxel. The mean SD (Gy) for all voxels receiving greater than 20% of the prescribed dose in the original plan was reported.

2.4. Delta[4] measurements
Positioning errors of the Delta[4] phantom (ScandiDos, Uppsala Sweden) were minimized by adjusting the treatment couch (six degrees of freedom) using cone beam computed tomography (CBCT) imaging and registration to a reference CBCT image of the phantom in optimal position.

All treatment plans were measured on the same day with the exact same phantom positioning to assure consistency within our results. The plans were recalculated in the Delta[4] phantom using AAA and prior to the delivery of the treatment plans, a half arc calibration field was delivered to correct for the output of the machine determined daily. The calculated and measured dose distributions were normalized to the maximum three-dimensional planned dose. The 3% global dose difference fail rate in the measurement points was evaluated in the Delta[4] software. Diodes receiving a dose less than 20% were excluded from the analysis.

3. Results and Discussion
The average EAM scores were 0.42, 0.62 and 0.78 for the simple, clinical and complex treatment plans respectively. The choice of ASC and MU objective affected the EAM score. For the MI_T, no obvious relation between the different plans was observed. (Figure 1A). The tools available in Eclipse to regulate complexity had higher impact on the aperture-shape than on the dynamical nature of the plan.

The mean SD due to delivery variations, EAM and measured dose difference fail rate were the highest for the plans designed to be most complex. The results for the simple and clinical plans were more diverse where EAM always resulted in lower scores for the simple plan while no obvious relations between simple and clinical plans were found for SD due to delivery variations or measured dose difference fail rate. The delivery simulations do not take into account uncertainties due to calculation errors which is included in both EAM and the measurement evaluations. One reason why the SD was lower for the clinical compared to the simple plan in two of the three cases is probably because of the increased number of arcs used in the clinical plan that will contribute to a blurring of the uncertainties related to the different arcs. Uncertainty estimations based on measurements with the Delta[4] includes several sources of uncertainties. Besides the fact that the measurement procedure includes uncertainties in itself and that the results are limited to a few measurement points located in the two detector planes of the Delta[4], the measurement result was based on only one delivery and does therefore show only one case of possible delivery variations. The mentioned uncertainties might lead to that the measurement evaluations cannot dissolve smaller dosimetric uncertainties. Whether the differences between EAM and dose difference fail rate are due to a non-optimal design of EAM or mentioned limitations in the measurement procedure needs further investigation.

In Figure 1(C-E) the 3D SD distribution of the dose is displayed for one of the prostate plans. For the simple plan the SD of the dose was higher on the edge of the PTV, especially in the caudal-cranial direction. For the complex plan the SD was higher for voxels at PTV borders but also inside the structure, in comparison to the clinical plan where the SD was lower. Hence clinical plans appear to be less susceptible to machine parameters variations.

Measurement points with a dose deviation higher than 3% were located mostly in the dose fall-off region for the simple and clinical plans. However, for the complex plans, deviations were spotted in several regions within the dose distribution. The SD analysis and measurement evaluations were the only methods in this work that could describe uncertainty in a 3D volume. The spatial information of uncertainties is important to be able to conclude on the clinical relevance. Furthermore, analyses that does not include 3D evaluation inevitably include some kind of averaging where important uncertainties can be evened out.

4. Conclusion

Different types of complexities are not necessarily related to each other. One plan can have a low complexity regarding aperture-shape but at the same time high complexity regarding the dynamic delivery or the delivery variations. In order to use specific complexity evaluation methods for taking clinical decisions it is essential to understand how different aspects of complexity are incorporated in each method.

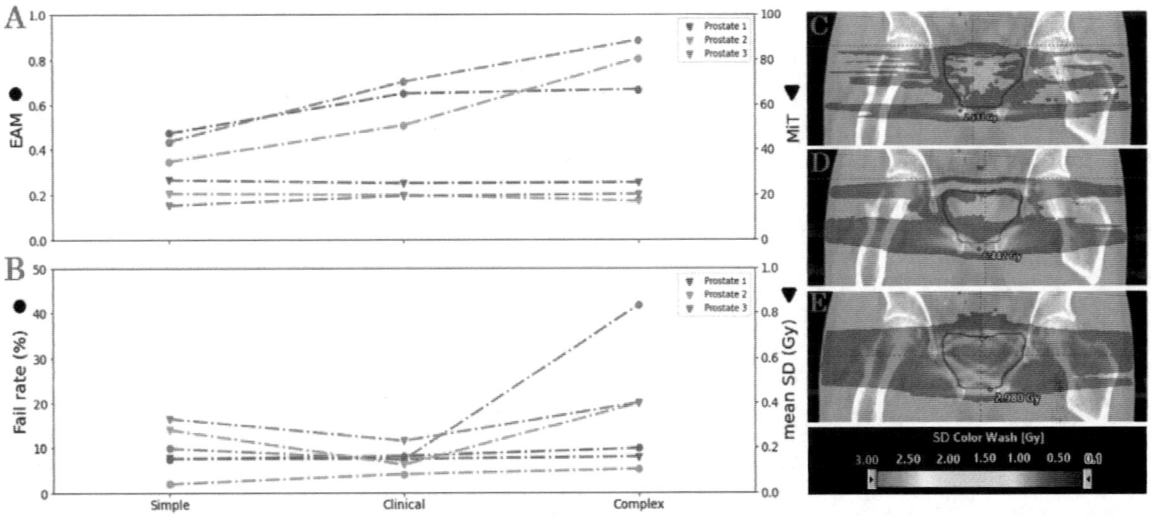

Figure 1. (A) EAM and MI_T values and (B) 3% dose difference fail rate (%) and mean SD in dose due to delivery variations (Gy) is shown for all nine prostate plans. Triangle points refer to the value of the MI_T and STD values while circle points to EAM and fail rate. Treatment plans originating from the same patient are colour coded. In C, D and E, the distribution of the SD of dose due to delivery variations is displayed in a central frontal slice for prostate case #2 for the clinical, simple and complex plans respectively. The red contour depicts the planning target volume and the colour wash scale in these distributions ranges from 0.1 up to 3 SD (Gy).

5. Acknowledgements

The authors gratefully acknowledge the King Gustaf V Jubilee Clinic Cancer Research Foundation, Varian Medical Systems and the Swedish Radiation Safety Authority for financial support.

6. References

[1] Hernandez V *et al* 2020 *Radiother. Oncol.* **53** 26-33
[2] Chiavassa S *et al* 2019 *Br. J. Radiol.* **92** 20190270
[3] Gotstedt J *et al* 2015 *Med. Phys.* **42** 3911-21
[4] Park JM *et al* 2014 *Phys. Med. Biol.* **59** 7315-40

[5] Varian Medical Systems 2015 Eclipse Photon and Electron Algorithms Reference Guide. Doc ID P1015026-001-A. https://jpneylon.github.io/ABR/PDFs/Add_052418/EclipseAlgorithms13.6_RefGuide.pdf
[6] Scaggion A *et al* 2020 *J. Appl. Clin. Med. Phys*. **21** 27-34
[7] Gotstedt J *et al* 2021 *Phys. Imag. Radiat Oncol.* **17** 124-9
[8] Webb S 2003 *Phys. Med. Biol*. **48** 2051-62

Journal of Physics: Conference Series **2630** (2023) 012032 doi:10.1088/1742-6596/2630/1/012032

On the comparison of two NIPAM gel dosimeters with high resolution 3D MRI sequences

M Cinq-Mars[1,2], J-D Jutras[3] and L Beaulieu[1,2]
[1]Département de physique, de génie physique et d'optique et Centre de recherche sur le cancer, Université Laval, Québec, Canada
[2]CHU de Québec – Université Laval et CRCHU de Québec, Québec, Canada
[3] Tom Baker Cancer Center - Alberta Health Services, Calgary, Canada

E-mail: melissa.cinq-mars.1@ulaval.ca; Luc.Beaulieu@phy.ulaval.ca

Abstract. Polymer gel dosimeters allow the measurement of a dose distribution in 3 dimensions. Irradiation induces polymerization reactions in the long chains of the gel molecules, which is characterized by a change in the transverse relaxation time T_2. Gel dosimetry is mainly advantageous for its high resolution, but can also be expensive. The main goal of this project is to compare two normoxic polymer gel recipes made with NIPAM as monomer. The analysis is based on the quality of the images obtained for a fixed MRI scan time and a comparative cost of analysis of each 1L phantom. The ΔT_2 maps are obtained by scanning the gels before and after irradiation with 3D sequences and the DESPOT2 technique is used to reconstruct the T_2 maps. The phantoms were irradiated with 8 photon beams of 6 MV with a fixed 250 MU, 4 at two opposite sides of the jar (AP-PA configuration), to obtain 4 dose distributions of 4.7 Gy, 8.9 Gy, 13.2 Gy and 17.3 Gy. The SNR is then determined in the ΔT_2 maps in function of the dose step and the concentration of NIPAM in the phantom. The first gel phantom is made with 15% NIPAM and has a total cost of $1561.65, including products and MRI scan costs. The SNR obtained for the 4 dose steps are 10.02, 31.18, 45.12 and 37.54. The contrast between the T_2 before and after the irradiation in the 4 regions of dose are 0.45, 0.66, 0.78 and 0.85. The second phantom contained 5% NIPAM, cost 1186.85$ and the SNR obtained are 16.80, 27.08, 30.11 and 26.17 and the ΔT_2 are of 0.61, 0.75, 0.79 and 0.77 for the 4 dose steps respectively. The comparison of the two recipes has shown that the increase of NIPAM concentration does not allow a significant increase in image quality for MRI. The recipe with 5% of NIPAM has a lower dynamic range but a better sensitivity at lower dose and is less expensive than the one with 15% of NIPAM. The results also show that by increasing the NIPAM concentration 3-fold, the cost of a 1L phantom increases by 30%.

1. Introduction

Gel dosimetry is a valuable tool for measuring 3D dose distributions and has many advantages for dosimetry: tissue equivalence, three-dimensional dose measurement, high spatial resolution and lack of energy dependence. There are two main categories of gel dosimeters: Fricke gels and polymer gels [1]. Polymer gels can characterize complex and non-uniform dose distributions with a large dose range and no diffusion effects, which could be interesting to study complex treatment plans, but their fabrication is difficult and the access to MRI is limited. However, their fabrication is more challenging, their reading via optical CT is limited by scatter conditions, while MRI-based reading may require long (costly) scan times given low contrast-to-noise efficiency. Polymer gel includes normoxic gels, which means that they are less sensitive to oxygen contamination and can be manufactured under standard atmospheric conditions [1]. One such type of gel is the NIPAM polymer gel which is useful for MRI analysis. During irradiation, a polymerization reaction is initiated in the long molecular chains, decreasing the transverse relaxation time (T_2) proportionally to the delivered dose. By measuring the change in the T_2 relaxation time, before and after the irradiation, a 3D dose map can be obtained with a calibration curve. The Driven Equilibrium Single Pulse Observation of T_2 (DESPOT2) technique is used to produce the T_2 maps from MRI sequences because of its high SNR efficiency per unit voxel size [2].

In 2006, Senden et al. [3] developed a polymer gel dosimeter using the NIPAM as monomer for his low toxicity and his high solubility in water. More recent studies have proposed new polymer gel recipes based on NIPAM monomer in significantly different concentrations. The goal of this work is to compare two new NIPAM gel recipes by analyzing the SNR and the ΔT_2 value in the dose distribution in function of the cost of each dosimeter. Indeed, NIPAM is the most expensive product in the manufacture of the gel and the goal of the project is to analyze the cost of the contrast-to-noise ratio for each dosimeter while maintaining a constant reasonable MRI scan time of 30 min pre- and 30 min post-irradiation.

2. Methods and materials

2.1. Gel preparation

Two phantoms are manufactured based on previous NIPAM recipes presented in the articles of Maynard et al. (2020) and Chiang et al. (2013). The recipes are shown in Table 1. The fabrication procedure is based on the one presented by Olding et al. (2010) [6]. The gelatine (300 type A, Sigma-Aldrich, Oakville, ON) is first dissolved in 80% of the distilled water and the solution is heated to 50°C. This temperature is maintained till the gelatine is completely dissolve. The BIS (N,N'-Methylenebis acrylamide, Sigma-Aldrich, Oakville, ON) product is then dissolved in the water-gelatine solution. At the same time, in another beaker the NIPAM (N-Isopropylacrylamide, Sigma-Aldrich, Oakville, ON) is dissolved in the 20% of distilled water remaining at ambient temperature without any heating. When NIPAM is completely dissolved in the water, the THPC (Tetrakis (hydroxymethyl) phosphonium chloride solution, Sigma-Aldrich, Oakville, ON) product is added, and the solution is stirred for about 1 to 2 minutes. Meanwhile, the water-gelatine-BIS solution is cooled to 34°C in running cold water. When the solution is cooled to the desired temperature, the water-NIPAM-THPC solution is added and the gel is then stirred for about 1 to 2 minutes and then is put in the 1L Polyethylene Terephthalate plastic jars (Modus Medical Devices Inc., London, ON).

Table 1. List of the ingredients (distilled water, gelatin, NIPAM, BIS and THPC) used in the two recipes tested.

	Recipe 1 Maynard et al. (2017)	**Recipe 2** Chiang et al. (2013)
Water (wt%)	75.5	87
Gelatin (wt%)	5	5
NIPAM (wt%)	15	5
BIS (wt%)	4.5	3
THPC (mM)	10	10

2.2 Treatment planning and irradiation

The two phantoms are irradiated on a Truebeam Linac with a 6 MV photon beam and 250 MU. Four photon beams of 5x10 cm^2, 5x7.5 cm^2, 5x5 cm2 and 5x2.5 cm^2 are used at gantry angles in an AP-PA configuration (90° and 270°) for a total of 8 photon beams for each phantom. A step-like distribution of dose from 4.2 Gy to 17.3 Gy calculated in RayStation is determined in 4 regions of interest (ROIs) in the phantom height.

2.3 T2 maps production

The two phantoms are first imaged with 3D sequences on MRI. The technique used is based on the method on T_2 maps reconstruction DESPOT2 [2]. Two sequences FLASH multi-echo with optimized flip angles combined with a B_1 map are used to obtained T_1 maps. Then, T_1 and B_1 maps, with four bSSFP sequences with multiple phase and flip angles are used to solve for T_2 voxel per voxel. The two phantoms were imaged before and after the irradiation to further improve uniformity. The $\Delta T2 = (T2_{\text{pre}} - T2_{\text{post}})/T2_{pre}$ map is obtained by the subtraction of the T_2 map before and after the irradiation and normalized by the T_2 pre-irradiation. The T_2 map pre-irradiation is also filtered via Gaussian anisotropic diffusion in 3DSlicer (conductance of 0.5 and 5 iterations) to reduce noise propagation in the final ΔT_2. The SNR in the ΔT_2 map (which is also essentially a contrast-to-noise ratio in this case) is measured in region of interest (ROIs) positioned at the center of each dose step.

3. Results and Discussion

3.1. ΔT_2 maps production

The figure 3 shows the ΔT_2 maps obtained for the two recipes. The 4 regions of dose of 4.7 Gy, 8.9 Gy, 13.2 Gy and 17.3 Gy can be seen in the sagittal and coronal views.

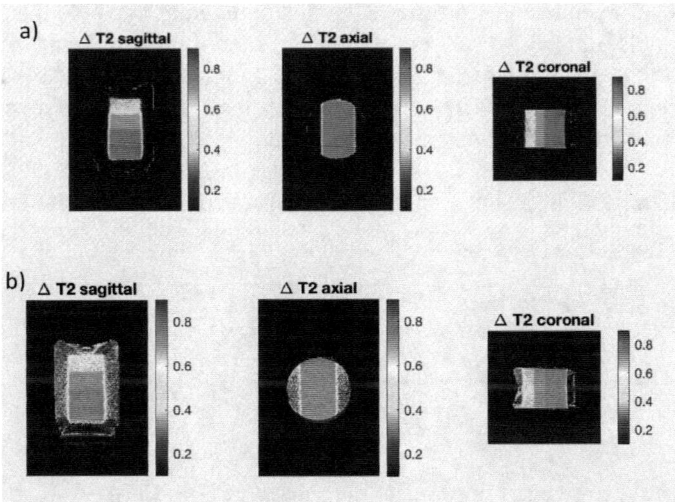

Figure 1. ΔT_2 maps in the three views for a) recipe 1 (15% NIPAM/ 4.5% BIS) and b) recipe 2 (5% NIPAM/ 3% BIS).

3.2. SNR and cost of gel phantoms

The signal to noise ratio (SNR) is calculated in 4 ROIs in the ΔT_2 maps obtained using the technique of Preibisch and Deichmann [7], and its value is related to the dose predicted in each step. The Figure 2a shows the 4 measures of SNR for the recipes vs the dose, while Figure 2b shows the contrast vs the dose. The SNR achieved for both phantom recipes looks very promising. However, both recipes appear to have saturated at lower dose than we anticipated (about 10 Gy for the 5% NIPAM and 15 Gy for the 15% NIPAM). More investigation will be required to evaluate the response of each recipe at lower dose of 0 to 5 Gy, and whether a threshold dose response occurs for the gels.

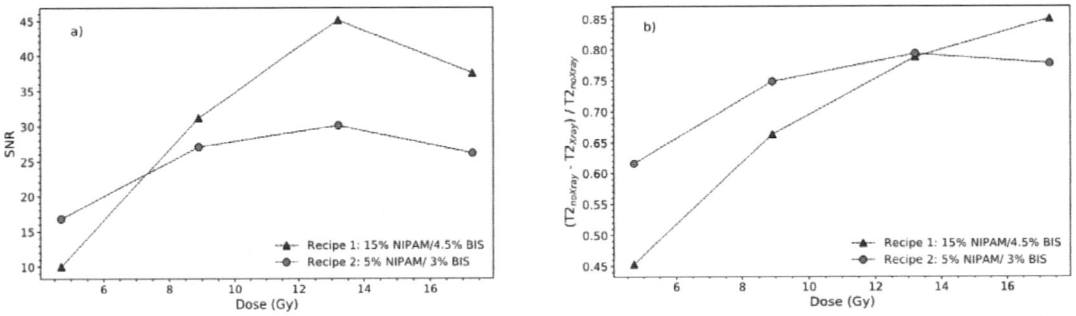

(a) (b)

Figure 2. a) Signal to noise ratio and b) contrast according to the estimated dose steps delivered to the phantom for the recipe made with 15% NIPAM (recipe 1) and the 5% NIPAM (recipe 2).

In Table 2, the cost of each of the 1L phantom is compared. This includes the cost of the products, the shipping, the jars and the MRI time scan of 32 minutes done before and after the irradiation [7]. The MRI scan time is kept constant and the main difference in the cost of each phantom is the NIPAM and BIS products. The cost of a 1L jars of gel is about 375$ higher for a NIPAM concentration of 3 times higher and a BIS concentration 1.5 times higher. However, the cost of the SNR at 13 Gy is higher for the 5% NIPAM (1186.85$/30=39.6$ per unit CNR) than for the 15% NIPAM (1561.64$/45=34.7$ per unit CNR), while the opposite is true at low dose of 4 Gy. **However, a fair assessment would require considering a range of cost in MRI scanning, as opposed to a mere average value of $500 per hour. We plan to do that in our final assessment.**

Table 2. Calculation of the cost of each phantom of 1L from recipe 1 and recipe 2 presented in Table 1.

	Recipe 1	**Recipe 2**
Products	604.28 $	159.84 $
Phantom jars	6.83 $	6.83 $
Total	**611.11 $**	**166.67 $**

4. Conclusion

A comparison between two gel dosimeters composed of NIPAM monomer allows to study the benefits of increasing NIPAM concentration in the gel by measuring the image quality for different doses for a fixed MRI time scan. The ultimate goal is to minimize the cost of the contrast-to-noise ratio at a dose of interest. A first recipe with a NIPAM concentration of 15% shows a higher SNR for high doses while the second gel has a better dynamic range and contrast at lower dose. The study of the cost of each phantom shows that a recipe with 3 times more NIPAM and 1.5 times more BIS is 1.3 more expensive and does not upgrade significantly the image quality. Gel dosimeters are known to be hard to manufacture but gel based on NIPAM monomer has shown progress and potential for increased clinical interest while they are less toxic. Even if MRI is the gold standard imaging, CT imaging can be used to characterize the change of opacity and density in the gel [1]. One of the main problems remaining for clinical application of gel dosimeters is the cost of the fabrication and their required imaging time at MRI. The next steps are to adjust the recipe by playing on the concentrations of the ingredients and increase or decrease the scan time in the MRI. Other properties of the gel are also important, including the dynamic range, dose rate response, whether there is a threshold dose response, linearity, etc. We are still in the early stage of optimizing our own recipe and the MRI pulse sequences.

5. References

[1] Schreiner L J *et al* 2009 in *Clinical Dosimetry Measurements in Radiotherapy: AAPM Summer School*. D W O Rogers and Joanna E Cygler (Eds). ISBN: 9781936366118
[2] Jutras J D *et al* 2016 *Mag. Res. Med.* **76** 1780-1804
[3] Senden R J *et al* 2006 *Phys. Med. Biol.* **51** 3301-3314
[4] E Maynard *et al* 2020 *Biomed. Phys. Eng. Express* **6** 035031
[5] Chiang C M *et al* 2013 *J. Radioanal. Nucl. Chem.* **298** 1435-1445
[6] Olding T *et al* 2011 *J. Med. Phys.* **36** 3-14
[7] Preibish C *et al* 2009 *Magn. Reson. Med.* **62** 240-246
[8] IRM Québec. 2021. https://irmquebec.com/

Iterative Image Reconstruction Methodology in Optical CT Radiochromic Gel Dosimetry

S Collins[1], A Ogilvy[1], D Huang[2], W Hare[2], M Hilts[3,1] and A Jirasek[1]
[1]Department of Physics, University of British Columbia - Okanagan campus, Kelowna BC V1V 1V7, Canada
[2]Department of Mathematics, University of British Columbia - Okanagan campus, Kelowna BC V1V 1V7, Canada[3]Medical Physics, BC Cancer - Kelowna, Kelowna BC V1Y 5L3, Canada

Email: andrew.jirasek@ubc.ca

Abstract. Modern advancements in radiation therapy require paralleled advancements in the dosimetric tools used to verify dose distributions. Optical computed tomography (CT) imaged radiochromic gel dosimeters provide comprehensive, tissue equivalent, 3D dosimetric information with high spatial resolution and low imaging times. Traditional CT image reconstruction methods (filtered backprojection) do not account for light refraction within the optical CT system reducing the image quality. Iterative reconstruction methods make use of a system matrix that describes this light refraction thus, improving the reconstructed image quality. However, use of iterative reconstruction methods is not widespread, largely due to the impractical storage size of the required system matrix. Furthermore, current iterative reconstruction methods do not address the issue of image degradation due to a single detector element collecting light from multiple raypaths. For optical CT radiochromic gel dosimetry to be used effectively as a radiation therapy treatment plan verification tool, the system must be both practical and accurate. Thus, this work has two main objectives: (i) reduce the size of system matrices by means of polar coordinate discretization in lieu of the traditional Cartesian coordinate discretization, and (ii) reduce image degradation due to multiple raypaths by a novel approach to populating the system matrix that accounts for multiple raypaths.

1. Introduction

Radiation therapy treatments plans with complex 3D dose distributions and steep dose gradients need dosimetric tools that can verify these complex treatment plans [1,2]. Traditional dosimeters are restricted to point or planar measurements, whereas, gel dosimeters can measure dose in full 3D geometry while being nearly tissue equivalent [3,4]. Furthermore, their radiation induced response can be measured with sub-millimetre spatial resolution [1,5]. Specifically, radiochromic gel dosimeters elicit an optical density change proportional to the delivered radiation dose, allowing for readout of the dose distribution via optical computed tomography (CT). For optical CT radiochromic gel dosimetry to be used in radiation therapy treatment plan verification the system needs to be both accurate and practical. That is, the optical CT images must give an accurate representation of the dose distribution as well as being practically viable for a clinical setting.

Journal of Physics: Conference Series **2630** (2023) 012033 doi:10.1088/1742-6596/2630/1/012033

1.1. System Matrix
Light is refracted as it traverses through different mediums in the optical CT system causing image degradation and artifacts when reconstructed using traditional filtered backprojection (FBP) methods. FBP assumes a straight raypath from source to detector, which is not the case when light undergoes refraction through the optical CT system. This loss of image quality can be remedied by utilizing iterative image reconstruction methods that account for this refraction. The path of each light ray as it travels through the optical system is compiled into what is known as a system matrix and is applied during iterative image reconstruction. These system matrices can become immensely large (>100GB) making them impractical for storage on a typical clinical workstation computer [6,7]. Here we develop a method to reduce the storage size of optical CT system matrices.

1.2. Single Detector Collection of Multiple Raypaths
As light is refracted through an optical CT system there arises the potential for a single detector to collect light from multiple (often very different) raypaths. This violates the single raypath assumption of the linear attenuation equation, which is utilized in the iterative reconstruction, causing artifacts in the resulting image. Engineering efforts can be made during the design of an optical CT scanner to minimize this occurrence by means of lenses and refractive index match materials. Nonetheless, the problem of multiple ray paths persists as all interfaces, including the dosimeter material, would be required to be perfectly index matched which is difficult to achieve in practice. Some rays may undergo severe refraction such that they are collected by detectors that also received a ray from a more direct raypath. These severely refracted rays are termed crossover rays and the rays that follow a more direct raypath are termed primary rays. Furthermore, light is partially reflected at each medium interface which allows for the potential that a single ray can undergo multiple reflections before being collected by the detectors and are termed reflected rays. Regardless of whether a detector element receives multiple raypaths due to crossover or reflected rays the problem remains the same; single detector collection of multiple raypaths. Here we aim to address the issue of multiple raypaths entering a single detector element and thereby reduce the induced image artifacts.

2. Materials and Methods
An in-house ray tracing simulator has been developed to track the path of each ray as it traverses the different mediums of the prototype tabletop solid-tank fan-beam optical CT scanner developed at the University of British Columbia – Okanagan (Figure1) [8]. The scanner consists of an acrylic (PMMA) block with a curved face and a hole bored through it in which a cylindrical radiochromic gel dosimeter is inserted. The dosimeter is submerged in a refractive index matched fluid to allow for the dosimeter to be rotated to acquire projections while the laser source and photodiode detector bay remain fixed. By discretizing the dosimeter into cartesian pixels, the simulation allows for the calculation of the length of every ray through every pixel and these calculated lengths were used to populate a system matrix. Furthermore, the ray tracing simulation was used to generate synthetic detector data (sinograms) mimicking the data collection of the actual optical CT system. These synthetic sinograms were used to evaluate the novel iterative reconstruction techniques under known conditions. Images were reconstructed with the Landweber simultaneous iterative reconstruction technique (SIRT) [9].

2.1. Polar Coordinate System Matrix Discretization
When populating the system matrix, the length of each ray through each pixel of the dosimeter must be calculated for every projection, since the resulting length will change as the ray is rotated in relation to the cartesian pixel grid. However, the dosimeter can be discretized into a polar pixel grid consisting of azimuthal lines and radial rings (Figure 2) [7]. If the azimuthal spacing is equal to the step size between projections, the length of each ray through each pixel is constant at each projection. This allows for the calculation of ray lengths through each pixel of just a single projection thus, reducing the required size of the system matrix by a factor of the number of projections. The radial rings were asymmetrically spaced such that the area of each polar pixel was constant ensuring each pixel was

Journal of Physics: Conference Series **2630** (2023) 012033 doi:10.1088/1742-6596/2630/1/012033

equally weighted during iterative reconstruction. For display purposes, the polar pixel images are converted back to Cartesian pixel images by means of oversampling. A 5x5 grid is created for each cartesian pixel to determine the contributions of the polar pixels to each Cartesian pixel. Each intersection of the 5x5 oversampling grid that falls within a polar pixel will give a weighting factor of one over the total number of intersections (1/25). This is used as an approximate calculation of the area of the polar pixel that overlaps the cartesian pixel (Figure 2). The value of each cartesian pixel is then calculated by the weighted sum of the contributing polar pixels. *Note: Ray tracing for the polar coordinate discretized system matrix is performed with the fan beam collimated such that there are no occurrences of a single detector collecting multiple raypaths, allowing for independent analysis.*

2.2. Single Detector Collection of Multiple Raypaths

As a first test case, simulations were performed with the fan beam collimated to allow rays on the outer edges of the beam span to be severely refracted and considered as crossover rays. Rays from the central region of the fan beam that follow a more direct raypath are thus considered the primary rays (Figure 1). To account for detectors receiving light from both primary and crossover raypaths, the multiple paths were treated as a single ray during the iterative reconstruction. The length through each pixel within the system matrix was then weighed based on the relative intensities of the rays to account for their relative contribution to the cumulative light intensity measured by the detectors. Synthetic sinogram data that included these crossover rays was generated for a simple ring phantom. Images were iteratively reconstructed with system matrices that accounted for these crossover rays and compared to images that were reconstructed when only primary rays were taken into account.

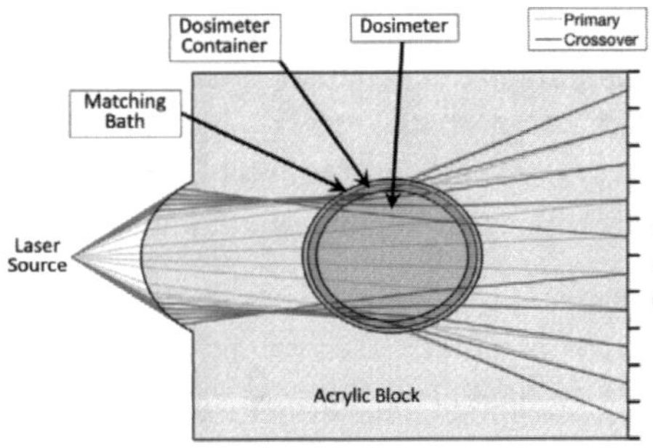

Figure 1. Ray tracing simulator produced visual representation of primary (green) and crossover (red) raypaths through the prototype optical CT system of [8].

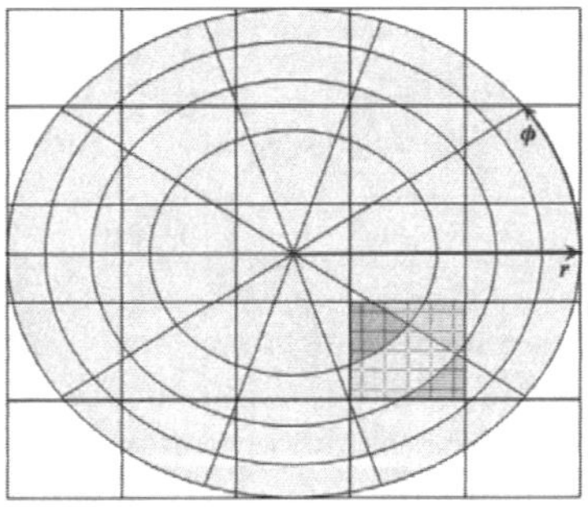

Figure 2. Polar coordinate discretization with asymmetrically spaced radial rings. Polar to Cartesian conversion is achieved through oversampling which approximates the area of polar pixel that overlaps a given Cartesian pixel.

3. Results and Discussion

3.1. Polar Coordinate System Matrix Discretization

Tracing 640 rays over 360 projections on a 512x512 Cartesian coordinate grid resulted in a system matrix size of 2.16 GB when stored in the sparse matrix file format. To produce a system matrix utilizing a polar coordinate pixel grid system with the same number of pixel elements as its 512x512 Cartesian coordinate counterpart, 728 radial rings and 360 azimuthal lines were used. This yielded the expected sparse system matrix size reduction of 360 times (6.0 MB). The asymmetrical spacing of the radial rings caused the central region of the image (where radial ring spacing is the largest) to have a 50% loss in spatial resolution (0.336 mm^{-1} at 10% MTF) (Figure 3). Although the number of azimuthal lines is fixed to the number of projections, the number of radial rings can be increased to reduce the size of the inner most pixels and thus reduce the loss of spatial resolution in the central region of the image. This increases the number of polar pixel elements and comes at a cost of increased system matrix size. Therefore, a balance between spatial resolution of the central region of the image and system matrix size must be found. Increasing the number of radial rings to 2048 provided an increase in spatial resolution to the central region of the image (0.546 mm^{-1} at 10% MTF) reducing the loss to only 23% and produced a sparse system matrix size reduction of 144 times (14.6 MB) (Figure 3).

3.2. Single Detector Collection of Multiple Raypaths

The occurrence of crossover rays within the optical CT data produce artifacts in the resulting iteratively reconstructed images. Specifically, when a simple ring phantom's synthetic sinogram is reconstructed a smaller radius, less intense ring artifact is produced. By merging the crossover rays with their primary ray counterpart within the system matrix, it was shown to produce a 26% reduction in the intensity of the induced ring artifact (Figure 4). Although the ring artifact persists, this first case test problem models crossover rays that are much higher in intensity than we would expect to see from reflected rays in our optical CT system. This first case test problem should therefore be looked at as an exaggerated case of the reflected rays problem. The application of merging multiple rays within the system matrix to reflected rays within the optical CT system is still under development. However, the reduction of artifacts under the exaggerated conditions described makes that application promising.

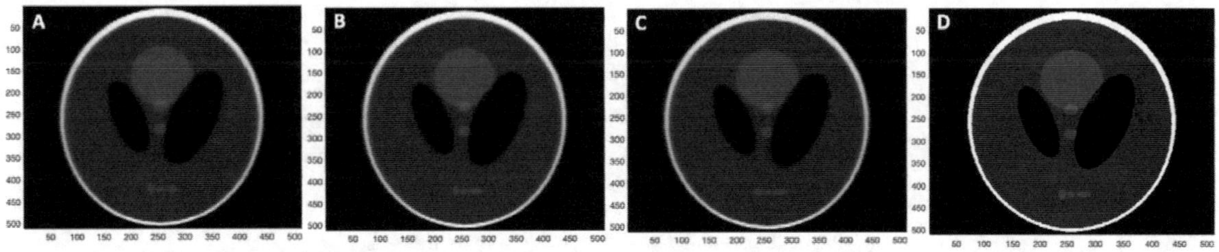

Figure 3. Reconstructed Shepp-Logan phantom synthetic data with system matrix discretized in: A) Cartesian coordinates, B) polar coordinates with 728 radial rings and C) polar coordinates with 2048 radial rings. D) The known Shepp-Logan phantom dose distribution.

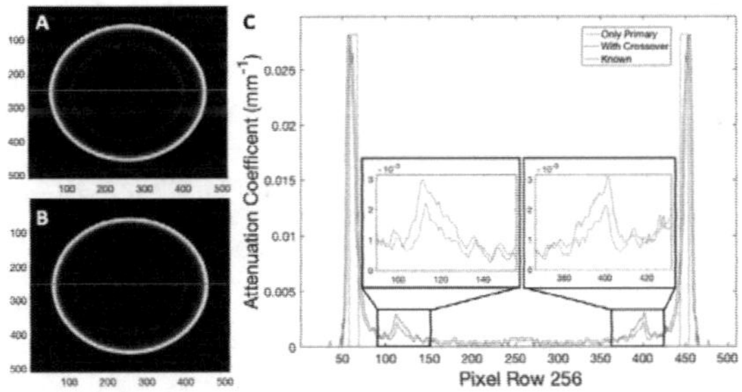

Figure 4. Reconstructed ring phantom synthetic data utilizing: A) a system matrix that only accounts for primary rays, B) a system matrix that merges crossover rays. C) Image profile comparison of the ring phantom images taken at pixel row 256. The highlighted regions show the slight reduction in intensity of the ring artifact.

4. Conclusions

A system matrix size reduction of 360 times (6.0 MB) was achieved by discretizing the image pixel elements in polar coordinates, allowing system matrix generation for only a single projection. However, this reduction in system matrix size came at the cost of decreased spatial resolution in the central region of the image. A more balanced approach found a system matrix size reduction of 144 times (14.6 MB) with only minimal reduction in spatial resolution.

Merging multiple rays within the system matrix was shown to produce a slight reduction in the intensity of image artifacts. If a more substantial correction for these crossover rays can be achieved, the correction can be applied to all sources of multiple ray paths including those from reflected rays.

These novel iterative image reconstruction methods make strides toward optical CT radiochromic gel dosimetry being a practical and accurate tool for radiation therapy treatment plan verification.

Journal of Physics: Conference Series **2630** (2023) 012033 doi:10.1088/1742-6596/2630/1/012033

5. References

[1] Oldham M *et al* 2003 *Med. Phys.* **30** 623–634
[2] Kairn T *et al* 2017 *J. Phys.: Conf. Ser.* **847** 012071
[3] Atiq M *et al* 2017 *Pak. J. Med. Res.* **56** 99-103
[4] De Deene Y *et al* 2015 *Phys. Med. Biol.* **60** 1543-1563
[5] Oldham M 2015 *J. Phys.: Conf. Ser.* **573** 012006
[6] Xu Y *et al* 2004 *Med. Phys.* **31** 3024-3033
[7] Thibaudeau C *et al* 2013 *Med Phys*. **40** 111904
[8] Ogilvy A *et al* 2020 *Phys. Med. Biol.* **65** 245012
[9] Hansen P C *et al* 2012 *J. Comput. Appl. Math.* **236** 2167–2178

Journal of Physics: Conference Series **2630** (2023) 012034 doi:10.1088/1742-6596/2630/1/012034

Polymer gel dosimetry for experimental verification of conformal small animal irradiation at a preclinical research platform

E Konradsson[1], M Blomstedt[1], C Jamtheim Gustafsson[2,3], S Å J Bäck[1,2,3], C Ceberg[1] and S Ceberg[1]

[1]Medical Radiation Physics, Department of Clinical Sciences, Lund University, Lund, Sweden

[2]Radiation Physics, Department of Hematology, Oncology and Radiation Physics, Skåne University Hospital, Lund, Sweden

[3]Medical Radiation Physics, Department of Translational Sciences, Lund University, Malmö, Sweden

E-mail: elise.konradsson@med.lu.se

Abstract. In translational research, there is a need for preclinical studies of high quality that corresponds to the conformal dose distributions conventionally delivered to humans in the modern radiotherapy clinic. To facilitate this need, novel preclinical systems consist of preclinical radiation platforms and small animal treatment planning systems. However, small-field dosimetry is challenging and requires dosimeters with high spatial resolution. In this study we demonstrate the feasibility of experimentally validating the dose distribution in small fields at a preclinical X-ray research platform using a polymer gel dosimeter and MRI-readout.

1. Introduction

Preclinical experiments with small animals are an important step between *in vitro* research and clinical implementation. To provide translatable results, preclinical experiments are required to follow the development of modern radiotherapy techniques with conformal dose delivery [1]. Radiotherapy platforms dedicated for preclinical research, such as XStrahl's XenX, as well as novel treatment planning systems (TPS) adapted for small animals, allow for precise irradiation of tumor and normal tissue [2,3]. For high quality studies, it is desirable to experimentally verify the dose distribution calculated by the treatment planning system, and complex treatment plans require precise dosimetry tools. A polymer gel dosimeter is a tissue equivalent dosimeter with a high 3D spatial resolution, high reproducibility, and the potential to provide accurate small field dosimetry [4].

The aim of this study was to 1) investigate the dose response and reproducibility of a polymer gel with MRI readout in small fields at a preclinical X-ray research platform, and 2) investigate if the polymer gel dosimeter can verify the conformal small field dose distribution calculated by the preclinical TPS μRayStation.

2. Materials & Methods

2.1. *Manufacturing of a polymer gel dosimeter*

The polymer gel N-isopropylamide (NIPAM), with low toxic monomer composition, was used [5,6]. Manufacturing of the polymer gel was performed according to the recipe of Waldenberg et al., under normoxic conditions in a fume hood. Water was heated and stirred using an IKA® C-MAG HS 7 control. After slowly adding 5% w/w gelatine from porcine skin (gel strength 300 type A, Sigma-Aldrich) and letting it dissolve, 3 % w/w NIPAM (97 % Sigma-Aldrich) and 3% w/w N,N'-methylene-bis-acrylamide (99 % Sigma-Aldrich) was added. After the mixture was cooled to 38°C, 21-26 mM of the antioxidant THPC (80 % solution in water, Sigma-Aldrich) was added. The gel mixture was dispensed in 15 ml vials. Before irradiation, the gels were stored in a dark room-temperature environment for 24h.

2.2. *Polymer gel irradiation*

Vials with gel were positioned one-by-one on top of solid water slabs at a source-to-surface distance (SSD) of 33 cm inside the preclinical research platform XenX (XStrahl Ltd, UK, Figure 1a), and irradiated with a 220 kV X-ray beam. To investigate the reproducibility in the gel response, seven vials were irradiated at each dose of 0.5, 2, 5 and 10 Gy (batch I). To investigate the dose response of the gel, vials were irradiated in the range 0.5-40 Gy (n=2/dose level, batch II). To investigate the feasibility to use the gel dosimeter for absorbed dose measurements in small field irradiations, vials were positioned in a narrow gutter in the XenX cabinet (Figure 1b) and irradiated at an SSD of 34 cm using three beams with gantry angles 135°, 180° and 225°, and field sizes of 5x5 mm^2 (batch III). A calibration curve was obtained by irradiating gels in the same batch with known absorbed doses.

(a) (b)

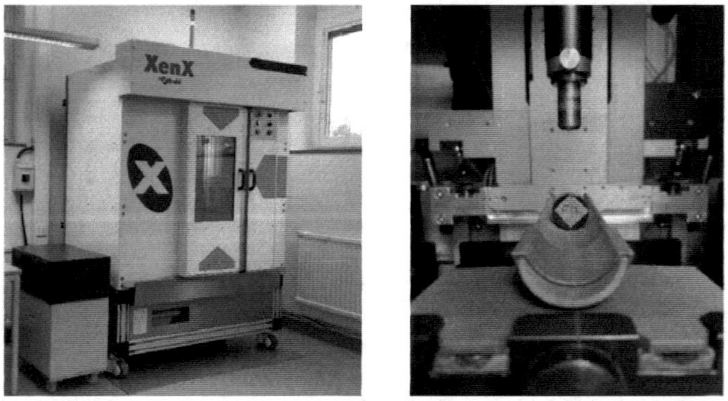

Figure 1. The preclinical research platform XenX (a)
and the setup used for irradiation of gel with 220 kV using 5x5 mm^2 beams (b).

2.3. *Polymer gel read-out*

At 24h after irradiation, vials were scanned at a General Electric Architect 3T MRI system using a fast spin echo (FSE) pulse sequence with a repetition time of 4000 ms and various effective echo times (TE_{eff}), to obtain T2 weighted MR images in the coronal plane (Figure 2). The signal (S) of the MR images were extracted using Mice Toolkit 2021.2.1, and the relaxation time (R2) was obtained by fitting the data to a monoexponential model $S = S(0)e^{-TE_{eff}*R2}$.

2.4. *Quantifying dose distribution*

The R2 map obtained from MR images was calibrated to absorbed dose and compared to calculated dose distribution from the Monte Carlo based preclinical TPS μRayStation (RaySearch, Sweden) commissioned for the 220 kV beam of the XenX irradiator. In the TPS, a vial (ρ=2.33g/cm^3) with gel

Journal of Physics: Conference Series **2630** (2023) 012034 doi:10.1088/1742-6596/2630/1/012034

(ρ=1g/cm^3) was built and the treatment set-up was reconstructed. A calculated dose profile was extracted along the beam axis of the 180° beam, and compared to the corresponding measured dose profile from the MR image.

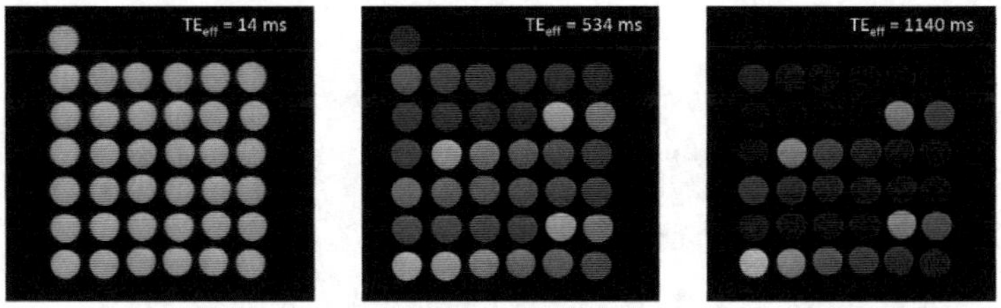

Figure 2. Coronal MR images of a gel batch. Images were acquired using a fast spin echo pulse sequence with TR = 4000 ms and sixteen different TE_{eff} ranging from 14 ms to 1140 ms. The MR images correspond to the shortest (left panel), middle (middle panel), and longest (right panel) TE_{eff} used.

3. Results & Discussion

3.1. Polymer gel dose response reproducibility and linearity

The T2 weighted signal from the MR images decreased with increasing dose. For the reproducibility assessment (batch I), the mean R2 ± SD for vials (n=7/group) irradiated with 0.5, 2, 5 and 10 Gy respectively, were 1.405 ± 0.016 s^{-1}, 1.477 ± 0.010 s^{-1}, 1.743 ± 0.015 s^{-1} and 2.231 ± 0.021 s^{-1}, showing a high reproducible gel response. For the dose response assessment (batch II), the R2 values increased linearly (at a sample correlation coefficient r^2 of >0.99) up to a dose of 12 Gy, after which the response saturated (Figure 3). For higher doses, a second-degree polynomial function could provide a more accurate description of the dose response. Based on the linear fit, the dose-R2 sensitivity of the gel dosimeter (i.e., the slope of the dose-response curve) in the dose range 0.5-40 Gy was 0.1 s^{-1} Gy^{-1}. In future work, attempts will be made to improve the sensitivity by refining the gel recipe, e.g., by increasing the percentage of NIPAM or adding acetone as a co-solvent to allow for increased concentration of N,N'-methylene-bis-acrylamide [7].

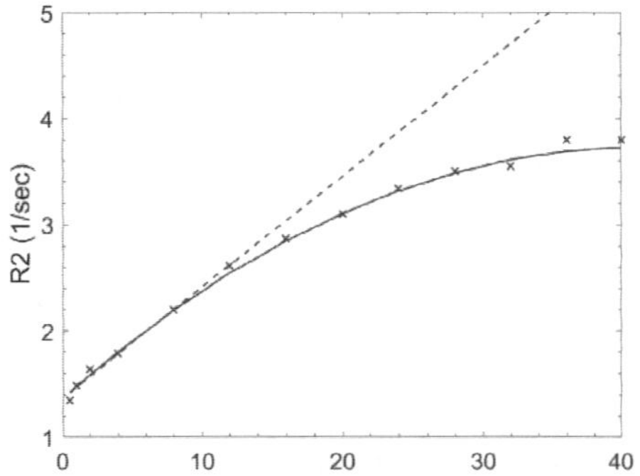

Figure 3. Dose response for the polymer gel irradiated with 220 kV X-rays. The R2 values are fitted with a second-degree polynomial function (solid line). For doses ≤12 Gy, the data are also fitted with a linear function (dotted line).

3.2. Polymer gel dosimeter to quantify dose disposition in preclinical experiments

As illustrated in Figure 4, the agreement between the absolute dose distribution measured with the polymer gel dosimeter and the dose distribution calculated with the TPS was good. In the central part of the profile (± 3 mm from isocenter) the absolute mean deviation was 2.9%. The mean absolute deviation over the entire profile was 6.3%. These results indicate that the polymer gel investigated can be useful for small field dosimetry.

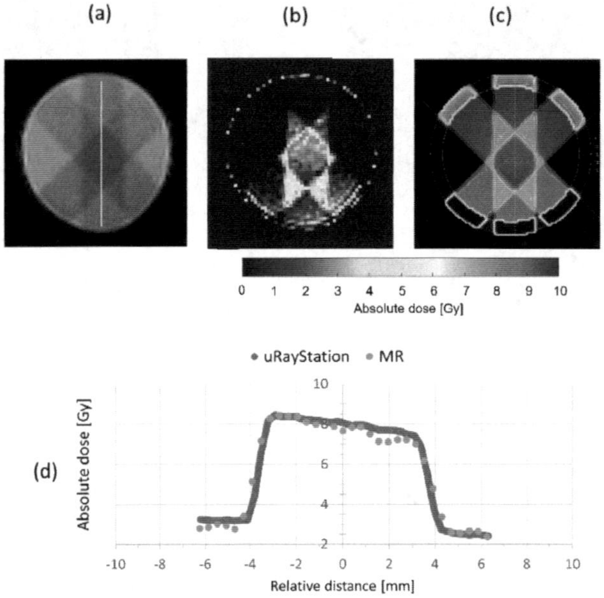

Figure 4. The MR image (a), MR dose image (b), and treatment plan (c) of the vial irradiated with three 5x5 mm^2 beams, as well as a comparison between an absolute dose profile extracted from the TPS (blue) and the MR dose image (orange) (d). The yellow line in the MR image indicates the position of the profile.

4. Conclusion

In this study we demonstrate the feasibility of experimentally validating the dose distribution of conformal small animal irradiations at a preclinical research platform using polymer gel dosimetry and MRI-readout. In future work we will refine the gel mixing procedure for increased sensitivity and optimize the MRI-readout to reduce noise.

5. Acknowledgments

This research was supported by the Swedish Cancer Society and Mrs Berta Kamprad Foundation.

6. References

[1] Verhaegen F *et al* 2011 *Phys. Med. Biol.* **56** R55-83
[2] Chiavassa S *et al* 2020 *Phys. Med. Biol.* **65** 035006
[3] Ghita M *et al* 2017 *Radiat. Oncol.* **12** 1-10
[4] Baldock C *et al* 2010 *Phys. Med. Biol.* **55** R1-63
[5] McAuley K B *et al* T 2013 *J. Phys.: Conf. Ser.* **444** 012001
[6] Waldenberg C *et al* 2017 *J. Phys.: Conf. Ser.* **847** 012063
[7] Merkis M *et al* 2021 *Med. Phys. Balt. States* **15** 183–86

Journal of Physics: Conference Series **2630** (2023) 012035 doi:10.1088/1742-6596/2630/1/012035

Characterization of a water-based liquid scintillator for use in megavoltage radiotherapy beams

D. Bernier-Marceau[1,2,3], É. Cloutier[1,2,3], M. Yeh[4], G. D Orebi Gann[5,6] and L. Beaulieu[1,2,3]
[1]Département de physique, de génie physique et d'optique, Université Laval, Québec, Canada
[2]Centre de recherche sur le cancer, Université Laval, Québec, Canada
[3]CHU de Québec – Université Laval and CRCHU de Québec, Canada
[4]Brookhaven National Laboratory, Upton, NY 11973-500, USA
[5]University of California, Berkeley, CA 94720-7300, USA
[6]Lawrence Berkeley National Laboratory, Berkeley, CA 94720-8153

Email:daphnee.bernier-marceau.1@ulaval.ca; beaulieu@phy.ulaval.ca

Abstract. The measurement of the dose delivered in radiotherapy treatments is carried out using dosimeters that are often expensive to produce and sometimes toxic to humans and the environment, which leads to more complex and rigorous clinical manipulations. It is in this context that it is necessary to provide new types of scintillators that would no longer have these problems while having properties equivalent to those of human tissues. Thus, the following study presents the performance of a water-based liquid scintillator used at radiotherapy energies. The characteristics studied include the proportionality of the scintillation signal to the dose, the scintillation efficiency at two different energies as well as the identification of the Cherenkov portion of the signal for photon beams of 180 kVp, 6 MV as well as 18 MV. Spectral measurements of the scintillation solution and a solution of distilled water were acquired in order to isolate the contribution of the scintillation signal from the spectrum obtained, and then compared to a commercial scintillator, Ultima Gold. The signal exhibits a linear dose relationship with a correlation coefficient of 0.999 and lower scintillation efficiency than Ultima Gold.

1. Introduction

The need to measure accurately the dose for radiation treatments is one of the reasons why there is various kinds of detectors available in clinical settings. However, in the photon beam energy range where Compton scattering is dominant, most of these detectors do not have interaction properties that are similar to those of human tissues [1]. Organic liquid compounds that are composed of a solvent and fluorophores possess this key advantage but often have a solvent that can be toxic and therefore require tedious manipulations [1]. Water-based liquid scintillators (WbLS) have tremendous potential since the solvent is mainly composed of water and exhibit similar light emission characteristics as a standard liquid scintillator. In this work, we focus on the characterizing of a WbLS composed of 80% of water with less than 20% of the organic solvent linear alkylbenzene (LAB) and 2,5-diphenyloxazole (PPO). More specifically, this work investigates the relationship between the dose and the scintillation response as well as the spectral response of the detector and its variation with the energy of the incident beam.

2. Methods and materials

2.1. Water based liquid scintillator description
The water based liquid scintillator (WbLS) was produced by the Neutrino and Nuclear Chemistry team at Brookhaven National Laboratory affiliated with the research team of Gabriel D. Orebi Gann. The liquid scintillator was given to our team for characterization. This dosimeter is composed of 80% of distilled water and 20% or less of an organic solvent named linear alkylbenzene (LAB). The fluorophores used are 2,5-diphenyloxazole (PPO) and represent less than 0.1% of the total composition of the liquid scintillator [3].

2.2. Irradiation of the scintillator and signal collection
All MV irradiations were performed with a medical linear accelerator TrueBeam (Varian, Palo Alto, USA). The scintillation signal was collected and guided through a clear optical fiber as seen in Figure 1. The spectra obtained from each irradiation was measured from the spectrometer QE-65 Pro (Ocean Optics, Dunedin, USA) with a fixed integration time of 25 s. It is to be noted that the spectra were not corrected for the wavelength dependencies of the fiber and the spectrometer. A field of 10x10 cm^2 was employed and the vial containing the solution to be analysed was placed at the linac isocenter. Water slabs were placed around the vial to act as a backscattering and build up medium and the vial was positioned at the depth of maximum dose for each energy used. For investigations using kV energies, the orthovoltage machine was used (Xstrahl, Camberley, United Kingdom).

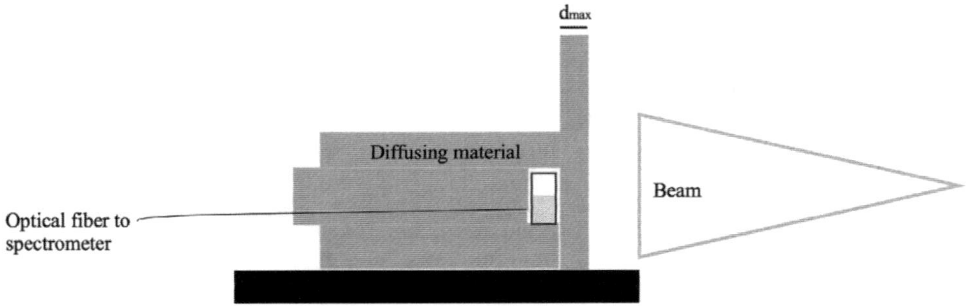

Figure 1. Experimental set-up for the scintillation collection at MV energies.

2.3. Linearity of the scintillation signal
The WbLS signal linearity was investigated for a photon beam energy of 6 MV. The dose rate was kept constant, only the number of MU changed to vary the dose at the isocenter. There were no corrections applied for the production of Cherenkov light in this section since it did not affect the linearity of the signal.

2.4. Energy dependence of the signal
Three distinct irradiations at 6 MV, 10 MV and 18 MV were performed with the vial of WbLS. The dose was kept constant for all of the irradiations. To correct for the Cherenkov emission in MV energy range, a vial containing water was exposed to the same beams under the same reference conditions. In water, only Cherenkov emission is collected. The minimum energy required to produce this parasite signal depends on the refractive index of the irradiated medium. In water, this energy is 264 keV [1]. Therefore, to obtain the signal without Cherenkov, irradiations at 180 kVp were performed and the signal without Cherenkov at MV energies was found using the hyperspectral method [2]. The liquid scintillator Ultima Gold (Perkin Elmer, Waltham, Massachusetts, United States) was irradiated in the same conditions as for the WbLS for comparison. A blank irradiation using an empty vial was performed for each energy to subtract the possible Cherenkov emission produced in the optical light collecting fiber.

3. Results and Discussion

3.1. Linearity of the scintillation signal

The scintillation signal obtained for the WbLS is shown in Figure 2 as a function of the dose delivered. The signal obtained follows a linear trend with doses ranging from 0 to 300 cGy. The correlation coefficient R^2 obtained is 0.999. The dose was determined knowing the number of MU programmed to the vial that was irradiated at reference conditions.

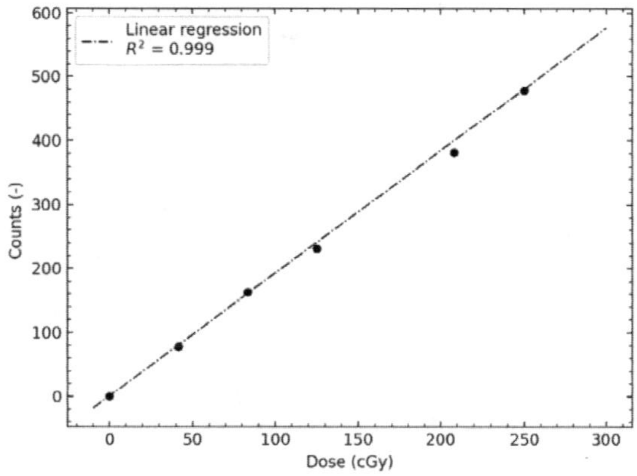

Figure 2. Linearity of the scintillation signal for the water-based liquid scintillator as a function of dose for 6 MV irradiation. Error bars are smaller than the symbols presented.

3.2. Energy dependence of the scintillation signal

The spectral response of the detector is represented on Figure 3. The signal is decomposed into its two components, the Cherenkov and the scintillation portion. There is a non negligeable production of Cherenkov scintillation in the signal that needs to be canceled for the analysis of the pure scintillation signal.

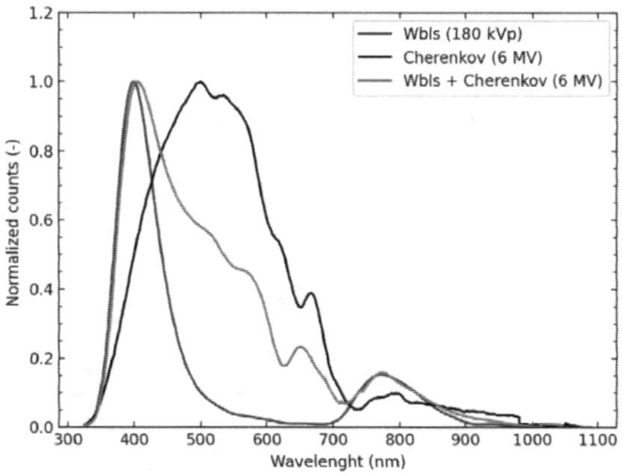

Figure 3. Spectral measurements of scintillation that illustrates the Cherenkov component created in the water-based liquid scintillator at 6 MV compared to the scintillation signal without the presence of Cherenkov at 180 kVp.

The scintillation signal of both the scintillators is shown in Table 1 as a function of the energy of the photon beam. Each value is normalized to the scintillation signal of the Ultima Gold at 6 MV. It is possible to see that the WbLS has a lower output than the liquid scintillator Ultima Gold at both energies. Ultima Gold has the highest energy dependance with an increased output of 5% at 18 MV compared to the signal obtained at 6 MV. WbLS presents a lower dependance with an output at 18 MV that is 0.6% more than the signal at 6 MV.

Table 1. Signal ratio of water-based liquid scintillator and Ultima Gold for beam energies of 6 MV, 10 MV and 18 MV normalized to the scintillation output of the Ultima Gold at 6 MV. The Cherenkov contribution was removed in all cases.

Beam nominal energy	6 MV	10 MV	18 MV
Scintillation ratio: WbLS	0.120	0.125	0.126
Scintillation ratio: Ultima Gold	1	1.021	1.051

4. Conclusion

This study illustrates the advancements in the characterization of a new type of water-based liquid scintillator. While the WbLS produced a lower emission yield than the reference commercial LS, the signal is sufficient for dosimetry purposes. The energy dependence needs to be explored further, including lower photon energies as well as clinical electron beams. It should be noted that the concentration could be modified in order to observe the effect of the %LS loading on the signal obtained. With an optimal concentration and known characterization, this scintillator could be a potential candidate for the development of a new dosimeter that is less toxic and more clinically safe than those existing on the market.

5. Acknowledgments

This work was supported in part by a research grant from the Natural Sciences and Engineering Research Council of Canada (NSERC). The authors would also like to thank the Brookhaven National Laboratory for providing us with the liquid scintillator.

6. References

[1] Beaulieu L *et al* 2016 *Phys. Med. Biol.* **61** 305-343
[2] Archambault L *et al* 2012 *Phys. Med Biol.* **57** 133-145
[3] Caravaca J *et al* 2020 *Eur. Phys. J.* **80** 1-10

Journal of Physics: Conference Series **2630** (2023) 012036 doi:10.1088/1742-6596/2630/1/012036

Preliminary commissioning of a prototype solid tank optical CT scanner for 3D radiation dosimetry

A Ogilvy[1], S Collins[1], M Guenter[2], D Huang[2], W Hare[2], M Hilts[3,1], and A Jirasek[1]
[1]Department of Physics, University of British Columbia - Okanagan campus, Kelowna BC V1V 1V7, Canada
[2]Department of Mathematics, University of British Columbia - Okanagan campus, Kelowna BC V1V 1V7, Canada
[3]Medical Physics, BC Cancer - Kelowna, Kelowna BC V1Y 5L3, Canada

E-mail: andrew.jirasek@ubc.ca

Abstract. Optical computed tomography (CT) is one of the leading modalities for imaging gel dosimeters. In previous research, it was shown that a design could significantly reduce the volume of the refractive index baths that are commonly found in optical CT systems. The proposed scanner has been manufactured and is in process of being commissioned. The rays refract through the system and their paths are estimated using a ray-tracing simulator. Reconstruction is preformed through algebraic reconstruction technique and iterated using the FISTA-TV algorithm. A Sylgard®184 edge phantom was created to commission the scanner's spatial resolution. The scanner is capable of a 0.929 mm^{-1} spatial resolution, although the spatial resolution is highly dependent on the number of iterations, and method of processing the edge spread function.

1. Introduction

Optical computed tomography emerged as a potential method for scanning three-dimensional dosimeters in 1996 when Gore *et al.* demonstrated its applicability in scanning PAG polymer gel dosimeters [1]. Since that initial work, a variety of scanner designs and methods have emerged [2-6]. Typically, these systems employ a standard filtered back projection (FBP) technique for reconstruction. Back projection reconstruction operates under the assumption of a direct path from signal source to the detector, which necessitates that these systems carefully index match the optical path. The requirement of a direct ray path can make researchers go to great length to reduce the refraction in optical CT systems. The primary method that most optical CT scanner systems use to minimizing refractive artifacts is an index matching bath [1, 7-10]. Managing a refractive index bath can be inconvenient, mess, and unappealing, as they often involve handling a relatively large volume (1-15L) of chemical mixtures (water, glycerol, propylene glycol, etc) that require constant maintenance and monitoring. The fluid can evaporate or separate over long scanning period, thus altering the remaining fluid's refractive index. In the event of the chemical mixture separating, Schlieren patterns may emerge in the viewing area as the result of motion in the fluid [11].

Previously published work has demonstrated that a solid tank scanner with a severely reduced need for index matching can be built and optimized given a set of design criteria [12]. The purpose of this work is to commission the design found from that previous research. The prototype scanner incorporates a solid acrylic block with minimal index matching fluid (5-30mL) and eliminates the need for rays to follow a direct path from source to detector. The scanner utilizes an altered version of the ray-tracing simulator, developed during design phase, to generate a ray-path system matrix. The *Fast Iterative Shrinkage-Thresholding Algorithm Total Variation* (FISTA-TV) algorithm is employed to perform reconstruction using the algebraic reconstruction technique. The commissioning of this scanner is still a work-in-progress.

2. Materials and methods

The rationale behind the design, as well as the procedure have been discussed in-depth in a previously published paper [12]. The prototype fan-beam scanner features a 635nm laser diode with a 60° line-generating Powell lens. The fan-beam enters a 290mm long solid acrylic (PMMA) block. The front face of the block is machined into a 61mm radius semi-circle to direct the beam towards a 104mm diameter hole that has been bored through the block (Fig. 1). A gel dosimeter is placed in the cavity of the bore, and through the use of an O-ring (CR39932, SKF; Toronto, ON), a very small volume of refractive index matching fluid is suspended in the gap between the dosimeter and the block (Fig. 2). The rear wall of the block was specially machined to allow five 64-element photodiode arrays (S8865 Series, Hamamatsu; Hamamatsu City Japan) to sit flush up against the acrylic without a collimator. Each of the 320 photodiode elements has an active area of 0.8mm by 0.7mm, with a 0.1mm spacer between each active area. The gel containers are mounted to a pair of motion controlling stages (ESP300, Newport; Irvine, CA), allowing for rotational and vertical movement.

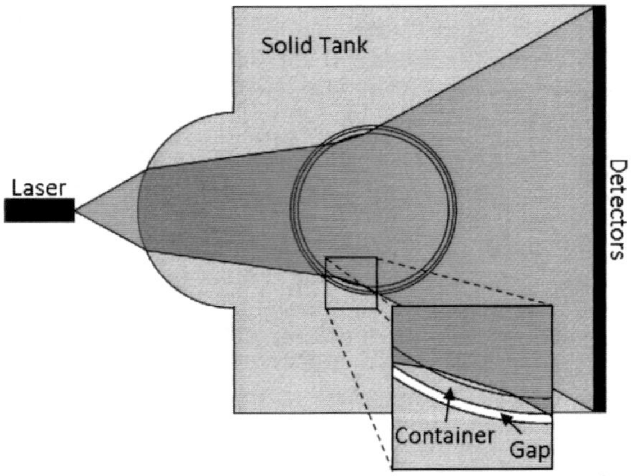

Figure 1. Shown is a bird's eye schematic representation of the scanner apparatus. The block, container, bore gap, and ray paths are all shown to-scale if a Sylgard®184 based dosimeter is held in the container.

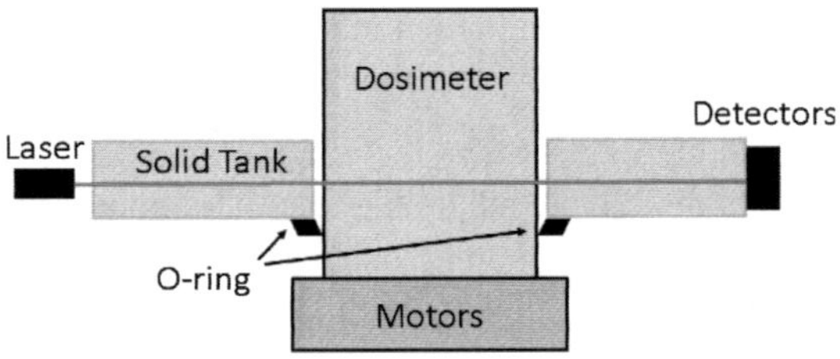

Figure 2. Shown is a not-to-scale cross-sectional schematic of the scanner apparatus. It can be seen that the dosimeter container is fixed to the motors from the base, and that an O-ring is used to allow the bore gap to be filled with index matching fluid.

This scanner is intended to be used on FlexyDos3D dosimeters, developed by De Deene *et al.* in 2015 [13]. The design geometry is specific to the refractive index of the base dosimeter material, Sylgard®184 which is made from a 10:1 ratio of Sylgard®184 Silicone Elastomer and Sylgard®184 Curing Agent respectively. If dosimeters with other optical properties were to be tested, a separate block geometry and apparatus setup would have to be calculated as outlined in the previously mentioned design optimization paper [12]. The overall design would remain relatively unchanged, outside of a new solid tank, and different laser set up.

Image reconstruction is performed using algebraic reconstruction technique (ART). The system matrix used during the reconstruction is a geometry matrix that was created using a retooled fork of the ray-tracer created for the scanner design optimization. Reconstruction iterations were performed using the FISTA-TV algorithm. Many other algorithms were tested on 24 synthetic test problems and 12 clinical test problems by Guenter *et al.* in 2021, and FISTA-TV was a top performer in both solution quality and low runtime [14].

3. Results and discussion

Spatial resolution was commissioned by creating a Sylgard®184 phantom, where a circular column was molded from a precisely machined rod. The mold was then back-filled with a dyed Sylgard®184 mixture (Fig. 3). After scanning the phantom and reconstructing the image, the centroid of the high-attenuation insert is found and the distance from the centroid is calculated for each pixel (Fig. 4). The edge spread function (ESF) of the region of circular attenuation with well-defined geometry is found by fitting a cubic spline to the edge data (Fig. 5), and the derivative of the ESF returns a line spread function (LSF). Applying a fast-Fourier transform to the LSF results in the modulation transfer function (MTF), and the 50% cut-off frequency is reported as the system's spatial resolution.

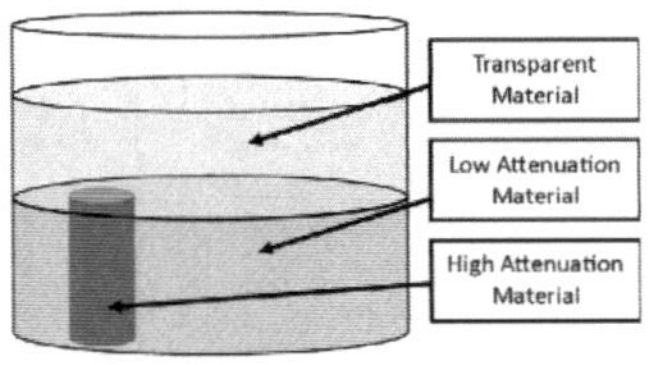

Figure 3. A schematic of a back-filled phantom used to measure spatial resolution.

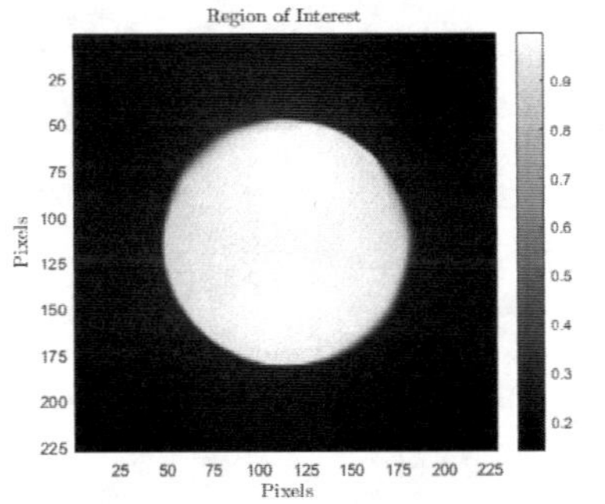

Figure 4. The reconstructed region of interest (200 iterations) of the back-filled spatial resolution phantom.

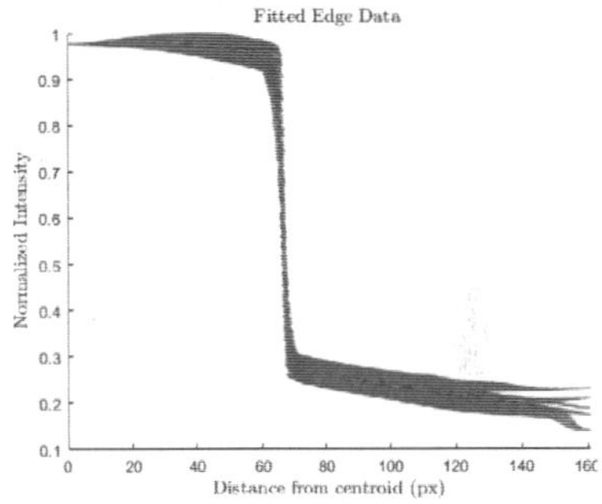

Figure 5. Edge data for the ROI in Figure 4 that has been fitted with a cubic spline.

With iterative reconstruction the number of iterations can greatly affect the clarity of an edge, as well as the noise in a system. During this experiment reconstructions were performed with a varied number of iterations (Table 1), and as expected a larger number of iterations corresponded to a higher spatial resolution up to 200 iterations. Due to the highly-varied nature of data on either side of the edge, some improvement to the spatial resolution can be found by cropping the data around the edge. The data was cropped using the KARI method, as outline in Viallefont-Robinet *et al.* in 2018 [15], resulting in significant improvements in spatial resolution. The highest spatial resolution result (MTF_{50}) was found to be 0.929 mm^{-1} after 200 iterations (Fig. 6). More iterations resulted in a loss of spatial resolution. Reconstructions with less than 80 iterations did not result in the conditions required for the KARI method of cropping, so those iteration numbers were disregarded.

Table 1. The MTF_{50} spatial resolution as the number of iterations increases.

Num. of Iterations	20	40	80	140	200	300
Uncropped	0.0337	0.0596	0.171	0.519	0.666	0.607
KARI Method	-	-	0.656	0.795	0.929	0.838

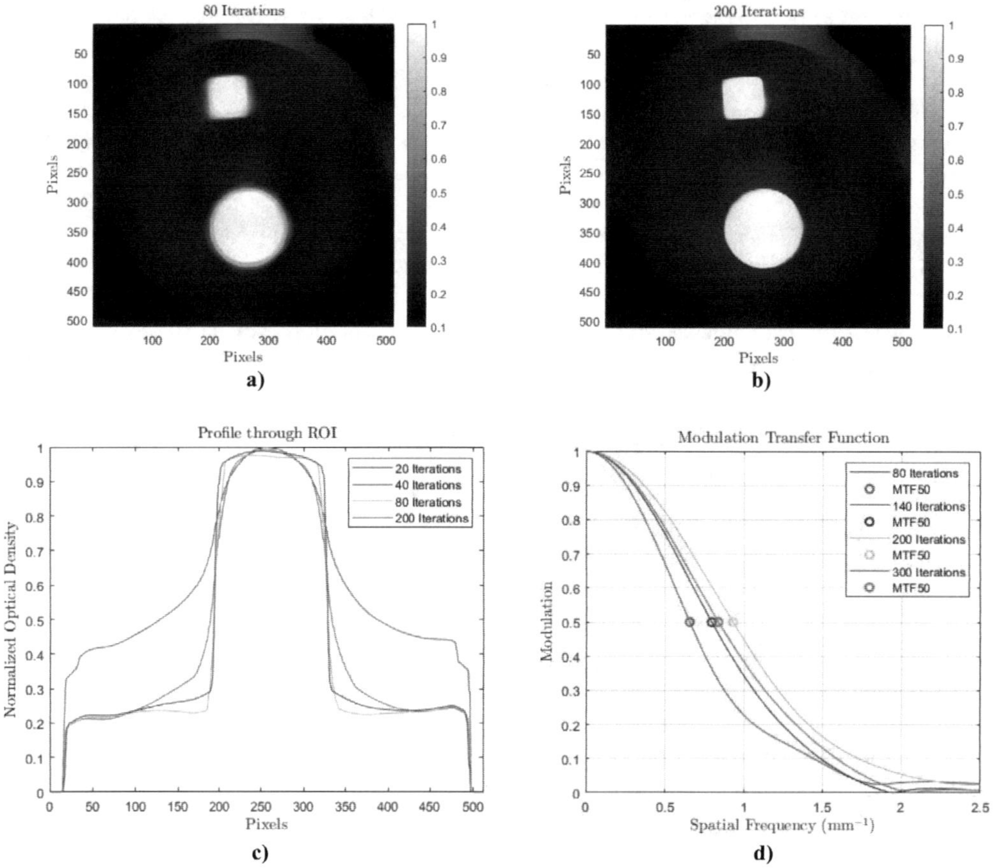

Figure 6. Shown are the reconstructed slices at a) 80 iterations, and b) 200 iterations. The square region of attenuation in the images was not used for this work. The c) profiles through the center of the circular spatial resolution phantom (row 342) shows a steeper edge for as iterations increase to 200. The d) MTF at 80, 140, 200 and 300 iterations from the spatial resolution phantom once ESF cropping has been implemented. Highlighted is the MTF_{50} for each data set.

4. Conclusion

In previous research, an optimal scanner was designed that would lessen the need for index matching and implement algebraic reconstruction technique with the FISTA-TV algorithm to reconstruct the images. That previously designed scanner has been manufactured, and commissioning is a work-in-progress. Currently, it was found that the scanner is capable of achieving a spatial resolution of 0.929 mm^{-1}. Best results were achieved at 200 iterations when cropping the data around the ESF. The future of this work is to finish commissioning the prototype scanner and characterize any artifacts that emerge during scanning or reconstruction.

5. References

[1] Gore JC *et al* 1996 *Phys. Med. Biol.* **41** 2695
[2] Oldham M *et al* 2001 *Med. Phys.* **28** 1436-1445
[3] Doran SJ 2001 *Phys. Med. Biol.* **46** 3191-3213
[4] Wuu C-S *et al* 2003 *Med. Phys.* **30**132137
[5] Krstajić N *et al* 2006 *Phys. Med. Biol.* **51** 2055-2075
[6] Sakhalkar HS *et al* 2008 *Med. Phys.* **35** 101-111

[7] Wolodzko JG *et al* 1999 *Med. Phys.* **26** 2508-2513
[8] Krstajić N *et al* 2007 *Phys. Med. Biol.* **52** 3693-3713
[9] Xu Y *et al* 2004 *Med. Phys.* **31** 3024-3033
[10] Campbell WG *et al* 2013 *Med. Phys.* **40** 061712
[11] Settles G S 2001 *Schlieren and Shadowgraph Techniques Visualizing Phenomena in Transparent Media* SpringerLink eBooks
[12] Ogilvy A *et al* 2020 *Phys. Med. Biol.* **65** 245012
[13] Deene YD 2015 *Phys. Med. Biol.* **60** 1543-1563
[14] Guenter M *et al* 2022 *Med. Phys.* **49** 1065-1082
[15] Viallefont-Robinet F 2018 *Opt. Express* **26** 33625-33648

Journal of Physics: Conference Series **2630** (2023) 012037 doi:10.1088/1742-6596/2630/1/012037

Investigation of the Dosimetry Characteristics of the GAFCHROMIC® EBT3 Film Response to Alpha Particle Irradiation

Mélodie Cyr[1], Victor D. Martinez[1], Slobodan Devic[1,3], Nada Tomic[1,3], David F. Lewis[4], Shirin A. Enger[1,2,3]
[1]Medical Physics Unit, Department of Oncology, Faculty of Medicine, McGill University, Montreal, Québec, Canada
[2]Research Institute of McGill University Health Centre, Montreal, Québec, Canada
[3]Lady Davis Institute for Medical Research, Jewish General Hospital, Montreal, Québec, Canada
[4]Rcfilmconsulting, 54 Benedict Rd. Monroe, CT, USA

E-mail: melodie.cyr@mail.mcgill.ca

Abstract. The purpose of this study was to investigate the dosimetric characteristics of the GAFCHROMIC® EBT3 film responding to alpha particle irradiation. Unlaminated GAFCHROMIC® EBT3 film pieces, were irradiated with a 30.055 kBq ^{241}Am alpha source, at eight different dose levels, between 0 and 509 Gy. The irradiations were performed inside an enclosed box. Epson Expression 10000 XL scanner in transmission mode was used to digitize irradiated films 24 hours post-irradiations as 16-bit RGB images in tagged image file format (TIFF). Optical density (OD) values were obtained by following the OD theorem. Raw and normalized pixel values (PV_{norm}) from the red, green, and blue colour channels were sampled from a 3 x 3 mm^2 region of interest. Calibration curves were created for both data sets (OD and PV_{norm}) and were fit accordingly. Monte Carlo simulations with the Geant4 toolkit were performed to establish a dose rate at the point within the sensitive layer of the film. An alpha dosimetry protocol for EBT3 films was obtained from the Monte Carlo calculated dose rate and dose calibration curves for alpha radiation were created. It is necessary to extend this study for different film types and compare to photon dosimetry calibration curves.

1. Introduction

GAFCHROMIC® EBT3 films are designed and used for the measurement of absorbed doses of ionizing radiation especially suited for high-energy photons. This film model has optimal dose ranges from 0.2 Gy to 10 Gy and can be developed in real time without post-exposure treatment [1]. EBT3 films need a color photo scanner to acquire 16-bit per channel of the red, green, and blue color components of light transmitted through the film. The EBT3 film model offers high spatial resolution and allows for depth dose measurements [2]. The EBT3 film has a structure comprised of an active layer (28 µm thick), sandwiched between two matte-polyester substrates (125 µm). The active layer contains an active component, a marker dye, stabilizers, and other elements [1]. The change in optical properties of GAFCHROMIC® films depends on the absorbed dose and the linear energy transfer (LET) of the ionizing radiation. The current experience of GAFCHROMIC® film dosimetry is based on low LET photon beams. Lately, the use of alpha emitting radionuclides such as the Alpha Tau Diffusing Alpha-emitters Radiation Therapy (DaRT) [3] for the treatment of cancer, has increased. This novel treatment modality relies on temporary or permanently implantable seeds impregnated with a small activity of ^{224}Ra, which are placed inside the tumours. Short-lived alpha-particle emitting atoms are released in the decay chain of ^{224}Ra and diffuse inside the tumour [3]. In this study, we investigate the GAFCHROMIC® EBT3 film model response to eight different alpha particle doses from an ^{241}Am source. However, since the range of alpha particles is very short (up to 100 micrometers), unlaminated films were used to directly expose the active layer to the alpha particles [4].

2. Materials and Methods

In this study, unlaminated GAFCHROMIC® EBT3 films were used for alpha particle dosimetry. The film was cut into 6.35 x 5.08 cm^2 pieces using a sharp cutter with a board for paper trimming. The films were cut longitudinally, and a landscape orientation was followed throughout the entirety of the study. The radioactive source emitting alpha particles, ^{241}Am was placed inside a metal box, mimicking a dark box environment. The source has a gold cover of approximately 51.8 nm to protect anything from being in direct contact with the source. The films were placed on the ^{241}Am source downwards, with the active layer in contact with the gold cover for varying exposure times to absorbed doses in the range of 0 Gy to 509 Gy. Figure 1 presents the experimental design. The absorbed doses were determined by calculating the dose rate of the experimental design using Monte Carlo Simulations performed with Geant4 simulation toolkit. The dose rate was calculated to be 28.26 Gy/h for irradiation at the level of the active layer of the film [5]. The film was scanned using Epson Expression 10000 XL scanner. The image was saved as a Tagged Image File Format (tiff). All scanned images were analysed using an inhouse python script in terms of pixel value (PV) for the red, green, and blue channels.

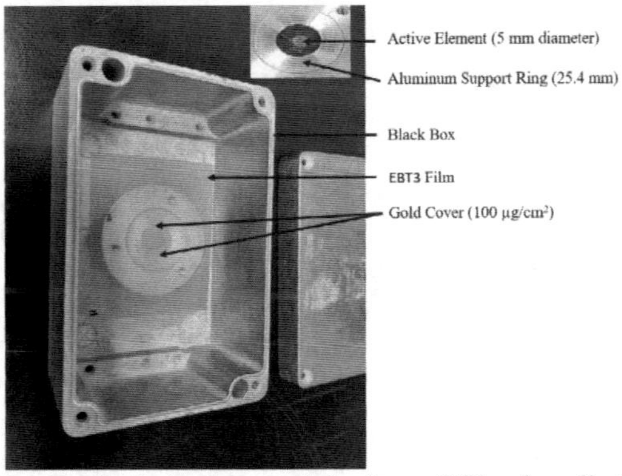

Figure 1. Experimental design of film irradiations.

The PV of the exposed region of the films were converted into average net optical densities (OD_{net}). The Equation 1 below indicates the conversion of 16-bit pixel channel numbers converted into OD_{net}. The film irradiations were repeated to lower error probabilities. It represents the EBT3 film at red, green, and blue colour channels. The calibration curves, i.e., the dependence delivered dose on the normalized PVs (Equation 2) and OD_{net}, were fitted to a power function (Equation 3) and exponential function (Equation 4) respectively.

$$OD_{net} = \left[\log_{10}\left(\frac{65535.0}{PV_{irradiated}}\right)\right] - \left[\log_{10}\left(\frac{65535.0}{PV_{background}}\right)\right] \quad (1)$$

Where normalized PV is calculated below:

$$PV_{norm} = \left(\frac{PV_{background}}{PV_{irradiated}} - 1\right) \quad (2)$$

And the normalized data was fit to:

$$Dose\ (Gy) = bx + c(PV_{norm})^n \qquad (3)$$

Where b, c, and n are fitting parameters, as a polynomial correction to a linear form. Finally, the fit for the OD_{net} data as mentioned, was fit to an exponential function:

$$Dose\ (Gy) = Ae^{(R*OD_{net})} + C \qquad (4)$$

where A, C, and R are fitting parameters. The plotted uncertainties were calculated based on total uncertainties from the relative fit parameter uncertainties and experimental uncertainties added in quadrature.

3. Results and Discussion

The dependence of the normalized PV and OD_{net} on the absorbed dose have been investigated. The result from the normalized PV concludes that the red colour channel has the highest sensitivity than the green and blue channels, however only useful for small doses (up to ~8 Gy). The higher sensitivity for the red channel is due to this channel having the highest absorption for low dose ranges up to approximately 8 Gy, however, the higher dose ranges up to 100 Gy, will saturate the red channel sensitivity curve. The green channel can provide a smaller uncertainty than the red channel when normalized PV is used as a response function for higher doses. The absorbed dose (Gy) as a function of normalized PV for all three colour channels were fit to the polynomial fit function, seen as equation 3 above. The absorbed dose (Gy) as a function of OD_{net} for all three colour channels, were fit to an exponential function. The errors bars associated with each plot are the calculated total uncertainties from the relative fit parameter uncertainties and experimental uncertainties, added in quadrature. Through further analysis, caution must be taken when looking at particular colour channels where the raw PV is less than 10 000. In this case, the capabilities of a flatbed optical scanner are interpreted rather than the function of the film. The measured response value is dependent on the light transmission through the film but for response values less than 10 000, it is corrupted by a signal from scattered light in the measurement system. This signal is now independent of the transmission of the film. Therefore, any dose above 50 Gy cannot represent response values appropriately for EBT3 film.

4. Conclusion

EBT3 film can measure the alpha radiation doses with high accuracy up to 50 Gy. This film model cannot be used for dosimetry higher than 50 Gy, therefore, above this dose limit, another film model such as HD-V2 film (high-dose) must be used. Further analysis is currently in completion with HD-V2 film to be used in combination with EBT3. In future studies, unlaminated EBT3 film will be used to benchmark autoradiography measurements for novel animal studies involving alpha emitting sources used for the treatment of cancer.

5. References

[1] Ng C *et al* 2016 *Radiat. Phys. Chem.* **124** 176-179
[2] Basset A *et al* 2019 *Nucl. Instrum. Methods Phys. Res. B* **454** 56-60
[3] Arazi L *et al* 2007 *Phys. Med. Biol.* **52** 5025-5042
[4] Mustaqim A S *et al* 2018 *J. Phys: Conf. Ser.* **1083** 12-19
[5] Agostinelli S *et al* 2003 *Nucl. Instrum. Methods Phys. Res. A* **506** 250-303

Developing a 3D-printing-based method to create anthropomorphic dosimeters for radiotherapy-delivery verification

S V Jensen[1,2], T B Erichsen[3], M B Jensen[1,2,4], E Worm[4], P R Poulsen[1,2], J B B Petersen[2,4], P Balling[5,6], D Q S Le[2] and L P Muren[1,2]

[1] Danish Centre for Particle Therapy, Aarhus University Hospital, Aarhus, Denmark
[2] Department of Clinical Medicine, Aarhus University, Aarhus, Denmark
[3] Faculty of Technical Sciences, Aarhus University, Aarhus, Denmark
[4] Department of Medical Physics, Aarhus University Hospital, Aarhus, Denmark
[5] Department of Physics and Astronomy, Aarhus University, Aarhus, Denmark
[6] Interdisciplinary Nanoscience Center, Aarhus University, Aarhus, Denmark

Email: sivije@au.dk

Abstract. Anthropomorphic three-dimensional (3D) dosimeters can be useful for verification of radiotherapy delivery. The aim of this study was to develop a 3D-printing-based method for creating anthropomorphic 3D dosimeters. Internal structures were created using water dissolvable 3D prints as negatives. External structures were 3D-printed, and a mould was produced using silicone rubber. Realistic liver and trachea dosimeters with both internal and external anthropomorphism were produced and subsequently irradiated with photons and protons. A 3%/3 mm gamma analysis resulted in 87% and 86% pass rates. The limiting factor to the resolution of the dosimeters was the 3D prints detail.

1. Purpose

Anthropomorphic phantoms play an important role for testing clinical equipment in conventional radiotherapy and proton therapy. They are usually constructed to imitate the head, thoracic, or abdominal region. They should ideally be compatible with both magnetic resonance (MR) and computed tomography (CT) scanners, be deformable and able to move in a respiratory manner [1-6].

Anthropomorphic phantoms cannot measure the dose themselves, but they can be combined with anthropomorphic 3D dosimeters. Other groups have developed head phantoms containing a 3D polymer gel dosimeter for proton therapy [7] and time-resolved 3D dosimetry using a combined MR linear accelerator system [8]. However, the regions of the body experiencing the highest degree of motion are the thorax and abdomen. In this region, considerable and partly respiratory driven organ deformations are common. Creating an anthropomorphic phantom and dosimetry system that can move and deform as the abdominal or thoracic region has the potential to enhance our understanding of how intrafractional and interfractional motion and deformations perturbs dose distributions.

In previous studies, a radiochromic silicone-based 3D dosimeter was proven suitable for both conventional radiotherapy [9-11] and proton therapy [12-14] even while being deformed [15,16]. The dosimeter can be cast into any shape with a high degree of detail [17] and has even been 3D printed directly [18]. If the dosimeters also contain air cavities complex air-tissue interfaces could be studied experimentally. Therefore, the ideal 3D dosimeter is deformable, anthropomorphic, can contain air cavities, be embedded in anthropomorphic phantoms, and work for both photon and proton irradiations.

The first aim of this study was therefore to develop a 3D-printing-based method for producing anthropomorphic silicone-based 3D dosimeters which can be deformed during irradiation, with and without air cavities. Particularly relevant are the liver in the thorax and the mediastinal volume surrounding the trachea in the upper abdomen. The second aim of the study was to test the performance of the dosimeters by irradiating them and compare the measured dose with the calculated dose distributions.

2. Materials and methods

2.1. *Internal structures*

Two different tracheas were 3D printed using a polyvinyl alcohol (PVA) water dissolvable filament (S1 in Figure 1a). The 3D prints were applied as "negatives" in their mould thereby creating internal structures. The first print was a simple trachea resembling a Y-shaped structure made to fit into a cylindrical 50 mm x 50 mm mould (Figure 1b). The second print was made from a 3D model of a trachea which was printed to fit into a 75 mm x 74 mm Ø cylindrical mould. The Y PVA print was immersed in boiling water for 20 seconds to smoothen its surface (S2).

The dosimeter material consisted of silicone elastomer (89.3%), curing agent (8.9%) and chloroform (1.5%), while the radiation-sensitive part of the dosimeter was leucomalachite green (0.26%). The dosimeters cured in a dark ventilated cupboard for 72 h at room temperature (S6).

The PVA prints were suspended in their moulds which was filled with the dosimeter mixture (S7). A weight was placed on top of the PVA 3D print to prevent it from floating. After curing, the PVA was punctured while submerging the dosimeter in warm water. Pliers were carefully applied to remove chunks of the dissolving PVA whereafter the PVA residue was rinsed off with water (step 8, Fig. 1a).

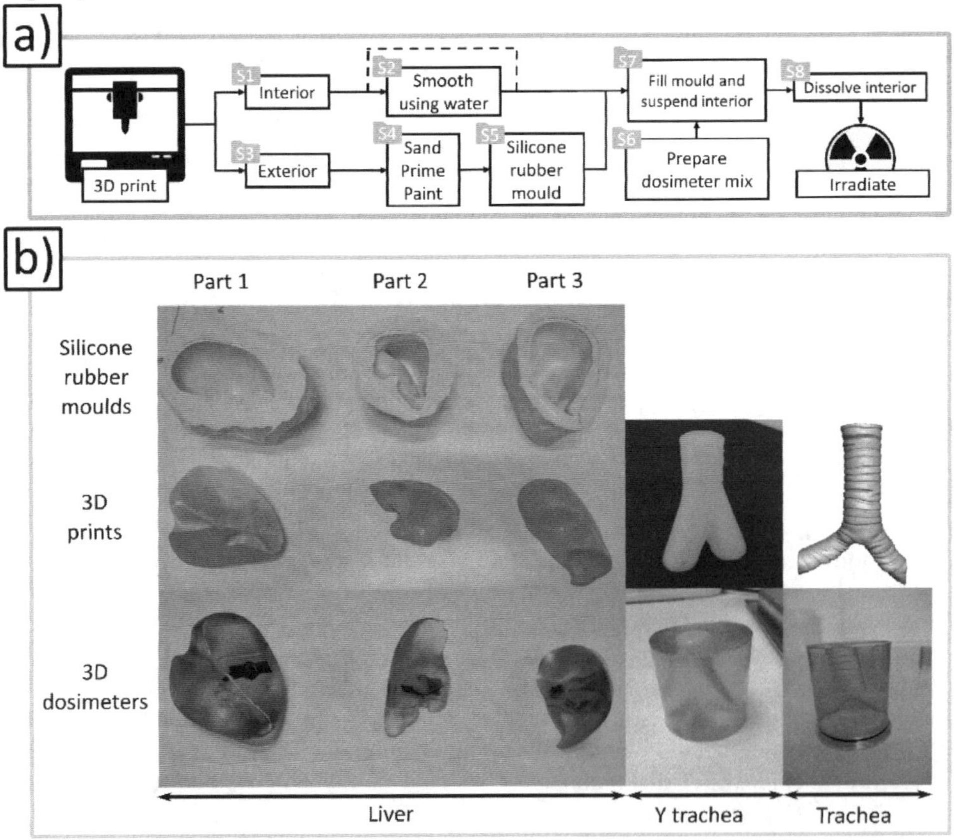

Figure 1. **a)** A flowchart depicting the steps included in producing the anthropomorphic dosimeters. **b)** Silicone rubber moulds for the dosimeters (top row), the corresponding 3D prints or models (middle row) and the final 3D dosimeters (bottom row).

2.2. *External structures*
The external structures were 3D printed using a polylactic acid filament (S3). The liver dosimeter print was divided into three parts to ensure that the parts could be read-out by the optical CT scanner, which could scan objects up to about 10 cm in size. Each 3D printed part was sanded down, coated with a layer of primer, and spray-painted to provide a smooth surface (S4). A casting basin was made from sculpting clay and filled with silicone moulding rubber. The 3D print was suspended in the basin and the moulding rubber was left to cure for 24 hours and thereafter removed (S5). The mould was filled with the dosimeter mixture (S6). A specially designed nut was suspended in the mixture during curing for later attachment in the optical scanner.

2.3. *Irradiation*
The dosimeters were CT scanned for treatment planning in Eclipse. The liver and Y dosimeter were irradiated with conventional photon beams using a Varian TrueBeam accelerator. The Y dosimeter received two opposing 16 mm x 16 mm fields to a maximum dose level of 10 Gy. The liver treatment plan consisted of six conventional 6 MV photon fields (15 mm x 15 mm) evenly spaced on a 180-degree arc with a prescribed maximum dose level of 5 Gy. A four-degrees-of-freedom couch correction was performed using a cone-beam CT. The trachea dosimeter received a two-field spot scanning proton plan using a Varian ProBeam system with a prescribed maximum dose level of 16 Gy.

2.4. *Read-out and data analysis*
The absorptivity of the dosimeters was read-out pre- and post-irradiation using an optical CT scanner with 1000 projections. The projections were reordered to have the same starting point in the pre- and post-read-out after which they were image registered in MATLAB. Using the ordered-subsets-convex total-variation reconstruction method led to 1 mm^3 voxels of the absorptivity.

The measured dose distributions (except the Y) were compared to the calculated distributions using a global 3%/3 mm 3D gamma analysis including all voxels in the datasets above 10% of the maximum calculated dose.

3. Results
Both methods developed to produce anthropomorphic dosimeters proved successful. The PVA filament was easy to remove without damaging the dosimeters. The detail of the dosimeters and silicone rubber moulds was fine enough that the layers of the 3D prints were visible on them.

Comparing the measured and calculated dose led to a gamma pass rate of 87% for part 2 of the liver dosimeter and 86% for the trachea dosimeter (Fig. 2).

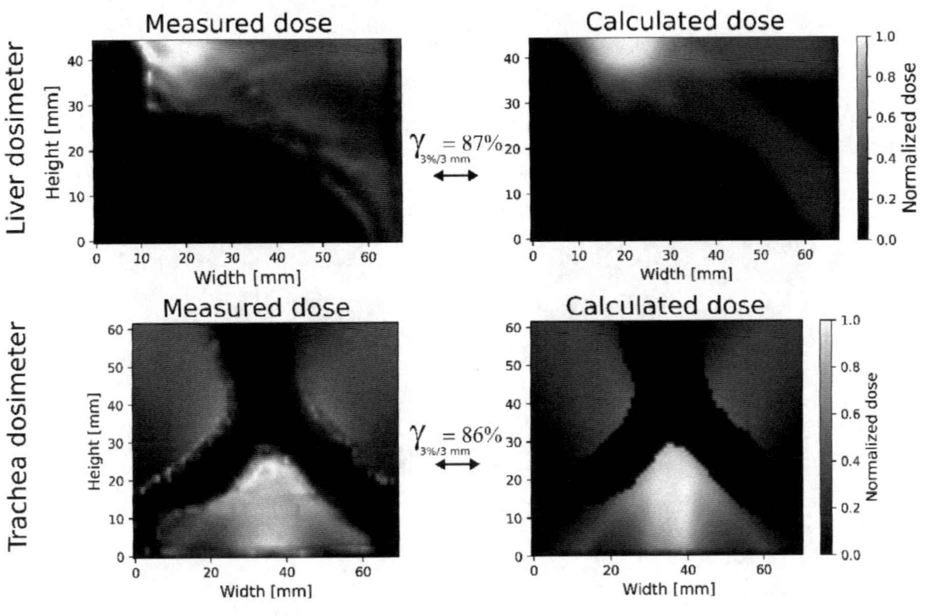

Figure 2: Central slices of the measured and calculated dose distributions. The top images are from part 2 of the liver dosimeter while the bottom is from the trachea dosimeter. The gamma analysis result is shown between the images.

4. Discussion

The developed method using 3D printing and silicone rubber was capable of producing anthropomorphic dosimeters and would work for structures with overhang of material and very intricate details only limited by the 3D prints. The dosimeters measured the expected dose distribution and could be useful in an anthropomorphic phantom.

For the photon irradiation the low gamma pass rate was primarily due to a low delivered dose. The read-out system has a limited dynamic range and if the dosimeter becomes too large then no light would reach the detector at the thickest projections which would impede the dose reconstruction. When using protons the measurements usually require a quenching correction due to dose-rate and linear energy transfer dependencies [13,14] which require a Monte Carlo simulation. For this the Hounsfield units of the dosimeter must be overwritten with -64 HU on the CT scans to get the correct stopping power [19]. However, the overwritten dosimeter structure in the treatment planning system could not be transferred to the Monte Carlo simulation and the quenching correction was therefore not applied. Furthermore, aligning the pre and post projections lead to some artefacts near the boundary of the dosimeters (see Fig. 2). If the dosimeter formulation could be altered to be tissue equivalent it would improve the applicability and transferability into clinical use. Also, the advantage of anthropomorphic dosimeters is limited if the size cannot exceed 10 cm since many organs are larger than that.

If the dosimeter could be printed directly with the dosimeter material as investigated by Wheatley *et al* [18] it would greatly decrease the production time. However, the method presented in this work does not require a specialized 3D printer and, in the case of the external moulds, these are reusable.

In future work, the 3D liver dosimeters will be used in proton spot scanning to investigate a protocol-specific plan during deformation in an anthropomorphic motion phantom [20]. Furthermore, replacing the radiochromic component with an optical stimulated luminescence component to make the dosimeters reusable is being investigated [21,22] which would likely increase clinical applicability.

In conclusion, we have successfully created a method to produce anthropomorphic dosimeters and shown they can measure the planned dose.

5. Acknowledgments

We kindly thank Mads Lykke Jensen for his skilled help with 3D printing. This project was funded by the Novo Nordic Foundation.

6. References

[1] Weidner A *et al* 2022 *Phys. Med. Biol.* **67** 045009
[2] Niebuhr N I *et al* 2019 *Phys. Med. Biol.* **64** 04NT05
[3] Ehrbar S *et al* 2019 *Med. Phys.* **46** 839-50
[4] Black D G *et al* 2021 *Med. Phys.* **48** 4205-17
[5] Tino R B *et al* 2022 *Med. Phys.* **49** 52-69
[6] Deene Y De *et al* 2022 *Phys. Medica* **104** 43-55
[7] Hillbrand M *et al* 2018 *Zeitschrift fßr Medizinische Phys.* **29** 162-72
[8] Deene Y De *et al* 2022 *J. Phys. Conf. Ser.* **2167** 012029
[9] Høye E M *et al* 2015 *J. Phys. Conf. Ser.* **573** 012067
[10] De Deene Y *et al* 2015 *Phys. Med. Biol.* **60** 1543-63
[11] Høye E M *et al* 2015 *Phys. Med. Biol.* **60** 5557-70
[12] Høye E M *et al* 2017 *J. Phys. Conf. Ser.* **847** 012021
[13] Høye E M *et al* 2017 *Phys. Med. Biol.* **62** N73-89
[14] Valdetaro L B *et al* 2021 *Phys. Imaging Radiat. Oncol.* **18** 11-8
[15] Kaplan L P *et al* 2017 *Phys. Med. Biol.* **62** 5612-22
[16] Jensen S V *et al* 2020 *Phys. Imaging Radiat. Oncol.* **16** 134-7
[17] De Deene Y *et al* 2015 *J. Phys. Conf. Ser.* **573** 012025

[18] Wheatley M J *et al* 2022 *J. Phys. Conf. Ser.* **2167** 012010
[19] Taasti V T *et al* 2016 *Med. Phys.* **43** 2780-4
[20] Abdominal phantom with liver dosimeter insert. https://youtu.be/X4VUM5mNnsk
[21] Sądel M *et al* 2021 *Sensors* **21** 6015
[22] Nielsen C L *et al* 2022 *J. Phys. Conf. Ser.* **2167** 012023

AUTHOR INDEX